Data-Driven Identification of Networks of Dynamic Systems

This comprehensive text provides an excellent introduction to the state of the art in the identification of network-connected systems. It covers models and methods in detail, includes a case study showing how many of these methods are applied in adaptive optics and addresses open research questions. Specific models covered include generic modelling for MIMO LTI systems, signal flow models of dynamic networks and models of networks of local LTI systems. A variety of different identification methods are discussed, including identification of signal flow dynamics networks, subspace-like identification of multi-dimensional systems and subspace identification of local systems in an NDS. Researchers working in system identification and/or networked systems will appreciate the comprehensive overview provided, and the emphasis on algorithm design will interest those wishing to test the theory on real-life applications.

This is the ideal text for researchers and graduate students interested in system identification for networked systems.

Michel Verhaegen is a professor at Delft University of Technology and a fellow of the International Federation of Automatic Control (IFAC). He co-authored *Filtering and System Identification: A Least Squares Approach* (Cambridge University Press, 2010).

Chengpu Yu is a professor at Beijing Institute of Technology.

Baptiste Sinquin is an algorithm engineer at SYSNAV.

Data-Driven Identification of Networks of Dynamic Systems

MICHEL VERHAEGEN
Delft University of Technology

CHENGPU YU
Beijing Institute of Technology

BAPTISTE SINQUIN
SYSNAV

CAMBRIDGE
UNIVERSITY PRESS

University Printing House, Cambridge CB2 8BS, United Kingdom

One Liberty Plaza, 20th Floor, New York, NY 10006, USA

477 Williamstown Road, Port Melbourne, VIC 3207, Australia

314–321, 3rd Floor, Plot 3, Splendor Forum, Jasola District Centre, New Delhi – 110025, India

103 Penang Road, #05–06/07, Visioncrest Commercial, Singapore 238467

Cambridge University Press is part of the University of Cambridge.

It furthers the University's mission by disseminating knowledge in the pursuit of education, learning, and research at the highest international levels of excellence.

www.cambridge.org
Information on this title: www.cambridge.org/9781316515709
DOI: 10.1017/9781009026338

First published 2022

A catalogue record for this publication is available from the British Library.

Library of Congress Cataloging-in-Publication Data
Names: Verhaegen, M. (Michel), author. | Yu, Chengpu, 1984– author. | Sinquin, Baptiste, 1991– author.
Title: Data driven identification of networks of dynamic systems / Michel Verhaegen, Chengpu Yu and Baptiste Sinquin.
Description: First edition. | New York : Cambridge University Press, [2022] | Includes bibliographical references and index.
Identifiers: LCCN 2021056490 (print) | LCCN 2021056491 (ebook) | ISBN 9781316515709 (hardback) | ISBN 9781009026338 (epub)
Subjects: LCSH: System analysis–Mathematics. | BISAC: TECHNOLOGY & ENGINEERING / General
Classification: LCC QA402 .V468 2022 (print) | LCC QA402 (ebook) | DDC 003/.85–dc23/eng/20220118
LC record available at https://lccn.loc.gov/2021056490
LC ebook record available at https://lccn.loc.gov/2021056491

ISBN 978-1-316-51570-9 Hardback

Contents

Preface

The identification of network-connected dynamic systems is currently a hot research topic within the community of systems and control. Researchers in other engineering areas, the social sciences and systems biology are putting a lot of effort into the study of network-connected systems. Modelling such networks and the identification of these models from acquired measurements is crucial in the analysis and understanding of the dynamics. Based on these models, synthesis to modify the behaviour of the network can also be performed. This book gives a unique overview of the state of the art of the research in the field of identifying networks of linear dynamical systems. This overview combines many of the pioneering contributions from the authors with those of other researchers who are playing a crucial role in the development of this new field.

This book is intended for researchers active in the field of network-connected systems. For researchers who are active in the field and would like to make a contribution, it gives an excellent overview of the state of the art that would save a lot of time and avoid the collection of scattered information from research papers, reports and unpublished work. Researchers active in the field of network-connected systems but not necessarily working on system identification will gain a good insight into what is possible. The book also contains a study on the application of novel identification methods to the real-life case study of adaptive optics. The book can be used as material for teaching the identification of network-connected systems at a master's or PhD level.

The book is organized as follows. Apart from an introduction (Chapter 1) stating the scope of the book, the book consists of three parts.

Part I: Modelling Large-Scale Dynamic Networks

This part gives an overview of the modelling of large-scale network-connected systems with an emphasis on the parametrization of the models. Different model structures are presented that balance the trade-off between the number of parameters and the accuracy of the model.

Part II: The Identification Methods

This part is the core part of the book and describes the new developments in the field of identification of large-scale network connections of linear dynamical systems.

Two families are presented. The first is the signal flow networks approach and is based on signal graph models. It focusses on the transfer function type of models. The second are networks that are a connection of linear state-space models, or, equivalently, structured state-space models in which the system matrices have a dedicated structure. For this class, subspace (like) identification methods are presented that allow one to find from data the size of the blocks of these matrices as well as estimates of the parameters in these blocks. The subspace-like nature is crucial from a user's perspective, as it takes the tedious work away from the user to a priori specify a model parametrization, something that is usually missing. For this second class, an important contribution is the identification of local systems in a large network using only local data. To improve the readability of the algorithmic derivations, a 'General Algorithmic Strategy' is given in each chapter. This structure outlines the general plan of solution for the considered identification problem.

Part III: Illustrating with an Application to Adaptive Optics

This part illustrates the use of many of the presented methods for an adaptive optics case study. Special attention is given to the trade-off between scalability of the algorithms versus the performance of the models and their use, for example in a model-based controller design.

The concluding chapter gives a brief overview of a number of research challenges and software developments in the use of these methods in practical scenarios.

Many of the results that have been documented in this book have been achieved in the course of an advanced grant of the European Research Council (ERC) under the European Union's Sevent Framework Programme (FP7-2007-2013, Grant Agreement No. 339681). These include mainly the research results in Chapters 7–10. In addition to these research efforts, this project has benefited from adjacent research activities and collaboration with colleagues. One such effort is research activity on the control of high-resolution imaging systems. This formed an interesting and stimulating application area in which to define, test and validate a number of the novel methodologies. Contributions in these adjacent research areas by Dr Reinier Doelman, Dr Pieter Piscaer, Dr Dean Wilding and Dr Paolo Pozzi are very much appreciated, not in the least for their help in expanding and consolidatiing the Smart Optics Lab of the Delft Center for Systems and Control. Platforms in this lab have served as testbeds for testing and validating a number of the new methodologies reported in this book. For the contributions that relied on collaborations with international experts, we acknowledge Professor Lennard Ljung and Professor Anders Hanson from the University of Lincköping, Professor Aleksander Haber from the City University of New York, Professor Paolo Massioni of the National Institut of Applied Sciences of Lyon, Dr Justin Rice of Harvard Medical School, Professor Kees Vuik of the Faculty of Mathematics and Computer Science of the Delft University of Technology, Professor Coen de Visser from the Faculty of Aerospace Engineering of the Delft University of Technology, Dr

Elisabeth Brunner of the Medical School of the University of Vienna and Dr Karel Hinnen of ASML.

Special thanks go to former MSc students who contributed in software development, method prototyping and validation. These include Peter Varnai, Guido Monchen and Wiegert Krijgsman.

erc European Research Council

Vocabulary

An explanatory vocabulary of frequently used terms related to the study of dynamical networks is introduced here.

centralized identification algorithm An identification algorithm is said to be centralized if the corresponding computations are carried out on a single computational unit. Standard Subspace Identification algorithms such as N4SID and MOESP identify state-space systems in a centralized manner. Also called 'global identification'.

computational complexity The computational complexity for a given mathematical operation, or algorithm, is the asymptotic growth rate of the computational cost.

data-sparse When a matrix is parametrized with far less parameters than its actual number of entries, it is said to be data-sparse. A sparse matrix is data-sparse. The reverse is not necessarily true: a data-sparse matrix is not necessarily sparse, for example a Toeplitz matrix.

dense matrix When a matrix has very few non-zero entries, it is said to be dense.

distributed identification algorithm An identification algorithm is said to be distributed when the calculation of the identified model can be distributed in separate parts. This may enable the distribution of these calculations over different cores with little communication required between each core.

dynamic network A dynamic network is a set of interconnected subsystems, and at least some of them have temporal dynamics.

general algorithm strategy A summary of the algorithmic plan of attack prior to detailing the solution for a particular identification problem.

global matrix representation The network dynamics may be represented at a local level, such as by writing a model for each subsystem, or at a global level by considering the network as one system. The underlying dynamics are then translated into special matrix structures. We will denote such matrices as global matrices.

heterogeneous set of subsystems This stands in opposition to an homogeneous set of subsystems. The subsystems need not be identical in a network.

homogeneous set of subsystems This is a synonym to express that the set of subsystems are all identical.

interconnected subsystems Two subsystems are said to be interconnected if at least one of them communicates with the other, thereby influencing the input-output behaviour of the subsystems. This interconnection may be physical due to the physical constraints these local systems have on one another, or logical due to the exchange of data these local systems communicate to one another.

large scale A system (or model) is said to be large scale when the dominant system (or model) dimension parameter, the dimension of its input or output vector (or the dimension of the state of the model), is large. Here the notion of 'large' depends on the intentions, but for this book we assume large-scale systems or models with a dominant dimension parameter exceeding a few hundred.

local model A large-scale system may be built as the assembly of interconnected subsystems. When it is spatially distributed, it is possible to define subsystems whose input-output behaviour is modelled using dynamical equations involving only localized input and output data. A local model is associated with a subsystem.

matrix structure A matrix structure is a matrix parametrization. For example, a matrix may be Toeplitz or banded. The parametrization may be affine or not. Exploiting these matrix structures is a main concern in the book in order to alleviate the computational complexity of identification algorithms.

multi-dimensional sensor or actuator A two- or three-dimensional sensor (actuator) array measures (actuates) a spatial phenomenon. One node in this (regular) array is defined by its spatial coordinates in the array or grid. Each node in this array measures (actuates) a local quantity of interest, such as a value of the wavefront at a particular location in space. An alternative terminology of such a collection of sensor (actuator) nodes is a 'sensor (actuator) network'.

multi-dimensional system A system is said to be multi-dimensional when both the sensors and actuators are regularly distributed on a multi-dimensional spatial array.

network-connected systems This is a set of interconnected subsystems that influence each other. The systems may exchange information between each other. The temporal dynamics of each subsystem cannot be described without knowledge of how its neighbours behave.

network topology The network topology is the interconnection pattern of the subsystems operating in the network.

scalable An algorithm is said to be scalable when its computational complexity and data storage scale favourably with the dominant dimension parameter of the problem it is trying to solve. A desirable favourable property is linear dependency of the algorithmic complexity on the dominant problem parameter.

sparse matrix A matrix is sparse when it has few non-zero entries relative to the total number of entries.

structured state-space model A state-space model is said to be structured when the state-space matrices are parametrized in terms of basis functions that can represent structures in geometrical space or other transformed domains.

subsystem A subsystem refers to a local part of a bigger system. It is associated with local input and output behaviour. It may be a single agent that is physically separated by other agents, such as in a platoon of satellites or cars, or it may be a virtual agent such as a discretization node that results in the spatial discretization of PDEs.

Notation and Symbols

$\mathbb{Z}$	the set of integers
$\mathbb{N}$	the set of positive integers
$\mathbb{C}$	the set of complex numbers
$\mathbb{R}$	the set of real numbers
$\mathbb{R}^n$	the set of real-valued n-dimensional vectors
$\mathbb{R}^{m \times n}$	the set of real-valued m by n matrices
∞	infinity
$\in$	belongs to
$=$	equal
$\approx$	approximately equal
$\square$	end of proof
$\otimes$	Kronecker product
$\odot$	Khatri–Rao product (the column-wise Kronecker product)
$x \circ y$	The outer product between two vectors x, y of length N is a matrix of size $N \times N$ with the (i, j)th element equal to $x_i y_j$.
I_n	the n by n identity matrix
$A_{i,j}$	the (i, j)th entry of the matrix A
$A(i, :)$	the ith row of the matrix A
$A(:, i)$	the ith column of the matrix A
$A(a : i : b, c : j : d)$	selects a submatrix from A composed of all entries with row index $a+ki$ (until it reaches b) and column index $c+kj$ (until d), where k is an integer starting at 0
A^T	the transpose of the matrix A
A^{-1}	the inverse of the matrix A
$A^{1/2}$	the symmetric positive definite square root of the matrix A
$\mathrm{diag}(a_1, a_2, \ldots, a_n)$	an $n \times n$ diagonal matrix whose (i,i)th entry is a_i
$\det(A)$	the determinant of the matrix A
$\mathrm{range}(A)$	the column space of the matrix A
$\mathrm{rank}(A)$	the rank of the matrix A
$\mathrm{trace}(A)$	the trace of the matrix A
$\mathrm{vec}(A)$	a vector constructed by stacking the columns of the matrix A on top of each other
$\|A\|_2$	the 2-norm of the matrix A
$\|A\|_F$	the Frobenius norm of the matrix A

$card(\mathrm{A})$	the number of non-zero elements of A
$[x]_i$	the ith entry of the vector x
$\|x\|_2$	the 2-norm of the vector x
$\|x\|_1$	the 1-norm of the vector x
lim	limit
min	minimum
max	maximum
sup	supremum (least upper bound)
$E[\,\cdot\,]$	statistical expected value
$E[xy^T]$	the covariance for two zero-mean vectors x and y
Tensors	Tensors are denoted with calligraphic letters.
$\mathcal{X} \in \mathbb{R}^{J_1 \times \ldots \times J_d}$	A tensor $\mathcal{X}$ with the entry at position $j_1, \ldots, j_d$ is denoted by $x_{j_1,\ldots,j_d}$.
$\mathrm{vec}(\mathcal{X}_{i,:,j})$	For a tensor of size $J_1 \times J_2 \times J_3, \mathrm{vec}(\mathcal{X}_{i,:,j})$ is equal to $\begin{bmatrix} x_{i,1,j} & \cdots & x_{i,J_2,j} \end{bmatrix}$.
$\mathrm{vec}(\mathcal{X})$	For a tensor $\mathcal{X}$, this denotes the vectorization using the vec operator such that $\mathrm{vec}(\mathcal{X}) = \begin{bmatrix} x_{1,\ldots,1} & x_{2,1,\ldots,} & \cdots & x_{1,2,1,\ldots} & \cdots & x_{J_1,\ldots,J_d} \end{bmatrix}$.
$\mathcal{O}(n)$	Floating-point operations (flops) finish in at most $c \cdot n$ flops, for some constant c.
$\mathcal{H}_{s,N}[z(n)]$	the block-Hankel matrix constructed from the sequence z starting in its top left position at time instance n, as defined in Definition (8.14)
$\mathcal{T}_s(A,B,C,D)$	the block Toeplitz matrix constructed from the quadruple of system matrices, as defined in Definition (8.15)
$\mathcal{O}_s(A,C)$	the extended observability matrix defined by the pair (A,C) given in (8.16)

Acronyms

1D	one-dimensional
2D	two-dimensional
ADMM	alternating direction method of multipliers
AIC	Akaike information criterion
ALS	alternating least squares
AO	adaptive optics
AR	auto-regressive
BCU	block-coordinate update
CCD	charged-coupled device
COSMOS	constrained subspace method for the identification of structured SSM
CPD	canonical polyadic decomposition
DARE	discrete algebraic Riccati equation
DC(P)	difference of convex (programs)
DM	deformable mirror
DNF	dynamic network function
DSF	dynamic structure function
EE	encircled energy
ELT	Extremely Large Telescope
FIR	finite-impulse response
GPU	graphical processing unit
K4SID	Kronecker-structured large-scale subspace identification
LKR	low Kronecker rank
LQG	linear quadratic Gaussian
LTI	linear time invariant
MIMO	multi-input multi-output
MLDS	multi-linear dynamical system
MOESP	multivariable output error state space
MSSM	matrix state-space model
MVM	matrix vector multiplication
N4SID	numerical algorithms for subspace state-space system identification
NDS	network of dynamic systems
NKP	nearest Kronecker product

PBSID	predictor-based subspace identification
PDE	partial differential equation
PSF	point spread function
QUARKS	Kronecker-based vector auto-regressive with exogenous inputs (KVARX)
RMS	root mean square
RMSE	root mean square error
SCAO	single conjugate adaptive optics
SH	Shack–Hartmann
SISO	single input single output
SNR	Signal-to-Noise Ratio
SOK	sums of Kronecker
SSARX	subspace identification method that uses an ARX estimation
SSM	state-space model
SSS	sequentially semi-separable
SVD	singular value decomposition
TSSM	tensor state-space model
VAF	variance accounted for
VAR(X)	vector auto-regressive (with exogeneous inputs)
VARMAX	vector auto-regressive moving average with exogeneous inputs
VARX	vector auto-regressive with exogeneous inputs
WFS	wavefront sensor

1 Introduction

1.1 Two Approaches: The Signal-flow Approach and the Network of Local LTI Systems Approach

The developments in hardware technology, wireless communication networks and computational technology contribute to the advancement of large-scale networks of interconnected systems. The large-scale nature of such networks has started to arouse curiosity within the systems & control community. A few cases may already be distinguished. The exchange of information between the systems interacting in a network can be known, partially known or be unknown. Examples of networks with known interconnections include networks of pipelines and electrical grids. Unknown and abstract connections occur in spatial-temporal systems that are governed by partial differential equations (PDEs). After discretization of the equations, the dynamics can be viewed as an array of local dynamical systems that exchange information via physical interactions between the local models that are each associated to local input and output signals. Examples include arrays of sensors for measuring the optical disturbance induced by a spatially and temporally varying windfield that hovers over a ground-based telescope, and in flow control to stabilize boundary layers and reduce the drag. A deformable mirror with non-negligible time response that may be used in optics also belongs to this class of spatial-temporal dynamical systems. In these cases, the sensor delivers at a high frequency rate a large number of measurements that will subsequently be processed for data-driven control.

The topic of this book is the system identification of interconnected systems operating in a network. System identification attempts to find a mathematical representation for the dynamics by which measured signals are related. Such a mathematical relationship is termed a model. Finding such a model requires a systematic approach towards designing experiments, specifying model classes, using mathematical methods to find estimates for the parameters in these models and validating these estimated models. These topics are well discussed in, for example, Ljung (1999); Verhaegen and Verdult (2007). This book focuses on the definition of model classes for networks of dynamical systems and presents some system identification methodologies to estimate such models from input-output measurements.

The main bottlenecks for the extension of classical identification methods to large-scale networks are the huge volume of data and the large number of parameters necessary to describe these models, even when considering compact local model dynamics.

A key challenge that will be addressed in this book is how to efficiently deal with this tremendous degree of complexity while obtaining models that are as accurate as possible in representing the system behaviour given, for example, by the input-output response.

Two major system identification approaches for identifying network connected systems are under full development. These developments occur very much along the lines of the two model classes that are in use for lumped parameter systems: transfer function models and state-space models. Many of the system identification approaches for networks have been inspired by particular applications.

The first main approach for identifying dynamic networks is based on the input-output transfer functions describing the network. This approach characterizes the network by employing dynamic signal flow diagrams and will be referred to as the *signal flow approach*. It starts from a global model representation that, in principle, can be formulated without knowledge of the way in which the local systems are connected to each other. If size allows, the global dynamics may be identified and a special parametrization in the global coefficient matrices may be looked for afterwards to find out or estimate a network topology. However, we usually strive for lean parametrizations of the coefficient matrices in these models in order to deal with the data complexity and overcome the computational bottleneck. Specifying such models and their consequences on the identification methods to find these models for data is a challenge to be pursued in this book. These parametrizations may come from physical insights such as localizability (each local system in the network has a very limited influence on its neighbours), spatial invariance (all local systems are identical) or separability of the system dynamics along the physical dimensions. Such models, which rely on transfer functions, are labelled in this book as *global* models. Typical applications are from biological or computer networks (see, for example, Gevers et al. [2019]; Gonçalves and Warnick [2008]).

The second main approach for identifying networks of dynamic systems is based on large state-space models with structured system matrices and are developed, among others, in the scope of designing data-sparse controllers for network connected systems. For examples of the latter we refer to D'Andrea and Dullerud (2003); Massioni and Verhaegen (2009); Rice and Verhaegen (2009). Typical areas of application of such controller design methods are formation flying, cross-directional control in paper processing, automated highway systems, micro-cantilever array control for massive parallel data storage and lumped parameter approximations of PDEs. We refer to the work in D'Andrea and Dullerud (2003) for more details on these applications. Local models may be related to a physical entity such as cars on an automated highway, or can be virtual as is often the case when modelling spatial-temporal dynamical systems. Key for these models is the interconnection pattern of the network, i.e. how the local dynamical systems are connected to each other. Priors from physics such as spatial invariance and separability may also be formulated here although at a local scale, that is, on the system matrices of the local models. These may eventually translate into parametrizations of the global matrices. This second approach is referred to as the *network of local LTI systems*-approach, and the mathematical methods derived in this context can be viewed as *identification for distributed control* methods that aim to be efficient for distributed control design. The efficiency envisaged is that the model structure of the identified model can be exploited to arrive at low-cost controller design, both in terms of calculating and implementing these controllers.

The examples given so far are illustrative and follow existing literature that uses either transfer functions or state-space models for subsequent control. Both approaches may nonetheless be applied succesfully to an application mentioned for the other approach.

An appropriate model structure is of prime importance for system identification as has been discussed in Ljung (1999); Verhaegen and Verdult (2007). If the model structure is not included within the set of candidates that accurately represents the true system, the estimates derived will be biased irrespective of the size of the dataset or the convergence properties of the dedicated algorithm. This is the case for small and medium size models as demonstrated by the bias-variance analysis given, for example, in Ljung (1999). For large-scale models, this will be even more prominent because the number of data points available for identification (and validation) is small compared to the number of parameters to be estimated. One often seeks (data-)sparse model parametrizations and the enforcing of structure in the system matrices of the model to be estimated. In general, this may help to reduce the variance of the estimates compared to the unstructured case for the same number of temporal samples. However, the other side of the coin is that the model quality may degrade: an incorrect parametrization of the matrices is likely to increase the bias, e.g. when the physical insights are too crude. Instead of conducting a bias-variance trade-off analysis like that in Ljung (1999), we develop identification methodologies to estimate model structures where structural parameters, such as the order of local models, are derived from the given data. These structural parameters offer the freedom to the user to get the best model within that model class. This will be the case for the subspace-like identification methods to be developed in this book.

In this chapter, we will start by providing motivating examples additional to those already mentioned that demonstrate how a network of dynamical systems arises through the use of multi-dimensional sensor grids. We will then briefly outline the remaining chapters of this book.

1.2 Examples of Networks of Dynamical Systems

Dynamical systems with multi-dimensional sensor/actuator arrays consisting of a large number of nodes appear, or will appear, in the near future in various engineering applications. These applications range from optics to flow control, but also in many other applications in economics, systems biology, social network analysis, etc. In this book, the examples considered are mainly restricted to engineering problems.

1.2.1 Data-Driven Predictive Control for Large-Scale and Extreme Adaptive Optics

In the current generation of large-scale adaptive optics (AO) and the forthcoming generation of extreme AO (XAO) being designed for the inspection of exoplanets, use will be made of (extreme) large multi-dimensional sensor/actuator arrays with thousands (ten-thousands) of nodes, each processing local signals at the kHz rate. The imaging of exoplanets using XAO is challenging due to the high contrast ratio and the small angular separation relative to the star; see Guyon (2018) for more details. In these AO systems, the turbulence induced in the atmosphere above the telescope induces a stochastic disturbance that greatly deteriorates the image quality and, without (real-time) compensation, reduces these extremely expensive telescopes to something comparable to Newton's telescope developed in the seventeenth century. The AO application is further discussed in Chapter 10. Here, we restrict ourselves to highlighting some identification challenges. Figure 1.1 illustrates the turbulence fields flowing over a telescope aperture. Such turbulence fields introduce optical aberrations that reduce the quality of the image recorded by the telescope. This would limit the science that can be done using such images such as investigating the atmosphere of exoplanets. Only the *wavefront* of the optical field aberrations is of interest as the intensity can often be assumed to be uniform. In order to improve the image resolution, use can be made of AO in order to reduce the effects of the optical aberrations. An important element in an AO system is the sensor that measures the optical aberrations.

The sensor measuring these wavefront aberrations in a ground telescope is a two-dimensional array, as depicted in Figure 1.2. The sensor, for example a Shack–Hartmann sensor, measures the combined effect of all the layers above the telescope, with each layer having a different wind speed and direction. A direct measurement of the optical wavefront aberration is not possible and a reconstruction always needs to be done. The reconstruction step may benefit from knowledge of model of spatial-temporal dynamics of the induced wavefront aberrations in order to predict these disturbances in time. Deriving such a model from the large number of measurements on an (extreme) large telescope is a challenging problem. The methods to be developed in this book serve as possible solutions for these types of problems. Based on the measured wavefront aberrations, a deformable mirror can then compensate the wavefront of the optical beam. Such a deformable mirror consists of a grid of actuators that operate under the mirror surface to deform that surface into a shape that compensates the overall wavefront distortation. Linking the sensor to this actuator is done by the controller. Modelling this link is also an identification problem that can be considered with the methods presented in this book.

Many instruments, such as a coronograph or a spectrograph, could benefit from scalable large-scale AO control algorithms. Examples of such instruments are HARMONI (high angular resolution monolithic optical and near-infrared integral field spectrograph) for the ESO's extreme large telescope (ELT) (Neichel et al., 2016), NFIRAOS (narrow field infrared adaptive optics system) for the Thirty Meter Telescope (Ellerbroek, 2011), and GPI (Gemini planet imager) (Poyneer et al., 2016).

Figure 1.1 A schematic representation of a telescope with two layers of the turbulence above the telescope. Each layer in this plot schematically represents the surface of the optical wavefront. In this figure we do not plot the isoplanatic angle between the star and the object of interest. For more details on the nomenclature refer to Chapter 10.

AO is not limited to astronomy. It has been widely used in other application areas, such as microscopy (Pozzi et al., 2020) and optical coherence tomography for in vivo high resolution imaging of the human retina (Pircher and Zawadzki, 2017). Although the control loop usually operates at a few hertz in these applications, there has not been much development regarding the data-driven minimum variance control of such systems.

1.2.2 System Identification for Wind Farm Control

Another example of the application of system identification of large-scale systems is the area of offshore wind farms (Gebraad, 2014). To illustrate, we refer to Figure 1.3

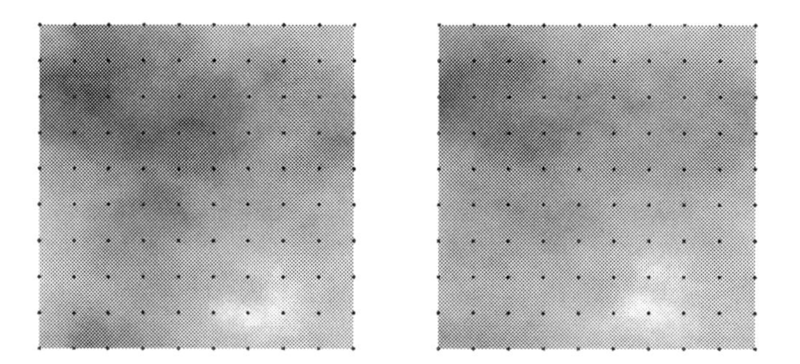

Figure 1.2 Example of a turbulence field consisting of the sum of two frozen layers propagating at different speeds: (left) $t = t_0$; (right) $t = t_1 > t_0$. The sensor array is of size 10×10 and the sensor nodes are represented by black dots. The correlations are not only spatial but also temporal. Reprinted/adapted from Sinquin (2019) with permission from TU Delft

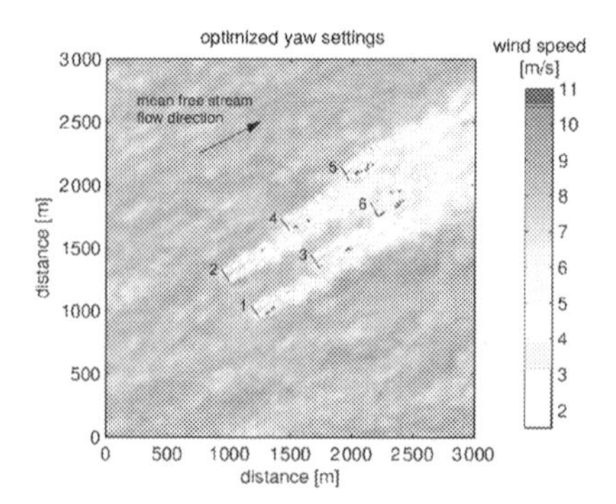

Figure 1.3 A 3×2 wind plant rotated 10° with respect to wind direction. Hub-height wind field at 800 s simulated time as calculated by the software SOWFA. The black lines indicate the rotor positions and yaw orientation of each turbine. Reprinted/adapted from Gebraad (2014) with permission from TU Delft

where six closely spaced wind turbines operate in a wind corridor. The control objective is to maximize the energy production of these turbines. This requires knowledge of the spatial-temporal interaction between them. Mathematical models are of use in the design of model-based controllers for maximizing the energy yield. Even though the sampling frequency for such applications is in the range of one Hertz (and much smaller than the AO applications considered in the previous subsection), the challenge for a mathematical model is that it should allow for an efficient controller design as well as its real-time operation. One approach to designing such a model is based on discretizing the Navier–Stokes equation (Boersma et al., 2018), and using this discretized model in ensemble Kalman filtering techniques (Doekemeijer et al., 2018). A second approach is to identify a dynamic model based on measurements of the wind velocity in time and space (e.g. on a 3D regular grid). Such models, can then be used to predict the future evolution of the wind field (such as its spatially distributed wind velocity and direction) in a predictive controller. The development of these

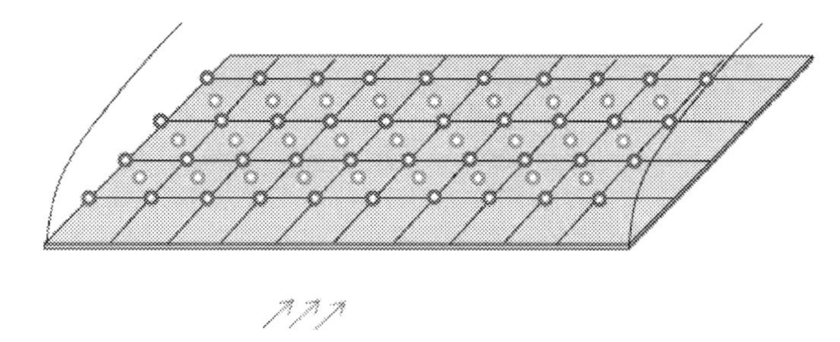

Figure 1.4 A 2D schematic of a multi-dimensional sensor and actuator array for active boundary control. The actuators are indicated by light circles, while the sensors by dark circles. The undisturbed flow is indicated by the arrows. The actuators and sensors are arranged in a regular grid. Reprinted/adapted from Sinquin (2019) with permission from TU Delft

models may be possible using the approaches outlined in this book, though the actual realization still needs to be done.

1.2.3 Active Boundary Layer Control

The goal of active boundary layer control on, for example, the wing of an airplane is to reduce the drag induced by the wind flow around the wing and fuselage by delaying the flow separation and preventing (as far as possible) the laminar flow (low drag) from transitioning into turbulent flow (high drag). This goal is achieved by using a multi-dimensional array of pressure sensors and a multi-dimensional array of actuators, such as blowing or suction actuators, to actively control the boundary flow around the surface. A possible configuration of these arrays is shown in Figure 1.4.

A trade-off in active boundary control is to determine a model for the temporal and spatial dynamics of the wind speed and vorticity that on one hand is compact, so that efficient control design and implementation becomes possible, and on the other that it is accurate enough to enable high performance control. In a number of occasions, such as the flow in a pipe, the dynamics of the flow close to the boundary of the pipe can be accurately described around a steady-state wind speed by the linearized Navier–Stokes equations. This yields a linear PDE for small variations around a steady state. The system identification methods developed in this book pursue the trade-off between model complexity and model accuracy using real-life measured data with multi-dimensional arrays of sensors and actuators.

For preliminary results on the development of compact models for boundary layer control via system identification we refer to Kim and Bewley (2007) and Inigo (2015).

1.2.4 Varied Applications Ranging from Weather Prediction to Sociology

Even without actuators to control, some datasets can be recast as a multi-dimensional networks, for example in weather prediction (Tsiligkaridis and Hero, 2013). There are many examples of geographic information systems that collect data via an array of sensors distributed geographically. The notion of dimension may also be more abstract, for example, in sociology and for studying relational networks (Hoff, 2015) as it does not feature a sensor grid. This area motivated the introduction of self-replicating patterns (Leskovec et al., 2010): the network is made up of different clusters that each replicate the same interconnection pattern, the clusters are themselves made of clusters of subsystems that again interact in a similar pattern. This may also occur in biology.

Adaptive optics systems, wind farms and flow control all rely on large arrays of sensors to measure wind field-induced aberrations and are potential applications of the system identification methods that are developed in this book. These examples may be recast into the signal flow approach or into the network of local LTI systems approach. In adaptive optics, identifying the spatial-temporal dynamics of the wave-front aberrations may be recast into a Signal Flow approach whereas identifying those of a deformable mirror would be seen as a set of interconnected subsystems with very limited neighbourhood allowing the use of the local approach. In such cases, there is already a strong a priori knowledge of the topology of the network that can be used to enhance the computational efficiency of the system identification. Both the signal flow approach and the network of local LTI systems may be used to represent discretized PDEs. This applies to the deterministic behaviour as well as the possible stochastic disturbances that are modelled by PDEs, such in modelling turbulence via the Navier–Stokes equations. The models derived may be used in the design of large-scale observers or Kalman filters.

In Section 1.3, we review the heat equation in the context of distributed systems to illustrate notions such as topology, subsystems, and how these arise from a discretized PDE.

1.3 The Spatio-Temporal Impulse Response

For particular type of waves, including the heat conduction or optical wave propagation in an empty medium, their behaviour is governed by a linear PDE featuring both spatial and temporal derivatives. At each discretization node of a PDE, a subsystem is assigned with its own input and output. In other words, the discretization and the interconnection pattern of the subsystems are highly related. Each subsystem shares information with its neighbours in time and space that can be pictured by its spatial-temporal impulse response.

A heat conduction example is used as an illustration. Here, we consider a thin metal plate with homogeneous material density and a known temperature distribution $T_0(\xi)$. In this notation, ξ represents the spatial variable and $t = t_0$ is the initial time point. Let the spatial boundary be denoted by Ω. The temporal propagation of the temperature over time $t > t_0$ is governed by thermal conduction principles and subjected to boundary conditions. Appropriate boundary conditions are the homogeneous Dirichlet boundary conditions. This results into the following PDE:

$$\begin{cases} \frac{\partial}{\partial t}T(t,\xi) &= -c\nabla^2 T(t,\xi) + u(t,\xi), \\ T(t_0,\xi) &= T_0(\xi), \\ T(t,\Omega) &= 0. \end{cases} \tag{1.1}$$

where c is a positive constant, ∇^2 is the Laplacian operator and $u(t,\xi)$ is a spatially (shown by its dependency on ξ) temporally (shown by its dependency on t) distributed input quantity. The discretisation of the PDE in Equation (1.1) can be obtained by finite-differencing in both time and space. This gives rise to a lumped parameter system model. When considering a uniform two-dimensional spatial grid of size $N \times N$, the discrete spatial-temporal model of Equation (1.1) becomes,

$$\begin{cases} T_{i_1,i_2}(k+1) &= (1+4\alpha)T_{i_1,i_2}(k) - \alpha \sum_{(\bar{i}_1,\bar{i}_2)\in\mathcal{N}_{(i_1,i_2)}} T_{\bar{i}_1,\bar{i}_2}(k) + \Delta t u_{i_1,i_2}(k), \\ T_{i_1,i_2}(0) &= T_0(i_1,i_2), \\ T_\Omega(k) &= 0. \end{cases} \tag{1.2}$$

where $\Delta t, \Delta\xi$ are the temporal and spatial discretisation steps, respectively, α is equal to $c\frac{\Delta t}{\Delta\xi^2}$ and $\mathcal{N}_{(i_1,i_2)}$ includes the four closest neighbours of node (i_1,i_2). This is a dynamic model (either in state-space form at or in difference equation format) that shows how local variables (temperatures) are modified by communication with its neighbours and the environment (input). Such dynamic models are local dynamic models, and the global model is a network between these local models. Their interaction can, for example, be displayed by considering the spatial-temporal impulse response of the dynamic systems. This is shown for two sets of model parameter values in Figure 1.5. Here, we restricted the spatial dimension to one in order to simplify the illustration.

In Figure 1.5, this spatial-temporal impulse response is displayed for a rod that is discretized spatially in 50 equidistant points (nodes). A non-zero initial condition is only provided at node 25 (while keeping the other initial conditions at zero).

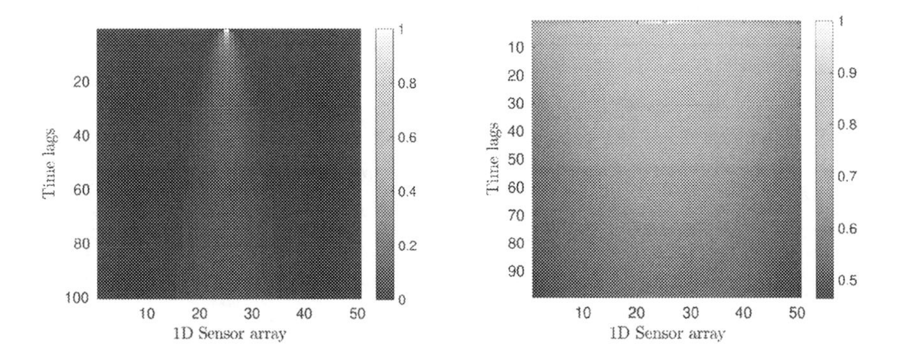

Figure 1.5 Spatial-temporal map for a one-dimensional string of subsystems $xi(k+1) = ax_i(k) + \alpha(x_{i-1}(k) + x_{i+1}(k))$, with (Left) $(a,\alpha) = (0.7, 0.15)$, and (Right) for a case with no spatial delay when propagating the state that corresponds to sequentially semi-separable systems that we will review in Chapter 3. Reprinted/adapted from Sinquin (2019) with permission from TU Delft

From Figure 1.5, the spatial and temporal dynamics become clear. This figure illustrates how fast information travels from one node to the other. The further away in both time and space, the less it matters to the local dynamics. The larger the value in the map (indicated by the lighter colour), the more the state of the neighbour (possibly in the past) contributes to the temperature of the system at node 25. The further away in time, the more neighbouring nodes contribute to the value of the state of the system at node 25.

To make the spatial-temporal connectivity analysis of this example more systematic, use can be made of the so-called funnel causality for a spatially-invariant system (that is when all subsystems in the grid are identical). This notion was introduced in Bamieh and Voulgaris (2005) and is a function defined for every possible distance between two nodes that is equal to the first time at which a node is affected by a change of another node located at a given distance. This distribution in time is fixed for the PDE in Equation (1.2) and resembles a cone. As the model in Equation (1.2) belongs to the class of sequentially semi-separable (SSS) systems (Rice and Verhaegen, 2009), Figure 1.5 shows that the wider the displayed cone (like in the figure in the right), the further away the local states are instantaneously shared with all the neighbours in that cone. This notion of funnel causality is closely related to the parametrization of the system matrices, some of which are detailed in Chapter 3, for the signal flow approach, and Chapter 4 for the network of local LTI systems approach.

As a conclusion to this introduction, the computational limitations of system identification algorithms in handling large input-output datasets (e.g. from large sensor arrays) have spurred the analysis of alternatives for deriving algorithms with linear computational complexity with respect to the number of sensor measurements for data-driven control. It is then relevant to introduce further parametrization on the matrices that stems from physical insights, for example whether the system is deterministic or stochastic, which is strongly linked to whether the network topology is known or not, or whether it has invariances, such that the best trade-off between scalability and model accuracy is found.

1.4 Organization of the Book

The core part of this book consists of nine chapters in addition to this introduction and the conclusion. There are three main parts. The first part (Chapters 2–5) discusses the models. The second part (Chapters 6–9) discusses identification methods. The third part (Chapter 10) illustrates the potential of some of the methods using the challenging case study of large-scale adaptive optics.

The chapters dedicated to the models start with a refresher, in Chapter 2, of different generic models that have been used for the identification of (classical) lumped parameter systems. These include input-output transfer function models and state-space models. As a preparation for network systems, a variant of input-output models indicated in the literature as dynamic or structure function (DNF or DSF) models is also discussed. Though these models have been introduced for the purpose of modelling dynamic networks, they can be used to model ordinary lumped parameter systems. The input-output models for dynamic networks from a *signal flow* perspective are presented in Chapter 3. Here, we start with sparse ARX models as a first simple model structure to deal with the complexity issue of large-scale networks. Along this line, sparse DNF/DSF models are discussed. An alternative way of reducing the number of parameters in the model is via tensor calculus. As an introduction to these models, a simpler variant is the use of Kronecker products. These model structures are introduced for VAR models and referred to as Kronecker-based VAR (Quarks) models. It should be noted that in this book we restrict ourselves to the VARX model class only. This is because the results for these model classes serve as an important platform for generalization towards variants such as VARMAX, Box–Jenkins, FIR, etc. Such generalizations are left as a problem for further research. The state-space modelling framework for networks of dynamical systems is presented in Chapter 4. To cope with the complexity of large network structures, various parametrizations of the system matrices of an LTI state-space model are discussed. Two general classes can be distinguished here. The first consists of state-space models where the system matrices are sparse. These include sparse parametrized state-space models, decomposable systems (Massioni and Verhaegen, 2009) and systems with block-tridiagonal system matrices. The second class includes state-space models where the system matrices are dense but sparsely parametrized. Examples are sequentially semi-separable matrices (Dewilde and van der Veen, 1998) or the use of sums of Kronecker products to model multi-dimensional state-space models. In Chapter 5, a brief analysis of the relationship between these different models and what happens when one tries to transform one model class into another is made. Of key importance is what happens, for example, with the sparsity structure of one model class, when transformed to another model class, as well as the consequences when discretizing continuous-time (structured) models. The latter topic matters when the a priori structural information about the model parametrization is given in continuous time.

The second part starts with the presentation of identification methods for the models discussed in Chapter 3. The approach taken in Chapter 6 is a global one: trying to identify the full models using all available data. For that reason, when interested in scalable algorithms, the sparsity in the model parametrization is of crucial importance. However, the consequence of sparsity constraints on the model coefficient matrices, which is a structural model constraint, is that the simplicity of estimating these coefficient matrices, such as with a simple linear least squares for lumped ARX models, is in general lost. The chapter focuses on the development of numerical methodologies that allow efficient data storage and calculation of the coefficient matrices. The spatial behaviour of the system is not viewed from the analysis of local subsystems that are related to each other according to a given topology. Instead use is made of the particular geometry or features of the sensor grid to induce structure in the coefficient matrices of the model. A key assumption in the models is that there are no unmeasured hidden quantities.

The identification of structured state space models is then considered in Chapters 7 and 8. These two chapters differ in the way sparsity is used. In Chapter 7, we develop methods for identifying the system dynamics under the assumption of separability in the horizontal and vertical dimensions. This is especially suited for large sensor grids, and then the state-space matrices are dense though data-sparse. In contast, in Chapter 8 sparsity is viewed in relation to the topology itself, for example when the subsystems are sparsely interconnected. The use of a state-space model naturally allows for the consideration of hidden state variables to be exchanged between the local models. For that reason, one of the key challenges in Chapter 8 is to identify a cluster of local models within a large network of dynamical systems.

The final chapter on identification methods is Chapter 9. Here, we present the identification of parametrized state-space models. In addition to the classical frameworks of maximum likelihood or expectation maximization methods, a new method is introduced that reduces a non-convex parameter optimization problem into a difference

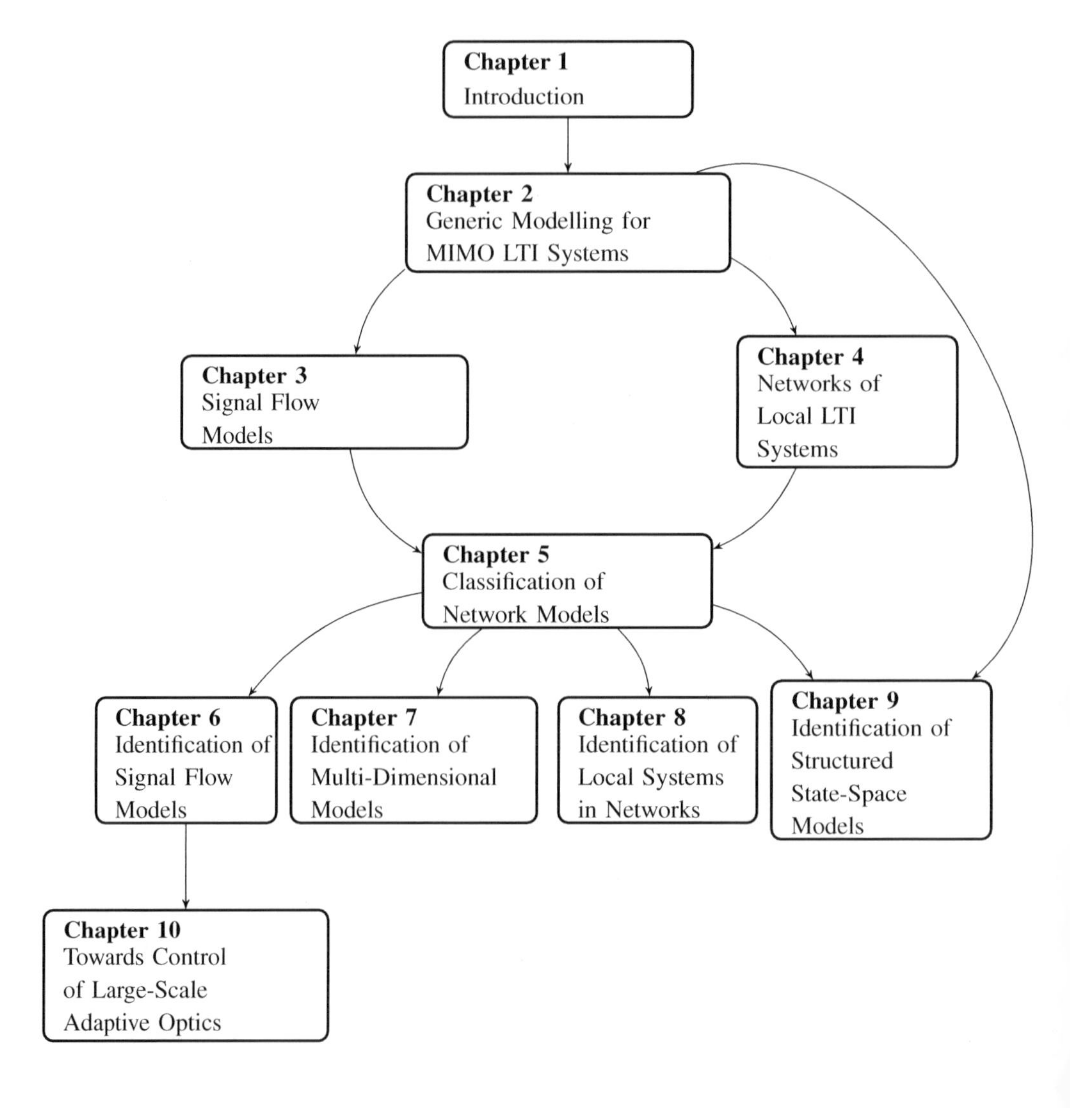

of convex program (DCP). The identification of structured state-space models, as described in Chapter 9, is the most general framework. It can be applied to every model structure discussed in Chapters 7 and 8, although definitely not with the same computational advantages as we have striven for in the previous chapters.

Chapter 10 considers in greater detail the application of some of the methods developed for the control of large scale AO.

Conclusions are presented in Chapter 11.

The organization of the book is presented in the flow chart above. The arrows indicate the suggested reading order for understanding particular topics in the book.

For example, when only interested in recent developments on estimating the parameters in parametrized state space models, treated in Chapter 9, the reader can, after the general outline in Chapter 2, directly jump to this chapter. When interested in the novel developments for identifying subsystems embedded in dynamic clusters, the reader best reads through the Chapters 4, 5 and 8. Chapters 3, 6 and 7 are dedicated to the derivations and discussions on the identification of multi-dimensional systems. The final concluding chapter is not listed in this chart.

Part I

Modelling Large-Scale Dynamic Networks

2 Generic Modelling for MIMO LTI Systems

2.1 Introduction

The dynamic relationship between (observed) input and output signals in a network can, at least on paper, be viewed as a dynamic relationship of a multi-input multi-output (MIMO) system. Within this book we restrict ourselves to LTI finite-dimensional MIMO systems. In principle, models that are used to represent these LTI finite-dimensional MIMO systems can also be used to model dynamic networks with linear dynamics. Though the large-scale nature generally prevents the construction/identification and use of such models with arbitrary dense coefficient/system matrices, in this chapter we give a brief overview of some widely used mathematical model representations for the generic MIMO setting. This review serves as a basis for later chapters, where these model classes are further detailed to represent networks of dynamic systems.

Two widely used modelling paradigms are presented for generic models of LTI finite-dimensional MIMO systems: transfer function models and state-space models. As prior information is often available in continuous time and this information can play a crucial role in identifying large-scale models, the identification problem is often addressed in continuous time. In this book we restrict ourselves mainly to the discrete-time case, and therefore the models presented in this chapter will be, apart from in the examples, in the discrete-time domain.

Section 2.2 presents general transfer function models. This includes the vector auto-regressive moving average models with exogenous inputs (VARMAX) and dynamic network/structure function models. The latter models, or simplifications thereof, will be used in the development of identification methods in Chapter 6. Section 2.3 presents state-space models in the so-called innovation form. Simplified versions of these models will be developed in Chapter 4.

2.2 Transfer Function Models

Transfer function models of finite-dimensional LTI single-input single-output (SISO) systems are defined based on rational functions of the form:

$$g(z) = \frac{b(z)}{a(z)},$$

where the numerator $b(z)$ and denominator $a(z)$ are polynomials in $z \in \mathbb{C}$. Such transfer functions are called *proper* when the degree of $b(z)$ is not greater than the degree of $a(z)$, and they are called *strictly proper* when the degree of $b(z)$ is strictly less than the degree of $a(z)$.

For multivariable systems we have to consider matrices with rational functions as entries. For the considered case, the transfer function matrix is then a rational function matrix $G(z)$. Such a rational function matrix is *proper* (or *strictly proper*) when all its entries are *proper* (*strictly proper*).

If we denote the rational function in the ith row and jth column of a matrix $G(z)$ by $G_{ij}(z)$[1] and given as:

$$G_{ij}(z) = \frac{ng_{ij}(z)}{dg_{ij}(z)}, \tag{2.1}$$

then the structure of this transfer function matrix is fully specified by all the orders of the numerators $ng_{ij}(z)$ and denominators $dg_{ij}(z)$. These orders are denoted, respectively, by $n_{ng}(i,j)$ and $n_{dg}(i,j)$, and they define the *structure* of the transfer function matrix $G(z)$. Such transfer functions can be used to represent the input-output relationship of LTI finite MIMO systems. Let q be the shift operator operating on time signals given by:

$$qy(k) = y(k+1), \tag{2.2}$$

and let the observable input be $u(k) \in \mathbb{R}^m$, the observable output be $y(k) \in \mathbb{R}^p$ and a (temporally) white-noise signal be denoted by $e(k) \in \mathbb{R}^p$ with covariance matrix Σ_e. Then, for system identification, we can denote such an input-output relationship as:

$$y(k) = G(q)u(k) + H(q)e(k). \tag{2.3}$$

The notation in the shift operator q specifies the model in the time domain, while, when taking $q = z$, the transfer functions are specified in the z-domain.

The full specification of the input-output model in terms of the general transfer functions requires the specification of the order of all the numerators and denominators of the entries of the rational function matrices $G(q)$ and $H(q)$. These are denoted by the following series of integers:

$$\{n_{ng}(i,j), n_{dg}(i,j)\}_{\substack{i=1:p \\ j=1:m}}, \quad \{n_{nh}(i,j), n_{dh}(i,j)\}_{\substack{i=1:p \\ j=1:p}}. \tag{2.4}$$

If normalizing the coefficient of $dg_{ij}(\infty) = 1$, then the number of parameters involved in describing these rational function matrices is:

$$\sum_{j=1}^{m}\sum_{i=1}^{p}\Big(n_{ng}(i,j) + n_{dg}(i,j) - 1\Big) + \sum_{j=1}^{p}\sum_{i=1}^{p}\Big(n_{nh}(i,j) + n_{dh}(i,j) - 1\Big). \tag{2.5}$$

[1] The same notation is used in the book to denote the matrix in the ith block row and the jth block column in a block matrix partitioning.

2.2.1 VARMAX Models

The rational matrix functions in Equation (2.3) can be written in factored form in terms of polynomial matrices. For that purpose, the following two definitions are introduced.

DEFINITION 2.1 (left coprime polynomial matrices) (Hippe and Deutscher, 2009) Two polynomial matrices $G_{D,L}(z)$ and $G_{N,L}(z)$ are left coprime if their greatest common divisor $U_L(z)$, such that

$$G_{D,L}(z) = U_L(z)\overline{G}_{D,L}(z), \quad G_{N,L}(z) = U_L(z)\overline{G}_{N,L}(z),$$

is a unimodular matrix; that is, the determinant of $U_L(z)$ is a non-zero constant independent of z.

DEFINITION 2.2 (left matrix fraction description) The left matrix fraction description (left MFD) of the $p \times m$ rational function matrix $G(z)$ is given as:

$$G(z) = G_{D,L}(z)^{-1}G_{N,L}(z),$$

where $G_{D,L}(z)$ and $G_{N,L}(z)$ are, respectively, a $p \times p$ and $p \times m$ polynomial matrices with $G_{D,L}(z)$ non-singular having a determinant that is not identical to zero, or equivalently the matrix $G_{D,L}(z)$ is non-singular for almost all values of z.

In general, it is required that in the left MFD, the factors $G_{D,L}(z)$ and $G_{N,L}(z)$ are left coprime. In that case the polynomial matrix $G_{D,L}(z)$ contains the information on the poles of the system, while $G_{N,L}(z)$ contains the information on the zeros.

From the generic input-output description in Equation (2.3), different special forms can be derived. A *first* special form is the so-called VARMAX model. When the compound rational matrix function $\begin{bmatrix} G(z) & H(z) \end{bmatrix}$ is given, we can define its left MFD as follows:

$$\begin{bmatrix} G(z) & H(z) \end{bmatrix} = A'(z)^{-1} \begin{bmatrix} B'(z) & C'(z) \end{bmatrix},$$

for $A'(z)$ and $\begin{bmatrix} B'(z) & C'(z) \end{bmatrix}$ left coprime. The existence of the inverse of $A'(z)$ makes this decomposition well posed with the matrix $A'(\infty)$ invertible. Exploiting this we may define the following alternative left MFD:

$$\begin{bmatrix} G(z) & H(z) \end{bmatrix} = \big(I - A(z)\big)^{-1} \begin{bmatrix} B(z) & C(z) \end{bmatrix}, \tag{2.6}$$

where $A(\infty) = 0$. Based on this left MFD, we can denote the generic input-output description in Equation (2.3) as:

$$y(k) = A(q)y(k) + B(q)u(k) + C(q)e(k). \tag{2.7}$$

The left MFD in Equation (2.6) can be considered as a *VARMAX realization* of the transfer function pair $\begin{bmatrix} G(z) & H(z) \end{bmatrix}$. In such a realization, relevant properties of the function pair, such as (strict) properness, may get lost and special care is necessary to preserve them. Two such properties are discussed next. First, the conditions on the polynomials $I_p - A(q), B(q), C(q)$ so that $(I - A(z))^{-1} \begin{bmatrix} B(z) & C(z) \end{bmatrix}$ is (strictly) proper. The second property is that of uniqueness.

To address the first issue, Lemma 2.1 is stated. Prior to that, we have two definitions standard in the analysis of polynomials. In this analysis (for convenience), we assume that both $(I - A(z))$ and $(B(z), C(z))$ only have positive or zero powers of z.

DEFINITION 2.3 (Row degree and row reduced) The degree of the ith row of a polynomial matrix $F(z)$ is the value of the degree of the highest power in z that occurs in that row. Let $\Gamma_r[F(z)]$ be the highest row degree coefficient matrix of which the ith row is zero, except in the jth column where j is the column number of that polynomial in the ith row with the highest power in z making the i, jth entry of the matrix $\Gamma_r[F(z)]$ equal to the coefficient of that highest power in z.

A square polynomial matrix $F(z)$ is row reduced if the matrix $\Gamma_r[F(z)]$ is nonsingular.

LEMMA 2.1 ((Strict) Properness of the right-hand side of Equation (2.6)) (Henrion and Šebek, 2000) *For a given left MFD as in Definition 2.2 with $G_{D,L}(z)$ row reduced, then the rational function matrix $G_{D,L}^{-1}(z)G_{N,L}(z)$ is strictly proper (resp. proper) if and only if each row of $G_{N,L}(z)$ has degree less than (resp. less than or equal) to the corresponding row degree of $G_{N,L}(z)$.*

The particular left MFD in Equation (2.6) is, by definition, already row reduced and therefore guaranteeing that its right-hand side is (strictly) proper requires the selection of the appropriate degrees of the polynomials in $B(z), C(z)$. From a system identification perspective, when n is the degree of $\det(G_{D,L}(z))$, or the order of the system, the degrees of the polynomials in $B(z), C(z)$ are chosen to be less or equal than n to guarantee properness of the right-hand side of Equation (2.6).

To make a (strictly) proper left MFD further unique, use can be made of the so-called Hermite form (Henrion and Šebek, 2000). This assumes that this form makes the polynomial matrix $(I - A(z))$ upper triangular. This shows that restricting $(I - A(z))$ to the Identity matrix multiplied by a scalar polynomial of degree n makes the parametrization unable to represent any pair of (strictly) proper matrices $\begin{bmatrix} G(z) & H(z) \end{bmatrix}$.

REMARK 2.1 VARMAX models have been used in the context of subspace identification of MIMO systems operating in a closed-loop (Houtzager et al., 2009a).

2.2.2 Dynamic Network/Structure Function Models

A second special form of the generic representation in Equation (2.3), which was recently proposed in Woodbury (2019), is the *dynamic network function* (DNF). Although this form has been studied mainly in the context of networks of dynamic systems, we will introduce this form as an alternative paradigm for modelling lumped parameter MIMO LTI systems in this chapter. Later on, in Section 3.4 of Chapter 3, we zoom in on using this modelling paradigm to model a dynamic network.

The DNF class models can be considered as a generalization of the VARMAX model by replacing the polynomial matrices $A(q)$ and $B(q), C(q)$ in Equation (2.7) by a strictly proper rational matrix $W(q)$ and proper rational matrices $V(q), T(q)$, respectively. This then yields the following DNF input-output model:

$$y(k) = W(q)y(k) + V(q)u(k) + T(q)e(k). \tag{2.8}$$

REMARK 2.2 In Woodbury (2019) the case for $W(q)$ to be proper is considered for dealing with a dynamic network containing feedback loops.

For the DNF model given by Equation (2.8) to be equivalent to the generic input-output model of Equation (2.3) special constraints are necessary. These include that the DNF model be well posed, meaning that it is necessary that the rational matrix function,

$$(I - W(z))^{-1}\begin{bmatrix} V(z) & T(z) \end{bmatrix},$$

exists. As shown in Woodbury (2019), this holds if and only if the rational matrix function $(I - W(z))$ is proper and invertible. Equivalently, the matrix $(I - W(\infty))$ should be invertible. The equivalence of the DNF model and the input-output model is then expressed by the following equality:

$$\begin{bmatrix} G(z) & H(z) \end{bmatrix} = (I - W(z))^{-1}\begin{bmatrix} V(z) & T(z) \end{bmatrix}. \tag{2.9}$$

To address the *DNF realization problem*, i.e. to derive the DNF triplet $(W(z), V(z), T(z))$ from a given the pair $(G(q), H(q))$, we write Equation (2.9) as:

$$\begin{bmatrix} W(q) & V(q) & T(q) \end{bmatrix}\begin{bmatrix} G(q) & H(q) \\ I_m & 0 \\ 0 & I_p \end{bmatrix} = \begin{bmatrix} G(q) & H(q) \end{bmatrix}. \tag{2.10}$$

The solution to this set of rational matrix equations in the frequency domain for $q = e^{j\omega}$ is discussed in Woodbury (2019). Here, the above equation is first transposed and then vectorized, yielding,

$$\underbrace{I_p \otimes \begin{bmatrix} G^T(e^{j\omega}) & I_m & 0 \\ H^T(j\omega) & 0 & I_p \end{bmatrix}}\text{vec}\left(\begin{bmatrix} W^T(e^{j\omega}) \\ V^T(e^{j\omega}) \\ T^T(e^{j\omega}) \end{bmatrix}\right) = \text{vec}\left(\begin{bmatrix} G^T(e^{j\omega}) \\ H^T(e^{j\omega}) \end{bmatrix}\right). \tag{2.11}$$

Based on Lemma 4.3.2 of Woodbury (2019), we can prove that the rank of the underbraced matrix in Equation (2.11) is equal to $p(p+m)$. The number of unknowns in this equation (for each ω) equals $2p^2 + pm$ and therefore we lack the p^2 equations required to make this set of equations have a unique solution. In Woodbury (2019), the number p^2 is called the *identifiability index*.

REMARK 2.3 (Network Identifiability) The identifiability index postulated in Woodbury (2019) is about identifiability of the network and not necessarily about the individual parameters in the individual transfer functions. Therefore we characterize this type of identifiability by *network identifiability* to distinguish it from parameter identifiability or structural identifiability as analysed, for example, in van den Hof (2002).

Therefore, to be able to define a bijective mapping between the DNF model matrices, $\begin{bmatrix} W(z) & V(z) & T(z) \end{bmatrix}$, and the given generic compound rational matrix functions, $\begin{bmatrix} G(z) & H(z) \end{bmatrix}$, further constraints on the DNF model are necessary.

This is far from trivial and, in general, depends on prior knowledge of the structure in the rational matrix function triplet $(W(z), V(z), T(z))$ of a DNF, such as an a priori known zero-pattern in these matrices. This topic is further discussed in Section 3.4. Here, we restrict ourselves to the definition of constraining DNF models to dynamic structure function (DSF) models (Gonçalves and Warnick, 2008).

These are derived by making the rational matrix function $W(z)$ of the rational matrix function triplet $(W(z), V(z), T(z))$ of a DNF 'hollow' (Woodbury, 2019), meaning to zero out the diagonal of $W(z)$. For computing this so-called hollow abstraction from a given DNF triplet $(W(z), V(z), T(z))$, we denote $W(q)$ as,

$$W(q) = W_O(q) + \text{diag}\big(W(q)\big). \tag{2.12}$$

When we consider the DNF triplet $(W(z), V(z), T(z))$ to be well posed, as stated in Section 2.2.2, it can be shown (see Theorem 3.4.7 of Woodbury [2019] that the matrix $(I_p - \text{diag}\,(W(q)))$ is proper and invertible. Hence, we can write Equation (2.8) as,

$$\begin{aligned} \big[I_p - \text{diag}\big(W(q)\big)\big]\, y(k) &= W_O(q)y(k) + V(q)u(k) + T(q)e(k) \\ y(k) &= \big[I_p - \text{diag}\big(W(q)\big)\big]^{-1}\, W_O(q)y(k) \\ &\quad + \big[I_p - \text{diag}\big(W(q)\big)\big]^{-1}\, V(q)u(k) \\ &\quad + \big[I_p - \text{diag}\big(W(q)\big)\big]^{-1}\, T(q)e(k) \\ y(k) &= Q(q)y(k) + P(q)u(k) + R(q)e(k), \end{aligned} \tag{2.13}$$

with the diagonal part of the proper rational matrix function $Q(q)$ being zero. Therefore, the DSF in Equation (2.13) adds p additional constraints to the problem of deriving the triplet $(Q(z), P(z), R(z))$ from a given pair $(G(q), H(q))$ compared to deriving the DNF triplet from that pair. Further constraints might be given for the special case $m = p$ by constraining the rational matrix function $P(z)$ to be diagonal. This would add a further $p^2 - p$ constraints.

An additional contribution that may help to formulate the necessary constraints to make the identifiability index zero is given when discussing sparse DNF/DSF in Section 3.4, and more precisely in Lemma 3.1.

2.3 State-Space Models

A finite-dimensional, discrete-time LTI state-space model in the so-called *innovation form* is given as (Verhaegen and Verdult, 2007):

$$\begin{aligned} x(k+1) &= Ax(k) + Bu(k) + Ke(k), \\ y(k) &= Cx(k) + Du(k) + e(k), \end{aligned} \tag{2.14}$$

for $x(k) \in \mathbb{R}^n$, $y(k) \in \mathbb{R}^p, u(k) \in \mathbb{R}^m, e(k) \in \mathbb{R}^p$ with n finite.

When the system matrix A is asymptotically stable, ensuring that there exist complex numbers $z \in \mathbb{C}$ outside the unit circle for which the expansion

$$(zI - A)^{-1} = z^{-1}I_n + z^{-2}A + z^{-3}A^2 + \cdots \tag{2.15}$$

is valid, then the transfer function pair $(G(q), H(q))$ in Equation (2.3) can readily be derived from the state-space model given in Equation (2.14) expressed in the shift operator q as

$$y(k) = \Big(C(qI - A)^{-1}B + D\Big)u(k) + \Big(C(qI - A)^{-1}K + I_p\Big)e(k). \tag{2.16}$$

The inverse problem of deriving a state-space model from the given pair $(G(q), H(q))$ is topic of the state-space realization problem of transfer functions. Such realization problems do not have a unique solution because Equation (2.16) is invariant under a state similarity transformation. One possible approach to such realization problems is to perform a Taylor series expansion of the given pair $(G(z), H(z))$ (similar to that given in Equation (2.15)) and to match these expansion coefficients to the so-called Markov parameters of the state-space model. The solution is based, for example, on Ho-Kalman realization theory and is outlined in Verhaegen and Verdult (2007).

2.4 Conclusions

The models presented in this chapter have been widely studied in an identification context, and powerful tools are available to identify such models from input-output data. In such identification methods, for example those presented in Ljung (1999) or Verhaegen and Verdult (2007), the coefficient matrices are, in general, unstructured, or their structure is not considered in the model identification. The structure of these coefficient matrices will turn out to become a crucial structural dimension that may complicate the identification solution drastically. In the next chapter, we discuss various approaches taken to define the structure of these coefficient matrices.

3 Signal Flow Models of Dynamic Networks

3.1 Introduction

The modelling paradigm presented in the previous chapter can, of course, be used for dynamic networks. However, when considering a large-scale network, a dimensionality problem arises when using full parametrizations for the coefficient matrices and we will therefore focus on special matrix structures of the coefficient/system matrices in these models. This chapter, which is a companion to Chapter 4, will address ways in which the number of parameters needed to represent the matrices of input-output models can be reduced. Two particular ways will be introduced to achieve this parameter reduction:

1. *Sparse coefficient matrices:* A reduction of the number of parameters necessary here is achieved by only considering the non-zero parameters to be estimated. A key challenge, of course, will be how to specify or find an appropriate non-zero pattern for the coefficient matrices.
2. *Dense but data-sparse coefficient matrices:* In this case, these matrices are completely full but can be specified by a reduced number of parameters. Examples from linear algebra are Toeplitz or Kronecker product matrices.

These two parameter reduction methodologies will be applied to the coefficient matrices of the transfer function models defined in Chapter 2. More precisely, we expand in this chapter on:

1. Sparse VARX models (Section 3.3).
2. Sparse dynamic/structure network models (Section 3.4).
3. Kronecker-product-based VAR models (Section 3.5).
4. Tensor VAR models (Section 3.6). For a brief introduction to tensor calculus, see Appendix B.

The first two model classes are examples of sparse coefficient matrix parameters, while the last two are examples of the dense, data-sparse way of parametrization. These model structures serve as a basis for defining a framework for defining input-output models for dynamic networks of LTI systems. In this book we will impose structural restrictions on the global network models as listed above. For these cases, identification methods will be developed later on in Chapter 6.

This Chapter is organized as follows. We start with a general introduction to how some basics of graph theory can be used to define structured matrices. Then, the above list of transfer function models are defined in the sections stated.

3.2 Graphs and Structured Matrices

To describe networks of all kinds such as in computer science, biology, social communication and matrix theory, frequent use is made of graph theory. For that reason, a few notions of graph theory will be summarized in this section. For more details, we refer to Diestel (1997).

A graph is visually represented by circles and connections between some of these circles. The circles are the *nodes* (or *vertices*) of the graph and the (connecting) lines are its *edges*. An example is given in Figure 3.1.

DEFINITION 3.1 (Graph) A simple graph consists of a non-empty set of nodes (or vertices) $\mathcal{V}$ and a set of edges $\mathcal{E} \subset \mathcal{V}^2$, which represents the connections between the nodes.

An element in the set $\mathcal{E}$ is a pair formed by elements of the set $\mathcal{V}$. When a and b are elements of that set, then an edge is, for example, denoted by (a,b). In a directed graph (or digraph) the order of elements in the pair defining an edge is important. Take the edge (a,b), then a is the initial node, or tail, while b is the terminal node, or head. In an undirected graph it is always assumed that when (a,b) is an edge, (b,a) is also an edge. Figure 3.1 is an example of an undirected graph. An example of a digraph is given in Figure 3.2.

DEFINITION 3.2 (Order of a Graph) The order of a graph defined by the set of nodes $\mathcal{V}$ and the set of edges $\mathcal{E}$ is the number of pairs in the set $\mathcal{E}$.

A common representation of a graph, apart from its graphical representation, is via its adjacency matrix.

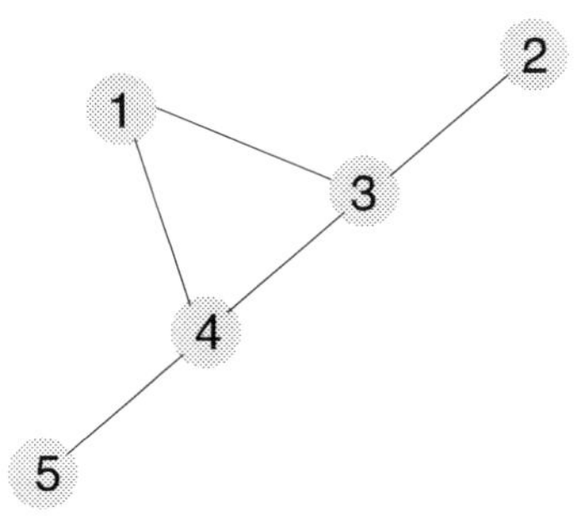

Figure 3.1 A simple undirected graph with nodes $\mathcal{V} = \{1,2,3,4,5\}$ and edges $\mathcal{E} = \{(1,3),(1,4),(2,3),(3,4),(4,5)\}$.

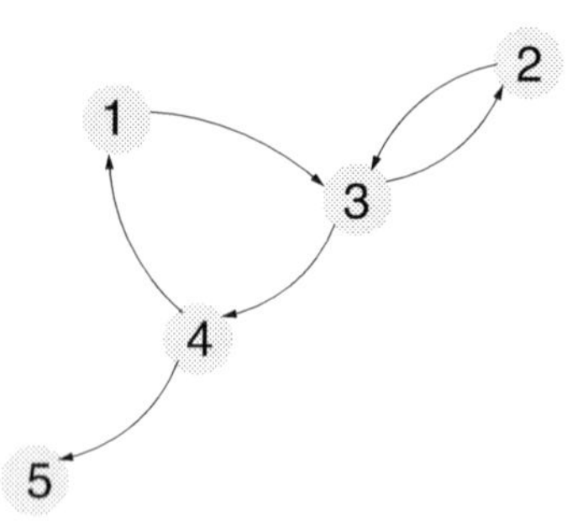

Figure 3.2 A directed graph, or digraph, with nodes $\mathcal{V} = \{1, 2, 3, 4, 5\}$ and edges $\mathcal{E} = \{(1, 3), (2, 3), (3, 2), (3, 4), (4, 1), (4, 5)\}$.

DEFINITION 3.3 (Adjacency matrix) The adjacency matrix A_G of a graph $\mathcal{G}$ with N nodes in the set $\mathcal{V} = \{v_i\}_{i=1}^N$ and edges $\mathcal{E}$ is defined as follows:

$$A_G(i, j) = \begin{cases} 1 & \text{if } (v_i, v_j) \in \mathcal{E}, \\ 0 & \text{otherwise.} \end{cases}$$

The adjacency matrices of the graphs in Figures 3.1 and 3.2 are on the left and right, respectively, below,

$$A_G = \begin{bmatrix} 0 & 0 & 1 & 1 & 0 \\ 0 & 0 & 1 & 0 & 0 \\ 1 & 1 & 0 & 1 & 0 \\ 1 & 0 & 1 & 0 & 1 \\ 0 & 0 & 0 & 1 & 0 \end{bmatrix}, \qquad A_G = \begin{bmatrix} 0 & 0 & 1 & 0 & 0 \\ 0 & 0 & 1 & 0 & 0 \\ 0 & 1 & 0 & 1 & 0 \\ 1 & 0 & 0 & 0 & 1 \\ 0 & 0 & 0 & 0 & 0 \end{bmatrix}. \tag{3.1}$$

The adjacency matrix will also be called the *pattern matrix*. The non-zero elements in such a matrix indicate a connection between the nodes (of a network) or the pattern of the network.

REMARK 3.1 (Ordering) The graph strongly depends on the ordering of the elements of the set $\mathcal{V}$ in Definition 3.1. A different ordering will result in a different adjacency matrix. This effect can be used for, example, to make the adjacency matrix banded.

There are many variants of the adjacency matrix, such as the normalized adjacency matrix or weighted adjacency matrix. The latter variant is used to represent edges in a digraph that have a weight. Such a weighted adjacency matrix is defined by the matrix A_G with entries:

$$A_G(i, j) = \begin{cases} w(v_i, v_j) & \text{if } (v_i, v_j) \in \mathcal{E}, \\ 0 & \text{otherwise.} \end{cases} \tag{3.2}$$

A weighted adjacency matrix is illustrated in Example 3.2.

Example 3.1 (Weighted adjacency matrix) Consider the digraph in Figure 3.3. Its weighted adjacency matrix is given in Equation (3.2).

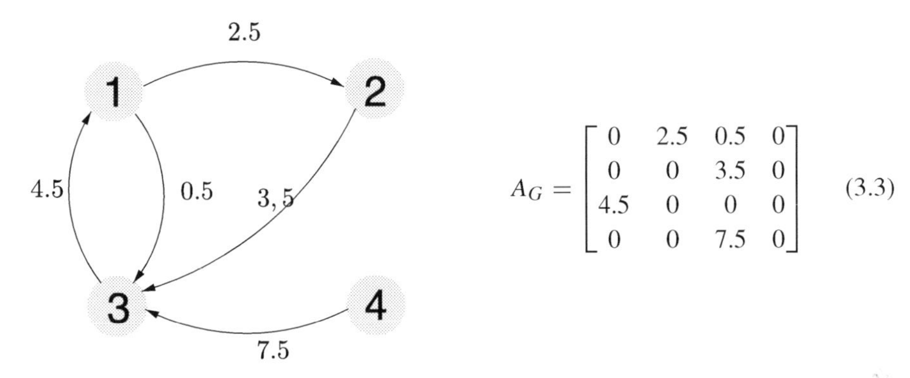

$$A_G = \begin{bmatrix} 0 & 2.5 & 0.5 & 0 \\ 0 & 0 & 3.5 & 0 \\ 4.5 & 0 & 0 & 0 \\ 0 & 0 & 7.5 & 0 \end{bmatrix} \tag{3.3}$$

Figure 3.3 A directed digraph with nodes $\mathcal{V} = \{1, 2, 3, 4\}$ and edges $\mathcal{E} = \{(1, 2), (1, 3), (2, 3), (3, 1), (4, 3)\}$.

These weighted adjacency matrices can be used to define sparse matrices or to parametrize such matrices. When in a graph, the node a has a *self-loop*, which is the set of edges containing the pair (a, a). Such self-loops define entries on the diagonal of the adjacency matrix.

3.3 Sparse VARX Models

A simplification of the VARMAX models, which is often used in system identification to initialize the VARMAX model parameters, is the VARX model (Houtzager et al., 2009a). This simplification with respect to Equation (2.7) is that $C(q) = I$. Although for small scale problems, the estimation of the coefficients of the polynomial matrices $A(q)$ and $B(q)$ can be done within the framework of ordinary linear least squares, this is no longer straightforward for large-scale problems. For that purpose, data-sparse parametrizations have been proposed in the literature to reduce the number of parameters in the coefficient matrices of the VARMAX model. Two such sparse model representations are presented in this book. In this section, we present sparse VARX models, while later, in Sections 3.5 and 3.6, VARX models are presented with dense coefficient matrices that are parametrized via a small number of parameters. These are so-called *Dense Data Sparse* model parametrizations.

Let the VARX model be denoted as,

$$y(k) = A(q)y(k) + B(q)u(k) + e(k), \tag{3.4}$$

where $e(k)$ again is a zero-mean white noise signal with covariance matrix Σ_e. Let the input and output vectors be explicitly denoted in terms of their components as:

$$u(k) = \begin{bmatrix} u_1(k) \\ \vdots \\ u_m(k) \end{bmatrix}, \quad y(k) = \begin{bmatrix} y_1(k) \\ \vdots \\ y_p(k) \end{bmatrix}. \tag{3.5}$$

There have been different attempts to use graph theory to represent the sparsity in VARX models. The difficulty to be overcome is that when considering the components of the signal vectors as nodes in the graph, these nodes become stochastic quantities rather than enumerated deterministic quantities.

We briefly discuss two generalizations using graph theory and restrict their definitions to VAR models only. Then, Equation (3.4) is explicitly denoted as:

$$y(k) = \sum_{i=1}^{n} A_i y(k-i) + e(k), \tag{3.6}$$

with the covariance matrix of $e(k)$, Σ_e, positive-definite.

DEFINITION 3.4 (Time Series (TS) chain graph for VAR) (Dahlhaus and Eichler, 2003) Let the VAR model in Equation (3.6) be given with the partitioning of $y(k)$ in its components as in Equation (3.5). Let the inverse of Σ_e be denoted by the matrix K_e, then the mixed graph (V_{TS}, A_{TS}, E_{TS}), with A_{TS} representing the directed edges and E_{TS} the undirected ones, is defined as follows:

$$\begin{aligned} V_{TS} = V \times \mathcal{Z} \quad & \text{for} \quad V = \{1, \ldots, p\}, \\ (i, t-\tau) \rightarrow (j,t) \in A_{TS} \quad & \Leftrightarrow \quad \tau \in \{1, \ldots, n\} \text{ and } (A_\tau)_{ji} \neq 0, \\ (i,t) \text{—} (j,t) \in E_{TS} \quad & \Leftrightarrow \quad K_{ij} \neq 0. \end{aligned}$$

The not defined edges are simply not there.

DEFINITION 3.5 (Granger Causality (GC) graph for VAR) (Dahlhaus and Eichler, 2003) Let the VAR model in Equation (3.6) be given with the partitioning of $y(k)$ in its components as in Equation (3.5). Let the inverse of Σ_e be denoted by the matrix K_e, then the mixed graph (V_{GC}, A_{GC}, E_{GC}), with A_{GC} representing the directed edges and E_{GC} the undirected ones, is defined as follows:

$$\begin{aligned} & V_{GC} = \{1, \ldots, p\}, \\ & (i,j) \notin A_{GC} \Leftrightarrow (A_\tau)_{ji} = 0 \ \forall \, \tau \in \{1, \ldots, n\}, \\ & (i,j) \notin E_{GC} \Leftrightarrow K_{ij} = 0. \end{aligned}$$

The main difference between the TS chain graph for VAR and the GC variant is that the first allows for a different sparsity pattern in each VAR coefficient matrix A_τ $(\tau \in \{1, \ldots, n\})$, while the second has a group sparsity pattern whereby all VAR coefficient matrices A_τ have the same sparsity pattern.

To represent both graphical models extended for the exogenous case, the following *generalized VARX* model is presented:

$$A_0^G y(k) = \sum_{i=1}^{n} A_i^G y(k-i) + \sum_{i=0}^{n} B_i^G u(k-i) + e_0(k), \tag{3.7}$$

for the matrix A_0^G lower triangular with unit entries on the diagonal. The covariance matrix of the white noise signal $e_o(k)$ is $\Sigma_0 > 0$, which is diagonal.

Transforming this generalized VARX into the original form like in Equation (3.6) for the non-exogenous case, the covariance matrix Σ_e can be expressed as:

$$\Sigma_e = \left(A_0^G\right)^{-1} \Sigma_0 \left(A_0^G\right)^{-T}, \tag{3.8}$$

yielding $K_e = \left(A_0^G\right)^T \Sigma_0^{-1} A_0^G$. The advantage of this generalized VARX model over the original one is that zeros in the matrix K_e, to express sparsity in the sense of the above two definitions, correspond to clear zero patterns in the matrix A_0^G. The same may hold for the ARX coefficients matrices via the relation $A_0^G A_i = A_i^G$.

3.4 Sparse Dynamic/Structure Network Models

The definition of sparse dynamic network models does not follow the lines of sparse ARMAX models. This is because the entries of $W(q), V(q), T(q)$ in Equation (2.8) are rational functions. Instead, it is common practice in the literature on DNF/DSFs to make use of signal flow graphs, see e.g. Gonçalves and Warnick (2008). This use is illustrated in Example 3.4.

Example 3.2 (Sparse DNF) Consider the signal flow graph depicted in Figure 3.4. In relation to the general DNF model in Equation (2.8), $p = m = 4$ and no noise is considered, i.e. $T(q) \equiv 0$. Considering the rules of signal flow graphs as, for example, summarized in section 3.9 of Ogata (2002), we can write this DNF as:

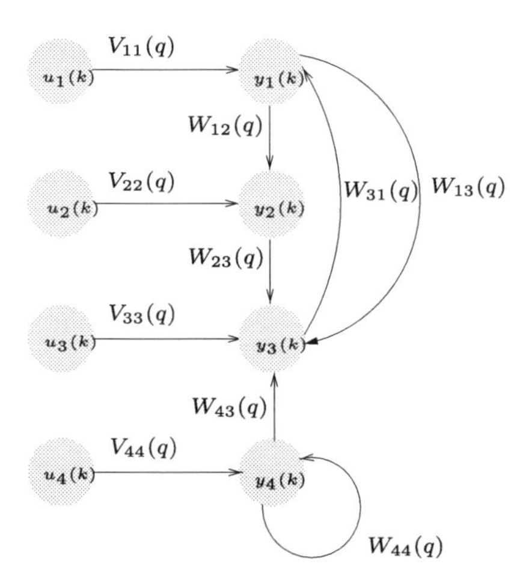

Figure 3.4 A signal flow graph of a dynamic network structure model.

$$\begin{bmatrix} y_1(k) \\ y_2(k) \\ y_3(k) \\ y_4(k) \end{bmatrix} = \begin{bmatrix} 0 & W_{12}(q) & W_{13}(q) & 0 \\ 0 & 0 & W_{23}(q) & 0 \\ W_{31}(q) & 0 & 0 & 0 \\ 0 & 0 & W_{43}(q) & W_{44}(q) \end{bmatrix} \begin{bmatrix} y_1(k) \\ y_2(k) \\ y_3(k) \\ y_4(k) \end{bmatrix}$$
$$+ \begin{bmatrix} V_{11}(q) & 0 & 0 & 0 \\ 0 & V_{22}(q) & 0 & 0 \\ 0 & 0 & V_{33}(q) & 0 \\ 0 & 0 & 0 & V_{44}(q) \end{bmatrix} \begin{bmatrix} u_1(k) \\ u_2(k) \\ u_3(k) \\ u_4(k) \end{bmatrix}, \tag{3.9}$$

where the non-zero transfer functions $W_{ij}(q)$ and $V_{ii}(q)$ are (strictly) proper, causal rational functions. From this equation, it becomes clear that the transfer matrix $W(q)$ can be considered as a generalized weighted adjacency matrix with weights given by rational transfer functions. The causal and strictly proper nature of the entries of these rational transfer functions displays the causal dependence of one quantity on the other.

In Section 2.2, it was shown that despite taking the transfer matrix $V(q)$ in Equation (2.8) to be diagonal, it was not sufficient to make the identifiability index 0. Introducing additional sparsity in the transfer matrix $W(q)$ in Equation (2.8) will further reduce the identifiability index. However, the strategic choices of introducing structure (for example, by placing zeros or known values) in the transfer function matrices so far is rather heuristic. A recent contribution (Gevers et al., 2019) on the unique identification of DSF models may be helpful. This result is stated in Lemma 3.1. This Lemma considers the following special case of the DSF (2.13):

$$y(k) = Q^0(q)y(k) + P^0(q)u(k) + v(k), \tag{3.10}$$

with all signals $y(k), u(k)$ and $v(k)$ p-dimensional, the matrix transfer function $Q^0(q)$ proper, hollow and satisfying $(I_p - Q^0(q))^{-1}$ being proper and stable. The noise term $v(k)$ is wide sense stationary with power spectrum $\Phi_v(z)$.

LEMMA 3.1 (Network Identifiability of DSF) (Gevers et al., 2019) *For the DSF satisfying the conditions stipulated in Equation* (3.10) *and, let from the related input-output model:*

$$y(k) = G^0(q)u(k) + w(k),$$

the transfer function $G^0(q)$ and the power spectrum $\Phi_w(z)$ be known. Further let the DSF be partitioned as:

$$\begin{bmatrix} y_1(k) \\ y_2(k) \end{bmatrix} = \begin{bmatrix} Q_{11}^0(q) & Q_{12}^0(q) \\ Q_{21}^0(q) & Q_{22}^0(q) \end{bmatrix} \begin{bmatrix} y_1(k) \\ y_2(k) \end{bmatrix} + \begin{bmatrix} P_{11}^0(q) & P_{12}^0(q) \\ P_{21}^0(q) & P_{22}^0(q) \end{bmatrix} \begin{bmatrix} u_1(k) \\ u_2(k) \end{bmatrix} + \begin{bmatrix} v_1(k) \\ 0 \end{bmatrix}, \tag{3.11}$$

with the power spectrum of $\Phi_v(z) = \begin{bmatrix} \Phi_{v_1}(z) & 0 \\ 0 & 0 \end{bmatrix}$, *and factorized as:*

$$\Phi_v(z) = H^0(z)\left[H^0(z^{-1})\right]^T. \tag{3.12}$$

Then, the triple $\left(Q^0(z), P^0(z), \Phi_v(z)\right)$ *can be uniquely derived from the pair* $\left(G^0(z), \Phi_w(z)\right)$ *provided that:*

$$\begin{bmatrix} P^0(z) & H^0(z) \end{bmatrix} \quad \textit{contains } L \textit{ known and linearly independent columns.} \tag{3.13}$$

Proof See proof of Theorem 4.1 of Gevers et al. (2019). □

REMARK 3.2 It is remarkable that the conditions for identifiability of DSF, apart from the hollow nature of the matrix transfer function $Q(q)$, are not further stipulated on this transfer function but on the input matrix transfer function $P(q)$ and the noise weighting matrix transfer function $R(q)$ in Equation (2.13). When $m = p$, Lemma 3.1 could be satisfied when either $P(q)$ or $R(q)$ is fully known. Taking, for example, $R(q) = I$, this means that the additive noise $v(k)$ in Equation (3.10) is zero-mean white noise with covariance matrix equal to the identity matrix. However, the Lemma allows for the additive noise $v(k)$ to have a singular spectrum, hence requiring a condition on the combination of $P(q)$ and $R(q)$.

3.5 Kronecker-Based VAR (Quarks) Models

To deal with large sensor arrays that give rise to a (stacked) vector $y(k)$ such as that in Equation (3.6) with large dimensions, it is crucial to reduce the number of coefficients to parametrize the coefficient matrices. Two such parametrizations that yield completely filled coefficient matrices but only require a small(er) number of parameters to describe these matrices are discussed in this section.

Example 3.3 (Regular 2D sensor array) In Chapter 10, an adaptive optics disturbance rejection problem is formulated that makes use of a Shack–Hartmann wavefront sensor to measure information about the wavefront disturbance in, e.g. ground-based telescopes (Roddier, 1999). Such a sensor consists of a 2D grid of lenslets and a CCD camera. Each lenslet, depicted by a gray spot in the schematic in Figure 3.5, measures the local gradient in the x and y directions of the wavefront (seen by the sensor). The depicted array consists of 32×32 sensors, each measuring two quantities. This amounts to 2 054 measurements at each time instance. This data may be stored in a matrix, for example, respecting the same column index and block-row index of the schematic in Figure 3.5. This then would yield at each time index a matrix of size 64×32.

Future telescopes will see an increase in the size of such a grid resulting in scalability issues of a minimum-variance controller. Although for telescopes the sensor measurements are usually only available on an annular grid, in this book we restrict ourselves to square grids. When the grid is not regular, or if there are missing measurements, interpolation techniques are often used to make the sensor array virtually regular.

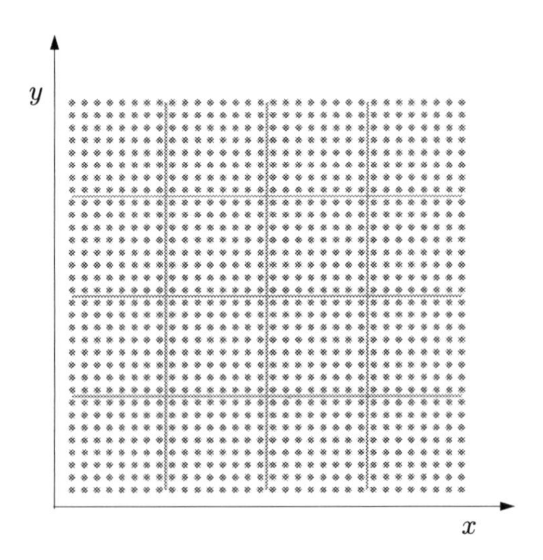

Figure 3.5 A schematic 2D sensor array where each dot represents a local sensor. The array consists of 32×32 sensors and the whole array is partitioned as indicated by the lines, with each partition consisting of 8×8 local sensors. Reprinted/adapted with permission from Sinquin and Verhaegen (2018) ©The Optical Society

Vectorizing the data matrix of a sensor array yields a data vector whose temporal dynamics may be described via the VAR model (3.6). The partitioning of the coefficient matrices of the VAR model can be exploited in representing such matrices by a sum of Kronecker products provided the sensor measurements stem from a regular grid and the system spatial dynamics are separable in the horizontal and vertical dimensions. This data-sparse representation is presented in the Lemma 3.2, where use is made of the following definition.

DEFINITION 3.6 (Re-shuffle operator (Loan and Pitsianis, 1992)) Let the indices $m_1, m_2, n_1, n_2 \in \mathbb{N}$, let $X \in \mathbb{R}^{m_1 m_2 \times n_1 n_2}$ and let $X_{i,j} \in \mathbb{R}^{m_2 \times n_2}$ partition X as:

$$X = \begin{bmatrix} X_{1,1} & \cdots & X_{1,n_1} \\ \vdots & \ddots & \vdots \\ X_{m_1,1} & \cdots & X_{m_1,n_1} \end{bmatrix}. \tag{3.14}$$

Then the reshuffling operator $\mathcal{R}(X) \in \mathbb{R}^{m_1 n_1 \times m_2 n_2}$ is defined as:

$$\mathcal{R}(X) = \begin{bmatrix} \text{vec}(X_{1,1})^T \\ \vdots \\ \text{vec}(X_{m_1,1})^T \\ \text{vec}(X_{1,2})^T \\ \vdots \\ \text{vec}(X_{m_1,n_1})^T \end{bmatrix}. \tag{3.15}$$

LEMMA 3.2 (Sum of Kronecker product (Loan and Pitsianis, 1992)) *Let $X \in \mathbb{R}^{m_1 m_2 \times n_1 n_2}$ be defined as in Definition 3.6, and let the SVD of $\mathcal{R}(X)$ be given as:*

$$\mathcal{R}(X) = \sum_{i=1}^{\min(m_1 n_1, m_2 n_2)} \sigma_i u_i v_i^T. \tag{3.16}$$

Let $U_i \in \mathbb{R}^{m_1 \times n_1}$ and $V_i \in \mathbb{R}^{m_2 \times n_2}$ be matrices such that $vec(U_i) = u_i$ and $vec(V_i) = v_i$, then,

$$X = \sum_{i=1}^{\min(m_1 n_1, m_2 n_2)} \sigma_i U_i \otimes V_i. \tag{3.17}$$

Proof The proof is straightforward based on the definition of the reshuffling operator and the existence of the SVD of a matrix (Golub and Van Loan, 1996). □

When the integer r satisfies,

$$r \leq \min(m_1 n_1, m_2 n_2),$$

we can approximate X by $\sum_{i=1}^{r} \overline{U}_i \otimes V_i$ for $\overline{U}_i = \sigma_i U_i$. The integer r will be called the Kronecker rank of a matrix in the class of matrices $\mathcal{K}_{2,r}$. The approximation of partitioned matrices with a sum of few Kronecker products has given rise to the definition of the *sum-of-Kronecker product matrices* class of matrices.

DEFINITION 3.7 (Sum-of-Kronecker Product Matrices (Sinquin, 2019)) The class of sum-of-Kronecker product matrices is defined as:

$$\mathcal{K}_{2,r} = \left\{ M \in \mathbb{R}^{m_1 m_2 \times n_1 n_2} \,|M = \sum_{i=1}^{r} M_{i,2} \otimes M_{i,1} \text{ for } M_{i,1} \in \mathbb{R}^{m_1 \times n_1}, M_{i,2} \in \mathbb{R}^{m_2 \times n_2} \right\} \tag{3.18}$$

The matrices $M_{i,1}$ and $M_{i,2}$ will be called the factor matrices.

DEFINITION 3.8 (Kronecker rank within $\mathcal{K}_{2,r}$) A matrix within $\mathcal{K}_{2,r}$ given by Equation (3.18) has Kronecker rank r when the following two matrices with r columns, with each column a vectorization of the matrices $M_{i,1}$ and $M_{i,2}$ respectively denoted as,

$$\begin{bmatrix} \cdots & \text{vec}\left(M_{i,1}\right) & \cdots \end{bmatrix} \quad \begin{bmatrix} \cdots & \text{vec}\left(M_{i,2}\right) & \cdots \end{bmatrix}$$

have full column rank r.

Lemma 3.2 shows that a low Kronecker-rank approximation of a matrix is equivalent to finding a low-rank approximation of the reshuffled matrix. The size of such rank is influenced by the size of the blocks of the matrix partitioning of the original matrix. Such partitioning is usually determined in engineering applications by the configuration of the used multi-dimensional sensor/actuator arrays.

REMARK 3.3 The class $\mathcal{K}_{2,r}$ contains the class of α-decomposable matrices. This class is discussed in Massioni (2014) as a generalization of the class of decomposable matrices defined in Definition 4.3.

In contrast to matrices belonging to the class of α-decomposable matrices, a sum-of-Kronecker product matrix does not require pre-knowledge of the adjacency matrix $\mathcal{P}$ (see Definition 4.3). Therefore, the topology of the network does not need to be known in advance. Moreover, the network can be composed of heterogeneous subsystems allowing for an arbitrary structure of the factor matrices. When describing large-scale networks, this structure may have the capacity to compress the data to represent the model in a significant manner. For example, when setting all the entries in quadruple (m_1, n_1, m_2, n_2) equal to N, N^4 entries are necessary to describe a generic matrix $M \in \mathbb{R}^{N^2 \times N^2}$. This compares to only $2rN^2$ elements when the matrix is described via a sum-of-Kronecker products.

Lemma 3.3 highlights the benefits of using the class of sum-of-Kronecker product matrices $\mathcal{K}_{2,r}$ to execute basic linear algebra operations efficiently. This efficiency is relevant both computationally and in terms of memory, and will be exploited in Chapter 6 for developing efficient identification methods for large-scale dynamic network systems.

LEMMA 3.3 (Computational Complexity in $\mathcal{K}_{2,r}$ (Loan and Pitsianis, 1992)) *Let $x \in \mathbb{R}^{N^2}$. Then, the orders of magnitude of the computational complexity for matrix-vector multiplication, matrix-matrix multiplication and matrix inversion are as follows:*

	$A, B \in \mathbb{R}^{N^2 \times N^2}$	$A, B \in \mathcal{K}_{2,r}$
Ax	$\mathcal{O}(N^4)$	$\mathcal{O}(rN^3)$
AB	$\mathcal{O}(N^6)$	$\mathcal{O}(r^2N^3)$
A^{-1} *(case: Kronecker rank of A is 1)*	$\mathcal{O}(N^6)$	$\mathcal{O}(N^3)$

The complexity obtained with the Kronecker-product parametrization considers the operations required for forming the factor matrices only.

REMARK 3.4 (Inversion of large-scale sum-of-Kronecker matrices) The approximation of the inverse of a large-scale sum-of-Kronecker matrix $A \in \mathcal{K}_{2,r}$ for the case where the Kronecker rank is larger than one is still an unresolved research topic for the general case. We refer to the work of Beylkin and Mohlenkamp (2005), Giraldi et al. (2014) and Varnai (2017) for more details.

Efficient linear algebra operations and compact data storage are possible when r is much smaller than N. The class of models with coefficient matrices that have this efficiency are the models of interest in this book. The coefficient matrices will then be referred to as *low Kronecker-rank matrices*.

Example 3.4 (Separability of the underlying function) Separability of the system function of spatial dynamics is essential for low Kronecker-rank matrices. Let a static mapping be represented by a smooth function from $\mathbb{R}^2$ to $\mathbb{R}$, and assume that this mapping is separable in both its coordinates. If this mapping describes the nature of a particular system mode, a basis for describing such a static system may be retrieved by concatenating columnwise the vectorized maps for all modes into

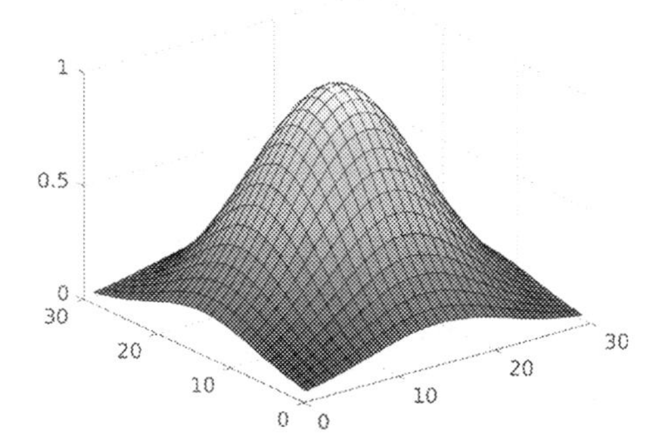

Figure 3.6 The depicted Gaussian function could represent the response of a single actuator (also called the influence function). The displacement on the vertical axis for a discrete 2D grid could be lifted into a column vector that determines a single column of the matrix on the left-hand side of Figure 3.7. When doing this for each actuator, we obtain the matrix representation in the latter figure. Reprinted/adapted from Sinquin (2019) with permission from TU Delft

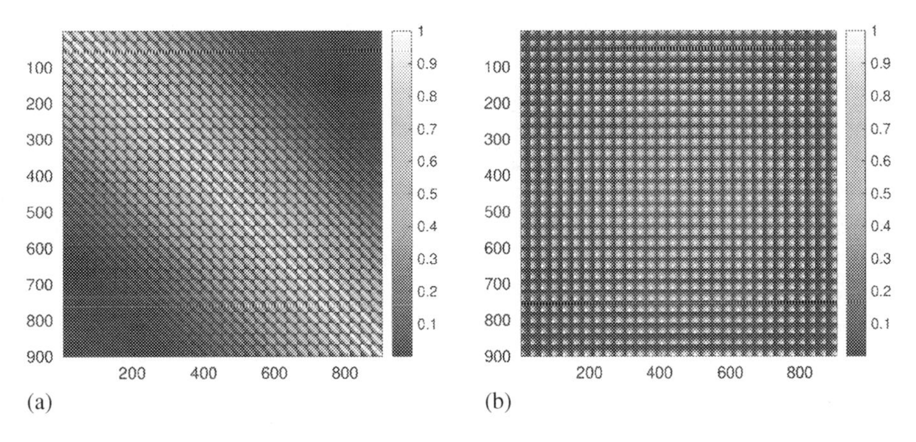

Figure 3.7 (a) A matrix with columns corresponding to the vectorization of a single mode of a deformable mirror. (b) The reshuffled version of the matrix on the left. Reprinted/adapted from Sinquin (2019) with permission from TU Delft

a matrix. Figure 3.6 illustrates a single mode of a deformable mirror. This figure represents the influence of deflecting one actuator by a unit step (and neglecting the temporal dynamics) on the wavefront aberation. An ideal Gaussian deflection is shown. The creation of the concatenated matrix with columns the vectorizations of all these modes (for all actuators) is displayed in Figure 3.7.

When there are temporal dynamics involved, the coefficient matrices of VARX models are parametrized with a sum of Kronecker products. This class of VARX models is indicated by Quarks models which is the prononciation of Kronecker VARX, although in the definition we restrict to VAR models only.

DEFINITION 3.9 (Quarks model (Sinquin and Verhaegen, 2019b)) Let the data vector $y(k) \in \mathbb{R}^{pN^2}$ consist of N column vectors of size pN stacked on top of each other. Then the Quarks model is:

$$y(k) = \sum_{i=1}^{n} \left(\sum_{j=1}^{r_i} M_{i,j,2} \otimes M_{i,j,1} \right) y(k-i), \tag{3.19}$$

with coefficient matrices $M_{i,j,1} \in \mathbb{R}^{pN \times pN}$ and $M_{i,j,2} \in \mathbb{R}^{N \times N}$.

The Quarks model can be presented in different ways. The first is by reversing the vectorization. For that purpose we assume that $y(k) = \text{vec}(Y(k))$ for $Y(k) \in \mathbb{R}^{pN \times N}$, and using the property of the vec operator that $\text{vec}(XYZ) = (Z^T \otimes X)\text{vec}(Y)$, we can write Equation (3.19) alternatively as,

$$Y(k) = \sum_{i=1}^{n} \sum_{j=1}^{r_i} \left(M_{i,j,1} Y(k-i) M_{i,j,2}^T \right). \tag{3.20}$$

When prior information is available about the factor matrices of the coefficient matrix of the VAR model, it may be used to parametrize them. This helps to either further reduce the computational complexity of the model identification step, or to cast the model into a structure useful for control.

Two examples of such coefficient matrix structures are given. The first example is that of banded, symmetric, Toeplitz and circulant matrices. Imposing such structures on the factor matrices is very attractive for reducing the number of parameters to represent these sum of Kronecker matrices. The Toeplitz block-Toeplitz structure arises, for example, when modelling 2D homogeneous spatially-invariant phenomena on a rectangular grid via PDEs. It should be said that the Kronecker and Toeplitz block-Toeplitz structures are related, but they are not equivalent.

LEMMA 3.4 (Relation between Toeplitz and low Kronecker-rank structures) *Let* $X \in \mathbb{R}^{N^2 \times N^2}$.

If X *is symmetric block-Toeplitz, then* X *has a Kronecker rank in* $\mathcal{K}_{2,r}$ *of at most* N.

If X *has a Kronecker rank of one, then this does not,* in general, *imply that* X *is block-Toeplitz nor that it has Toeplitz-blocks.*

Proof The first proposition is proved by using the reshuffling operator $\mathcal{R}$ defined in Definition 3.6 (for fixed dimensions). It is then observed that the Toeplitz-block structure is responsible for the rank constraint.

For the second point, the factor matrices may be, for example, randomly generated. □

The second example of structure in the coefficient matrices is that of a sparse matrix (with an unknown pattern of non-zero entries) or the sequentially semi-separable (SSS) matrix structure (defined in Definition 4.5). The SSS structure is more general compared to the Toeplitz matrix structure as it allows the modelling of spatially varying dynamics. Such parametrization is not affine. This may further complicate the computational aspects in algorithms for such systems.

3.6 Tensor VAR Models

Linear maps between signals that are static, such as in the case of the influence matrix in Figure 3.7 relating the actuators of a deformable mirror to the wavefront induced by that mirror and neglecting the temporal dynamics, are often highly structured. An illustration hereof is shown in Figure 3.7 (Left). In the general case, when f is a function that maps a set of variables $(x_1, x_2, \ldots, x_d)$ to $\mathbb{R}$, we may approximate this function with a sum of products between d functions ϕ_j^ℓ as:

$$f(x_1, \ldots, x_d) \approx \sum_{\ell=1}^{r} \prod_{j=1}^{d} \phi_j^\ell(x_j). \tag{3.21}$$

When considering indexed dimensions, such a mapping can equivalently be rewritten as a tensor F as follows:

$$F(i_1, \ldots, i_d) \approx \sum_{\ell=1}^{r} \prod_{j=1}^{d} v_j^\ell(i_j), \quad i_j = 1, \ldots, I_j, \tag{3.22}$$

where v_j^ℓ is a vector of appropriate size. The multivariate function approximation problem in Equation (3.21) and the tensor approximation problem in Equation (3.22) are equivalent; see Mohlenkamp (2013). The tensor F is able to approximate any static mapping f with finite error ℓ_2-norm when sufficiently increasing r in Equation (3.22). Such a tensor F is actually a generalization of the reshuffled matrix, which we formed by using the reshuffling operator $\mathcal{R}$ in Lemma 3.6. For example, when f is the Gaussian function depicted in Figure 3.6, which maps $\mathbb{R}^2$ to $\mathbb{R}$, the matrix F has rank one. Consequently, the parametrization of the system matrices with a sum of Kronecker products is a tensor approximation problem.

To make the above assertion explicit, we consider a function f that is separable in all its coordinates. Then this function f can be written as the Kronecker product with d factors, and the input and output signals are reshuffled into tensors of order d. When the sensor nodes are distributed over d spatial dimensions on a grid of d dimensions each with N (equidistant) nodes, the output vector is reshuffled into a tensor according to the grid dimensions. Alternatively, when the system is two-dimensional, the output signal may also be reshuffled into a tensor of order d such that the product of its dimensions is equal to pN^2. The trade-off between the bias and variance of the approximation can be regulated by selecting the Kronecker rank r.

The second reformulation of Equation (3.19) is to use the n-mode matrix tensor product formalism as defined in Definition B.4. This also prepares the generalization of the Quarks model within the framework of tensor calculus. Using this n-mode product, we can equivalently write Equation (3.19) as:

$$Y(k) = \sum_{i=1}^{n} \sum_{j=1}^{r_i} \Big(Y(k-i) \times_1 M_{i,j,1} \times_2 M_{i,j,2} \Big), \tag{3.23}$$

where the product between brackets is illustrated in Figure 3.8.

The Quarks model may have a reduced number of parameters as the model is *bilinear* in the coefficient matrices, as illustrated in Lemma 3.3. Therefore, the reduction in parameters will come at a cost of solving more difficult estimation problems than just linear regression. The full parametrization of such Quarks models in Equation (3.19) includes:

1. The temporal order index n.
2. The spatial order indices $\{r_i \mid i = 1 : n\}$.
3. A parametrization of the matrices $M_{i,j,1}$ and $M_{i,j,2}$ for $i = 1 : n$ and $j = 1 : r_i$. Such parametrization may, for example, be the specification of an affine dependence of these matrices on some parameter vectors $\theta_{i,j,1}$ and $\theta_{i,j,2}$ to define a (block) Toeplitz structure or banded/sparse matrix structure (Sinquin and Verhaegen, 2019b).

REMARK 3.5 (Multi-linear regression) Static variants of the temporally dynamic models that have been presented in this section, which only considers multilinear regression problems, have been proposed in the field of statistics, see e.g. Hoff (2015).

It will now be shown that the partitioning of a data matrix recorded by a sensor network influences the number of parameters required to parametrize models of DNS. The framework that is used for this purpose is *tensor calculus*. The method is first illustrated on an example to familiarize us with tensor notation, and the general tensor VAR model is then stated.

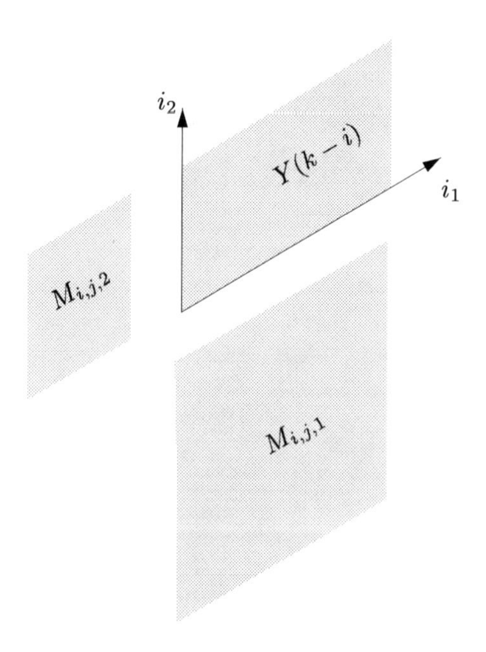

Figure 3.8 Visualization of the 2-mode product of a matrix $Y(k-i)$ (considered as a tensor) with two other matrices $M_{i,j,1}$ and $M_{i,j,2}$ as given in Equation (3.23). Here, i_1 is the 'ordinary' row index of the matrix and i_2 is the 'ordinary' column index of the matrix $Y(k-i)$.

Example 3.5 (Tensor VAR model) Consider the 12×9 sensor array in Figure 3.9 (Left). This array is partitioned as indicated by the straight lines (without dots) in this figure with each partitioning containing 3×3 local sensors. Each sensor provides at each time instance p data points, and the data matrix $Y(k)$ is constructed from these sensor read-outs as follows: $Y_{x,y}(k)$ is a p vector that contains the read-out of sensor with coordinates (x, y) in the sensor coordinate system (x, y) displayed in Figure 3.9 (Left). We now partition this data matrix according to the partitioning indicated by the straight lines in that figure. It should be remarked that such partitioning of a data matrix is not unique and might be the result from (engineering) insight in the data array. The sub-matrices that are defined by this partitioning can be stored in a tensor $\mathcal{Y}(k) \in \mathbb{R}^{J_1 \times \cdots \times J_4}$ with the four index set given by:

$$J_1 = J_2 = J_3 = \{1,2,3\}, \quad J_4 = \{1,2,3,4\},$$

such that,

$$n_1 \in J_1, \quad n_2 \in J_2, \quad n_3 \in J_3, \quad n_4 \in J_4.$$

The part of the tensor $\mathcal{Y}(k)_{:,:,n_3,1}$ that is formed of the first block-column of sensors is shown in Figure 3.9 (Right). A generalization of Equation (3.23) to describe the temporal dynamics of the tensor $\mathcal{Y}(k)$ uses the n-mode matrix tensor product defined by Definition B.4:

$$\mathcal{Y}(k) = \sum_{i=1}^{n} \sum_{j=1}^{r_i} \left(\mathcal{Y}(k-i) \times_1 M_{i,j,1} \times_2 \cdots \times_4 M_{i,j,4} \right). \tag{3.24}$$

The operation of the n-mode product is illustrated in Example B.4.

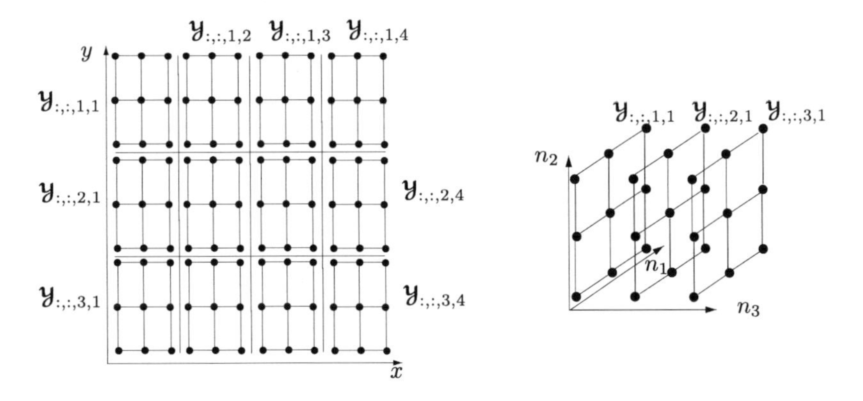

Figure 3.9 (Left) A schematic of a 9×12 2D sensor array where each dot represents a local sensor. The array is partitioned, as indicated by the lines connecting the dots, with each partitioning consisting of 3×3 local sensors. (Right) Illustration of part of the 4D tensor made of the first block column of the 2D sensor array. (The time dependence on the index k being neglected for the sake of clarity.)

The general format of the tensor VAR model, excluding the noise input, models the temporal dynamics of a tensor $\mathcal{Y}(k) \in \mathbb{R}^{J_1 \times \cdots \times J_d}$ time sequence as:

$$\mathcal{Y}(k) = \sum_{i=1}^{n} \sum_{j=1}^{r_i} \Big(\mathcal{Y}(k-i) \times_1 M_{i,j,1} \times_2 \cdots \times_d M_{i,j,d} \Big). \tag{3.25}$$

The matrices $M_{i,j,1}, \ldots, M_{i,j,d}$ are also called the factor matrices.

Using Proposition B.1, we can rewrite this equation as:

$$\mathrm{vec}\Big([\mathcal{Y}(k)]_{(1)}] \Big) = \sum_{i=1}^{n} \sum_{j=1}^{r_i} \Big(M_{i,j,d} \otimes \cdots \otimes M_{i,j,1} \Big) \mathrm{vec}\Big([\mathcal{Y}(k-i)]_{(1)}] \Big). \tag{3.26}$$

This shows the generalization with respect to the Quarks model formalism in Equation (3.19). Compared to this model, the tensor variant has the additional index d that depends on the way the tensors have been constructed (see the above Example 3.6). The effect of the Kronecker rank r_i and the number of factor matrices in approximating a coefficient matrix on the accuracy is discussed in Example 3.6 for a static regression problem.

Example 3.6 (Impact of Kronecker rank and partitioning on the modelling accuracy (based on example 6.1 of Sinquin [2019])) Let us consider the following matrix $M \in \mathbb{R}^{N^2 \times N^2}$ for $N = 32$:

$$m_{p,q} = e^{-\frac{(i-k)^2 + (j-\ell)^2}{\sigma^2}}, \tag{3.27}$$

for $p = (i-1)N + j$ and $q = (k-1)N + \ell$, with $(i, j, k, l) \in \{1, \ldots, N\}^4$ and $\sigma = 0.54$. The latter parameter is chosen to avoid the matrix M being dominantly sparse. The matrix M represents a frequently used static model of the influence of a deformable mirror on the wavefront in an adaptive optics problem (Roddier, 1999). To approximate this matrix with a sum of Kronecker matrices and a particular tensorization of the data vectors, the following multi-linear regression problem is defined:

$$\min_{M_{j,n}} \sum_{k=1}^{N_t} \| y(k) - \sum_{j=1}^{r} (M_{j,d} \otimes \cdots \otimes M_{j,1}) u(k) \|_2^2. \tag{3.28}$$

Here, $y(k)$ is generated as $Mu(k)$ – where M is the true model defined in Equation (3.27) – and $u(k)$ is randomly generated for $k = 1 \ldots N_t$ (with $N_t = 500$). Determination of the factor matrices $M_{j,n}$ requires the solution of a multilinear optimization problem (Shen et al., 2016). Here, use is made of alternating least squares as outlined in more detail in Section 6.3.1. A total of 20 iterations were performed for each test. The estimated factor matrices are denoted by $\widehat{M}_{i,j,n}$, and the estimated global matrix by $\widehat{M}$.

Table 3.1 Influence of the Kronecker rank r on the relative root mean square error between vec(M) and vec($\widehat{M}$).

Kronecker rank, r	1	3	5
32×32	$7.07 \cdot 10^{-15}$	$1.34 \cdot 10^{-12}$	$2.66 \cdot 10^{-13}$
$32 \times 8 \times 4$	$1.50 \cdot 10^{-1}$	$1.5 \cdot 10^{-1}$	$1.95 \cdot 10^{-3}$
$16 \times 16 \times 2 \times 2$	$3.64 \cdot 10^{-1}$	$2.78 \cdot 10^{-1}$	$2.67 \cdot 10^{-1}$
$8 \times 8 \times 4 \times 4$	$2.70 \cdot 10^{-1}$	$1.18 \cdot 10^{-1}$	$9.18 \cdot 10^{-2}$

Table 3.2 Akaike-like information criteria for Equation (3.29) for the different model structures (in *log*10).

Kronecker rank, r	1	2	3
32×32	0.61	0.91	1.09
$32 \times 8 \times 4$	3.84	2.09	1.05
$16 \times 16 \times 2 \times 2$	4.61	4.37	3.86
$8 \times 8 \times 4 \times 4$	4.35	3.62	3.31

The relative root mean square error (RMSE) is computed between vec(M) and vec$(\widehat{M})$ and is displayed in Table 3.1. The exponential function in Equation (3.27) is separable in both its coordinates (horizontal and vertical). As a consquence, this matrix M can be decomposed exactly into a single Kronecker product of two matrices. This explains why the algorithm converges globally. In practice, such exactness does not hold in general due to, for example, the non-separability of the function f or the fact that the entries of the associated matrix are derived from noisy measurements.

When increasing the tensor order, we are looking for a trade-off between model accuracy and model complexity. Such compromise is illustrated in Table 3.2. Here we use an Akaike-like information criteria (AIC) for comparing these different approximations (Ljung, 1999),

$$\mathrm{AIC} = \frac{1}{N_t}\left(R + dr\sum_{i=1}^{d} J_i^2\right), \tag{3.29}$$

where R is the residual of the cost function in Equation (3.28) and J_i is the size of the ith dimension of the tensor in $\mathbb{R}^{J_1 \times \cdots \times J_d}$. The smaller the value of AIC, the better the model approximation. Owing to the separability of Equation (3.27), the optimal model structure corresponds to $(r,d) = (1,2)$. Among the decompositions for $d > 2$, the configuration $(r,d) = (3,3)$ is best from an AIC perspective. In Figures 3.10 and 3.11, we display the matrix $\widehat{M}$ derived from solving Equation (3.28) for different choices of the dimensions of the tensor. A particular block structure can be observed

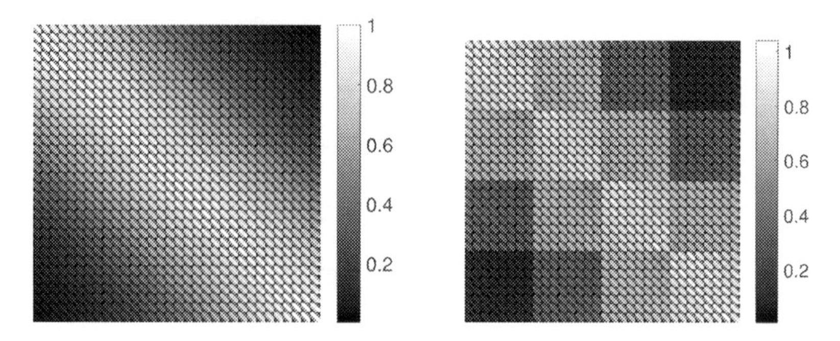

Figure 3.10 Entries of the estimated matrix $\widehat{M}$ with: (Left) $d = 2$ and $\mathcal{Y}(k) \in \mathbb{R}^{32\times 32}$; (Right) $d = 3$ and $\mathcal{Y}(k) \in \mathbb{R}^{32\times 8\times 4}$. Reprinted/adapted with permission from Sinquin and Verhaegen (2018) ©The Optical Society

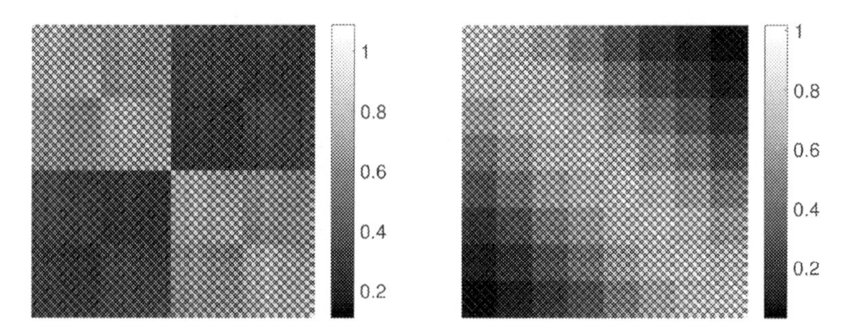

Figure 3.11 Entries of the estimated matrix $\widehat{M}$ with: (Left) $d = 4$ and $\mathcal{Y}(k) \in \mathbb{R}^{16\times 16\times 2\times 2}$; (Right) $d = 4$ and $\mathcal{Y}(k) \in \mathbb{R}^{8\times 8\times 4\times 4}$. Reprinted with permission from Sinquin and Verhaegen (2018) ©The Optical Society

for all the matrices when $d > 2$ and $r = 1$. The size of these blocks gets smaller when increasing the Kronecker rank.

The parametrization of a dense data-sparse VAR(X) model does not require the specification of a network topology, e.g. in terms of a network adjacency matrix or pattern matrix. Example 3.6 has, however, shown that these dense data-sparse models may reveal the sparsity pattern in the network as it models the dynamic connection between different signals (that are each associated to a node).

3.7 Conclusions

In this chapter, we have proposed to cope with the large-scale nature of the dynamic networks by enforcing structure in the coefficient matrices of the models in the signal flow class.

The measurements signals are vectorized, yielding large dynamical input-ouput equations that were either formulated with transfer functions or VAR models.

A parametrization is then enforced to ensure scalable representations in terms of data storage, that later on will be exploited for developing scalable estimation methods. Prior knowledge of how the subsystems interact with each other may be exploited, as is the case for sparse VARX models or sparse DNF. When this is unknown, a dense data-sparse representation is preferable. One example of this is the class of low Kronecker-rank matrices which offers high data compression for large sensor grids. This was generalized to higher dimensions by using a Kronecker product between more than two matrices. Such generalization yielded higher compression rates. Such generalization is obtained for 2D arrays by tensorizing the sensor data, allowing the establishment of a trade-off between compactness of the model on the one hand, and its accuracy on the other hand.

Another example, that is not further investigated here, includes the Toeplitz structure (Songsiri and Vandenberghe, 2010). Reducing the number of parameters in the system matrices may narrow down the set of model candidates, which would no longer be able to represent the temporal dynamics accurately. It is then useful to expand these parametrizations. This is, e.g. the case for invariant systems with boundaries that translate into a *Toeplitz plus Hankel* matrix structure (Rojo, 2008) (Epperlein and Bamieh, 2016). The *low Kronecker-rank plus low-rank* matrix structure presents interesting computational properties. For example, it is useful for multi-dimensional systems that have some dynamics best represented on a zonal basis, for which the Kronecker format is suited; and have other dynamics governed by only a few modes corresponding to low spatial frequencies.

4 Models of Networks of Local LTI Systems

4.1 Introduction

We start here by developing model structures for a particular network configuration that consists of a set of interconnected local systems, each governed by an LTI model description. The connection between these local systems represents the signal exchange between these networks. It will be assumed that (a major part of) this local exchange between networks is not measured. These local systems and their interconnections can then be lifted into a global model. In this way, this modelling approach is bottom-up, starting from the local system towards the global model structure; whereas the models discussed in the previous chapter may be considered as a top-down modelling approach.

The combination of local models and missing information about the exchange in between these local models make the modelling approach highly suitable for using state-space model structures. The challenge lies in devising generic parametrization methodologies (for interesting classes of dynamical networks) that allow us to substantially reduce the number of parameters for the system matrices. As in the Chapter 3, the same two categories will be studied:

1. *Sparse system matrices:* Three types of sparse system matrices for state-space models will be considered. First, an a priori sparse system matrix specified e.g. based on first principle information. Second, the class of so-called decomposable systems is reviewed that is of interest in controlling the formation of dynamical systems introduced in Massioni and Verhaegen (2009). Third, state-space models where the exchange is limited to two neighbouring systems only. We restrict ourselves to strings of subsystems, so that the system matrices have a block-tridiagonal structure.
2. *Dense but data-sparse coefficient matrices:* The two types considered are system matrices represented by the matrix class of sequentially semi-separable (SSS) matrices and multidimensional systems where the system matrices are sums-of-Kronecker product matrices. The first class is again narrowed down to interconnected strings as system identification methods will only be developed for these.

The organization of the sections is that the above two parts are treated in Sections 4.2 and 4.3, respectively.

4.2 Sparse State-Space Models for Networks of Dynamic Systems

Three different ways are presented to arrive at sparse system matrices in a state-space model of a network connection of dynamic systems. These are indicated as 'a priori parametrized state-space models', 'decomposable systems' and 'state-space models with block-tridiagonal system matrices'.

4.2.1 A Priori Parametrized State-Space Models

Attempts to find structure in LTI models of dynamical systems often rely on state-space models derived from first principles, such as conservation laws or constitutive equations. Systematic computer-aided design tools are readily available; see for example Fritzson (2011). In general, mathematical models derived in this way and formulated as state-space models are sparse. This is illustrated in the following example.

Example 4.1 (DC Servomotor) This is a classical example in the modelling of dynamical systems. It consists of an electrical part governed by the electrical constitutive equations to model the relationship between the voltage $u(t)$ and the motor shaft angle $\phi(t)$. Following the notation introduced in Figure 4.1 and modelling the electromotive force $V_b(t)$ as $K\frac{d\phi(t)}{dt}$, this electrical part is modelled as:

$$u(t) = Ri(t) + L\frac{di(t)}{dt} + K\frac{d\phi(t)}{dt}, \tag{4.1}$$

where K is the motor's back electromotive force constant. When the electrical current $i(t)$ causes an applied torque to the shaft with moment of inertia J equal to $K_T i(t)$, Newton's second law for rotational motion describes the mechanical part of the DC motor as:

$$J\frac{d^2\phi(t)}{dt^2} + f\frac{d\phi(t)}{dt} = K_T i(t) - T_\ell(t), \tag{4.2}$$

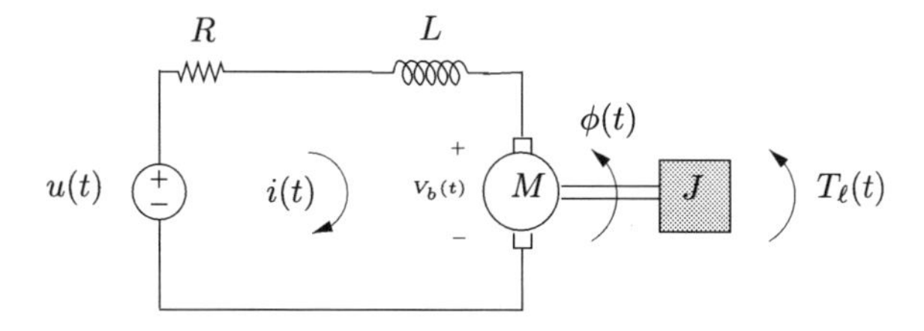

Figure 4.1 Electrical scheme representation of a DC motor connected to a rotating mass with inertia J.

where $T_\ell(t)$ is the torque exercised by the load of the DC motor. The resulting state-space equations can be derived by combining Equations (4.1) and (4.2) as:

$$\begin{cases} \frac{di(t)}{dt} = -\left(\frac{R}{L}\right)i(t) - \left(\frac{K}{L}\right)\frac{d\phi(t)}{dt} + \frac{1}{L}u(t) \\ \frac{d^2\phi(t)}{dt^2} = -\left(\frac{f}{J}\right)\frac{d\phi(t)}{dt} + \left(\frac{K_T}{J}\right)i(t) - \frac{1}{J}T_\ell(t) \end{cases} \tag{4.3}$$

When the state is defined as:

$$x(t) = \begin{bmatrix} i(t) \\ \phi(t) \\ \frac{d\phi(t)}{dt} \end{bmatrix}, \tag{4.4}$$

and the output is a measurement of the rotation angle $\phi(t)$, we can write the combination of Equations (4.1) and (4.2) as the following linear continuous time state-space model:

$$\begin{cases} \dot{x}(t) = \begin{bmatrix} -\frac{R}{L} & 0 & -\frac{K}{L} \\ 0 & 0 & 1 \\ \frac{K_T}{J} & 0 & -\frac{f}{J} \end{bmatrix} x(t) + \begin{bmatrix} \frac{1}{L} \\ 0 \\ 0 \end{bmatrix} u(t) + \begin{bmatrix} 0 \\ 0 \\ -\frac{1}{J} \end{bmatrix} T_\ell(t), \\ y(t) = \begin{bmatrix} 0 & 1 & 0 \end{bmatrix} x(t) + v(t), \end{cases} \tag{4.5}$$

where $v(t)$ represents the measurement noise. When there is no load, i.e. $T_\ell(t) \equiv 0$, and defining the parameter vector $\theta^T = \begin{bmatrix} -\frac{R}{L} & -\frac{K}{L} & \frac{K_T}{J} & -\frac{f}{J} & \frac{1}{L} \end{bmatrix}$, we can write this state-space model compactly as:

$$\begin{cases} \dot{x}(t) = A(\theta)x(t) + B(\theta)u(t), \\ y(t) = C(\theta)x(t) + v(t). \end{cases} \tag{4.6}$$

For this SISO case, the minimal number of parameters for the input-output transfer function is $2n + 1$ where n is the order of the state-space model (which is 3 in this case). Therefore, the size of the parameter vector θ is smaller than the minimal number required to model the transfer from $u(k)$ to $y(k)$.

It should be remarked that the system matrices are affine in the components of θ, but not in the physical parameters R, L, K, K_T, J, f. However, the special relationship with the coefficients of θ and these physical parameters allow us (in this special case) to find the physical parameters once the parameter vector θ has been estimated.

Neglecting the external influences, such as the load torque in Example 4.1, may allow a first principles model (when the dynamics are linear) to yield parametrized state-space models like that of Equation (4.6). Whether such parameters can be found uniquely via system identification is analysed within the context of *structural (or parameter) identifiability*. This is, in general, a complex problem that has mainly been addressed for specific model classes. For example, for the analysis of the identifiability of the parameters in compartmental models to be derived from the Markov parameters of the state-space model we refer to van den Hof (2002). The general representation of a structured state-space model may then read like Equation (4.6) with the system matrices $\big(A(\theta), B(\theta), C(\theta)\big)$ each defined as a weighted adjacency matrix corresponding to a particular graph.

So far, models have been presented in discrete time. However, the first principles models often used in retrieving a priori information about the structure of network models are specified in continuous time, like that in Example 4.1. Their translation to discrete time may lead to additional difficulties. For example, it may lead to ambiguities such that two different continuous-time models discretized by the same discretization rule give rise to the same discrete time model, see e.g. in section 4.6 of Ljung (1999).

Furthermore, the sparsity of the continuous-time model may get lost by discretizing the model. These problems may be alleviated by two additional measures in system identification. First, is that the selection of a sufficiently small sample period such that simple integration rules like Euler's integration yields suficiently small approximation errors. Second, is that by developing dedicated continuous-time identification methods, such as those in Yu et al. (2017). For that reason, we mainly continue in a discrete-time setting.

A more systematic way of finding a structured state-space model for a *network of dynamic systems* (NDS) makes use of block diagrams to represent the network of systems. These are discussed, for example, in section 2.3 of Ogata (2002). Their focus is on the connection of systems rather than signals themselves, like in the signal flow graph. As the boxes in a block diagram contain system descriptions, they may also be more amenable for modelling (structured) state-space models wherein typically the state (signal) is not completely measured. An example of such a NDS model giving rise to a structured state-space model is given in the next example.

Example 4.2 (Block diagram versus signal-flow diagram) Consider the network of three systems S_i for $i = 1, 2, 3$ in Figure 4.2(a). The three boxes represent these systems and edges in this figure represent the signals by which these boxes interact. This would be opposite of a signal flow graph that is depicted for this same network in Figure 4.2(b). Here the nodes (circles) represent the signals, such as y_i for $i = 1, 2, 3$ while the edges represents the dynamic link between these signals. These dynamic links are then the systems in the network. This is illustrated for the example in Figure 4.2(a) using a state-space framework. Let the system indicated by S_1 be described by the local state-space model:

$$\begin{cases} x_1(k+1) = A_1 x_1(k) + B_1 y_2(k), \\ \quad\;\; y_1(k) = C_1 x_1(k) + u_1(k) + v_1(k). \end{cases} \tag{4.7}$$

Similarly, the other two local systems S_2 and S_3 can be represented by a state equation using the triplets (A_2, B_2, C_2) and (A_3, B_3, C_3), respectively. When we lift the local state vectors $x_i(k)$ into the global state vector $x(k) = \begin{bmatrix} x_1(k)^T & x_2(k)^T & x_3(k)^T \end{bmatrix}^T$, and in a similar way define the global input $u(k)$ as $u_1(k)$, the global noise $w(k) = \begin{bmatrix} v_1(k)^T & v_2(k)^T \end{bmatrix}^T$ and the global output $y(k) = \begin{bmatrix} y_1(k)^T & y_2(k)^T & y_3(k)^T \end{bmatrix}^T$, then we obtain the following (global) state-space model:

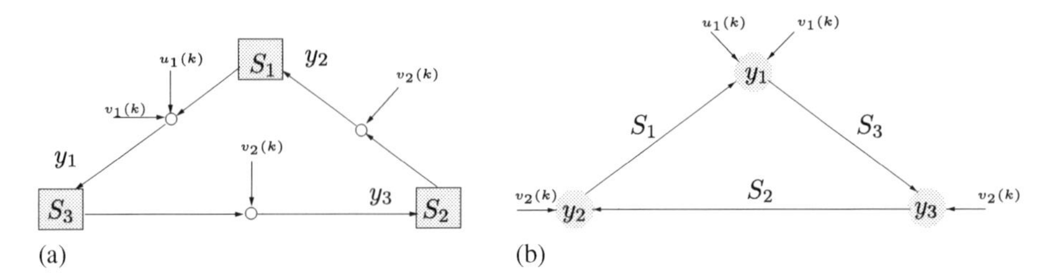

Figure 4.2 Two schematic representations of the same network of dynamic systems S_i for $i = 1, 2, 3$. (a) The block diagram representation with the blocks representing the systems S_i and with the edges representing the interconnecting signals. The small circles $\circ$ represent summations, with its output the sum of its input signals. (b) Signal flow diagram representation whereby the circles now represents the nodes or signals y_i for $i = 1, 2, 3$ and the edges represents the interconnecting systems. The incoming signals to the nodes are added together. The arrows in both network representations depict the direction of data transfer. Reprinted/adapted with permission from Verhaegen (2019) ©2019, Springer-Verlag London Ltd., part of Springer Nature

$$
\begin{cases}
x(k+1) = \begin{bmatrix} A_1 & B_1C_2 & 0 \\ 0 & A_2 & B_2C_3 \\ B_3C_1 & 0 & A_3 \end{bmatrix} x(k) + \begin{bmatrix} 0 \\ 0 \\ B_3 \end{bmatrix} u(k) + \begin{bmatrix} 0 & B_1 \\ 0 & B_2 \\ B_3 & 0 \end{bmatrix} w(k) \\
y(k) = \begin{bmatrix} C_1 & 0 & 0 \\ 0 & C_2 & 0 \\ 0 & 0 & C_3 \end{bmatrix} x(k) + \begin{bmatrix} I \\ 0 \\ 0 \end{bmatrix} + \begin{bmatrix} I & 0 \\ 0 & I \\ 0 & I \end{bmatrix} w(k)
\end{cases}
\tag{4.8}
$$

The system matrices of this lifted state-space model are structured in the sense that these matrices contain a number of zero matrix blocks. However, it is not yet a parametrization of the state-space model. This would require parametrizations of the system matrices in the individual models.

The lifting approach illustrated in Example 4.2 can be used to find a global state-space model for complex NDS. Such resulting global state-space models can be compactly denoted as:

$$
\begin{cases}
x(k+1) = \mathcal{A}x(k) + \mathcal{B}u(k) + \mathcal{H}w(k), \\
y(k) = \mathcal{C}x(k) + \mathcal{D}u(k) + \mathcal{G}w(k).
\end{cases}
\tag{4.9}
$$

In general, this lifting results in a state vector of extremely large dimensions and sparse system matrices. The structuring and/or parametrization of these system matrices is summarized in the following two definitions.

DEFINITION 4.1 (Full Topology NDS for state-space models) Consider the (global) state-space model in Equation (4.9) for a NDS. Then the full topology of this state-space model comprises a full parametrization of both the zero and non-zero matrix blocks, their sizes and an individual parametrization of the non-zero matrix blocks.

Remark that the topology does not provide information about the noise quantity $v(k)$ in Equation (4.9). In the context of subspace identification however, the following definition will be more relevant.

DEFINITION 4.2 (Partial Topology NDS for state-space models) Consider the (global) state-space model in Equation (4.9) of a NDS. Then the partial topology of this state-space model comprises the specification of the pattern of zero-matrix blocks in the system matrices **but** not their sizes.

Example 4.3 (Block diagram versus signal-flow diagram (continued)) The partial topology of the state-space model given in Equation (4.8) comprises the specification of just the location of the zero-matrix blocks in the system matrices. It does not require the specification of their sizes, nor those of the non-zero matrix blocks and nor a parametrization of the latter. For specifying this partial topology, use can be made of graph theory in the form of the concept of a 'pattern' matrix introduced, for example, in Massioni and Verhaegen (2009). For the network depicted in Figure 4.2(a), a possible pattern matrix $\mathcal{P}$ is:

$$\mathcal{P} = \begin{bmatrix} 0 & 0 & 1 \\ 1 & 0 & 0 \\ 0 & 1 & 0 \end{bmatrix}.$$

Here, the ones and zeros indicate the non-zero and zero matrix blocks, respectively, in the A-matrix of Equation (4.8). This particular pattern matrix is the adjacency matrix of the network in Figure 4.2(a) with the blocks S_i as nodes of a digraph and only considering the edges between the nodes. For a definition of a digraph we refer to Section 3.2. Such a pattern matrix does not, however, specify the sizes of the zero or non-zero matrices in the corresponding matrix nor their parametrization. A full topology is then necessary.

To obtain such a full topology for the example, we can, for the special case where all the input and output signals of the local systems S_i for $i = 1, 2, 3$ are scalar, use the so-called observer canonical form parametrization (Kailath, 1980). This would yield an affine parametrization of all system matrices in Equation (4.8).

4.2.2 Decomposable Systems

Decomposable LTI systems are state-space models of the form in Equation (4.9) with their system matrices belonging to the set of *decomposable matrices*. This set is defined next.

DEFINITION 4.3 (Decomposable matrix) (Massioni and Verhaegen, 2009) For a given pattern matrix $\mathcal{P} \in \mathbb{R}^{N_n \times N_n}$, and matrices $M_a, M_b \in \mathbb{R}^{n_\ell \times n_r}$, the class of decomposable matrices contains all matrices $\mathcal{M} \in \mathbb{R}^{n_\ell N_n \times n_r N_n}$ of the following form:

$$\mathcal{M} = I_n \otimes M_a + \mathcal{P} \otimes M_b, \tag{4.10}$$

where $\otimes$ represents the Kronecker product.

The parametrization of decomposable systems only requires to provide the pattern matrix $\mathcal{P}$ and a parametrization of the matrices (M_a, M_b) for each system matrix in Equation (4.9).

REMARK 4.1 (Non-uniqueness) Inspecting Equation (4.10), the same matrix $\mathcal{M}$ can be generated by different pattern matrices and by modifying the matrices M_a and M_b appropriately. Therefore, when defining the topology of a decomposable matrix the pattern matrix is not unique.

DEFINITION 4.4 (Decomposable system) A decomposable system is a state-space model of the form taken in Equation (4.9) where the system matrices $\mathcal{A}, \mathcal{B}, \mathcal{C}, \mathcal{D}, \mathcal{H}, \mathcal{G}$ belong to the class of decomposable matrices with the same pattern matrix $\mathcal{P}$.

A slight geneneralization of Definition 4.4 has recently been proposed in Massioni (2014) for efficient control design. This generalization is called α-decomposable and considers α different decomposable systems and also allows the pattern matrix to be time varying (Massioni, 2014).

Although the system matrices of decomposable sytems have a very specific form, the class of decomposable systems frequently occurs in NDS when the dynamical systems in the network are *identical*. This includes identical interaction between these subsystems. Examples include formation flying with identical satellites in a circular orbit or in deep space, paper machines and adaptive optics (Massioni, 2015). The class of decomposible systems is illustrated for an adaptive optics application in Example 4.4. Interest in decomposable systems stems from their usefulness in the efficient analysis and design of distributed controllers (Massioni and Verhaegen, 2009). It is an example of *network modelling for control* where the class of models (for identification) allows for (extreme) efficient design of distributed controllers. An example is the design of a distributed H_2-controller (Fraanje et al., 2010) that inherits the decomposable structure of the model of the deformable mirror for adaptive optics discussed in the next example. Such a controller can be implemented in a distributed manner, yielding a performance close to the centralised controller (in the H_2 norm sense) and a performance that is two times better than a decentralised PI controller that has additional notches for suppressing the high-resonance frequencies of the deformable mirror (Fraanje et al., 2010).

Example 4.4 (Deformable mirror in adaptive optics) In adaptive optics, as discussed in more detail in Chapter 10, use is made of deformable mirrors. In this example, we discuss a linear model of such a deformable mirror. It is based on Fraanje et al. (2010). A deformable mirror consists of a membrane or a reflecting surface that is deformed by actuators. Using thin plate theory, the deflection of the membrane

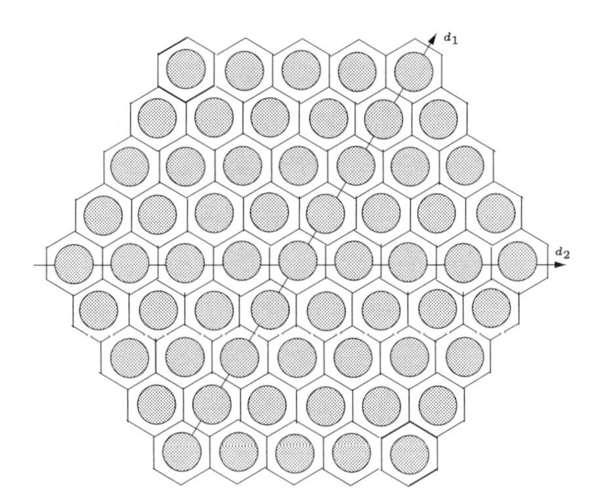

Figure 4.3 Schematic representation of the hexagonal actuator grid where each circle represents a local actuator that can apply a pressure on the membrane above the actuators (not depicted).

$w(x, y, t)$ – where x, y are the spatial 2D coordinates and t represents time – can be described by a partial differential equation (Timoshenko and Woinowsky-Krieger, 1959):

$$\left(EI\ \nabla^4 \left(1 + \eta\frac{\partial}{\partial t}\right) + \rho h\frac{\partial^2}{\partial t^2}\right) w(x, y, t) = p(x, y, t), \tag{4.11}$$

where $\nabla^4 = \left(\frac{\partial^4}{\partial x^4} + 2\frac{\partial^4}{\partial x^2 \partial y^2} + \frac{\partial^4}{\partial y^4}\right)$ is the biharmonic operator, E is Young's modulus, h is the thickness of the membrane, ρ the density, η the material damping parameter, $I = \frac{h^3}{12}$ the moment of inertia and $p(x, y, t)$ is the pressure applied by the actuators at position (x, y) at time instant t. These actuators may be distributed below the membrane in various geometrical configuration. One such configuration, as shown in Figure 4.3, is the widely used hexagonal configuration, but the modelling methodology presented in this example also holds for other grid configurations where the actuator spacing is equidistant. Linear actuators are assumed in this example, and these are modelled by putting a spring with stiffness c in parallel with a damper with damping constant $d = \zeta c$. The controller output is the force $f(i, j, t)$. The indices i, j indicate the spatial location of the actuator in the 2D grid (see Figure 4.3). We assume that the stiffness and damping ratio are equal for all actuators.

Equation (4.11) is spatially discretized. Here, we assume that the spatial distance Δ between the actuators is so small that the following finite difference approximation of the biharmonic operator yields a valid model approximation (Southwell, 1946):

$$\begin{aligned}\nabla^4 &\approx \mathcal{G}(D_1, D_2)\\ &= \frac{4}{9}\Big(42 - 10\big(D_1 + D_1^{-1} + D_1 D_2^{-1} + D_1^{-1} D_2 + D_2 + D_2^{-1}\big)\\ &\quad + 2\big(D_1^2 D_2^{-1} + D_1^{-2} D_2 + D_1 D_2 + D_1^{-1} D_2^{-1} + D_1 D_2^{-2} + D_1^{-1} D_2^2\big)\\ &\quad + \big(D_1^2 + D_1^{-2} + D_1^2 D_2^{-2} + D_1^{-2} D_2^2 + D_2^2 + D_2^{-2}\big)\Big),\end{aligned} \tag{4.12}$$

where D_1 and D_2 are the unit spatial shift operators along the two principal axis of the mirror depicted by d_1 and d_2 in Figure 4.3.

For this spatial discretization, we represent (using the same symbols) the displacement at the point with coordinates index (i,j) by $w(i,j,t)$, and the force in that location by $f(i,j,t)$. The (spatial) discretization of Equation (4.11) is:

$$\left(\frac{AEI}{\Delta^4}\mathcal{G}(D_1,D_2)\left(1+\eta\frac{\partial}{\partial t}\right)+c\left(1+\zeta\frac{\partial}{\partial t}\right)+\rho Ah\frac{\partial^2}{\partial t^2}\right)w(i,j,t)=f(i,j,t). \tag{4.13}$$

Here, the area of one element of the membrame corresponding to one actuator is equal to $\frac{3\sqrt{3}}{8}\Delta^2$ and is denoted by A. It will be assumed that the mirror is clamped at its boundary, yielding $w(i,j,t)=0$ for the grid coordinates (i,j) that fall outside the geometry of the deformable mirror membrane.

The decomposable system can be derived from this finite difference equation by writing it in the following state-space form (for the given expressions for A and I, and denoting $\frac{\partial w(i,j,t)}{\partial t}$ by $\dot{w}(i,j,t)$):

$$\frac{\partial}{\partial t}\begin{bmatrix} w(i,j,t) \\ \dot{w}(i,j,t)\end{bmatrix}=\begin{bmatrix} 0 & 1 \\ -\left(\frac{Eh^2\mathcal{G}(D_1,D_2)}{12\rho\Delta^4}+\frac{8c}{3\sqrt{3}\rho\Delta^2h}\right) & -\left(\frac{\eta Eh^2\mathcal{G}(D_1,D_2)}{12\rho\Delta^4}+\frac{8\zeta c}{3\sqrt{3}\rho\Delta^2h}\right)\end{bmatrix}\begin{bmatrix} w(i,j,t) \\ \dot{w}(i,j,t)\end{bmatrix}+\begin{bmatrix} 0 \\ \frac{8}{3\sqrt{3}\rho\Delta^2h}\end{bmatrix}f(i,j,t), \tag{4.14}$$

$$y(i,j,t)=\begin{bmatrix} 1 & 0\end{bmatrix}\begin{bmatrix} w(i,j,t) \\ \dot{w}(i,j,t)\end{bmatrix}+v(i,j,t), \tag{4.15}$$

where $v(t)$ represents local measurement noise. We now denote the local state $\begin{bmatrix} w(i,j,t) \\ \dot{w}(i,j,t)\end{bmatrix}$ by $x(i,j,t)$ and the local system matrices by:

$$A_{a,c}=\begin{bmatrix} 0 & 1 \\ -\frac{8c}{3\sqrt{3}\rho\Delta^2h} & -\frac{8\zeta c}{3\sqrt{3}\rho\Delta^2h}\end{bmatrix},\quad A_{b,c}=\begin{bmatrix} 0 & 0 \\ -\frac{Eh^2}{12\rho\Delta^4} & -\frac{\eta Eh^2}{12\rho\Delta^4}\end{bmatrix},$$

$$B_{a,c}=\begin{bmatrix} 0 \\ \frac{8}{3\sqrt{3}\rho\Delta^2h}\end{bmatrix},\qquad C_{a,c}=\begin{bmatrix} 1 & 0\end{bmatrix}.$$

Furthermore, we stack the local state, inputs and outputs (again with some abuse of notation) into the vectors $x(t)\in\mathbb{R}^{2N}$, $f(t)\in\mathbb{R}^N$ and $y(t)\in\mathbb{R}^N$, respectively, when considering n grid points. Then the (global) model of the deformable mirror model can be writen as the following decomposable system:

$$\begin{cases} \dot{x}(t)=(I_N\otimes A_{a,c}+P_N\otimes A_{b,c})x(t)+(I_N\otimes B_{a,c})f(t), \\ y(t)=(I_N\otimes C_{a,c})x(t)+v(t). \end{cases} \tag{4.16}$$

The pattern matrix P_N is a sparse matrix that has only non-zero entries determined by the coefficients of the operator $\mathcal{G}(D_1, D_2)$ in Equation (4.12) that defines the coupling between the different sub-elements.

4.2.3 State-Space Models with Block-Tridiagonal System Matrices

A generalization of decomposable systems is presented in this subsection. This generalization allows us to remove the restriction of identical systems and identical connection dynamics but retains the very sparse structure of the system matrices for large networks. Also, it retains the state-space formalism to allow for missing quantities (represented by state variables).

The *block-tridiagonal state-space model* is of the general form of Equation (4.9) with system matrices having a block-triangular structure. This structure is explicitly defined for the A-matrix $\mathcal{A}$ of Equation (4.9):

$$\mathcal{A} = \begin{bmatrix} A_1 & A_{1,r} & & \\ A_{2,\ell} & A_2 & \ddots & \\ & \ddots & \ddots & A_{N-1,r} \\ & & A_{N,\ell} & A_N \end{bmatrix}. \tag{4.17}$$

All matrices on the main diagonal are square but can be different while the off-diagonal blocks can further be rectangular. A similar block structure can be defined for the other system matrices in Equation (4.9).

A full topology definition of these system matrices would then need to specify the size of all block matrices as well as a full parametrization of them. In Chapter 8, a class of subspace-like identification methods will be presented that only require a partial topology specification, including the number N. Such identification methods allow us to extract the size of the block matrices from data. Two examples will be given next to illustrate this class of systems.

Example 4.5 (1D Heterogeneous NDS) Consider the 1D-DNS in Figure 4.4. Here, the local systems in the network are indicated simply by their index.

Now consider the local systems Σ_i for $i = 1, \dots, N$ to be represented as,

$$\begin{aligned} \Sigma_1 : x_1(k+1) &= A_1 x_1(k) + A_{1,r} x_2(k) + B_1 u_1(k), \\ y_1(k) &= C_1 x_1(k) + e_1(k), \\ \Sigma_i : x_i(k+1) &= A_i x_i(k) + A_{i,\ell} x_{i-1}(k) + A_{1,r} x_{i+1}(k) + B_i u_i(k), \\ y_i(k) &= C_i x_i(k) + e_i(k), \\ \Sigma_N : x_N(k+1) &= A_N x_1(k) + A_{N,\ell} x_{N-1}(k) + B_N u_N(k), \\ y_N(k) &= C_N x_N(k) + e_N(k). \end{aligned} \tag{4.18}$$

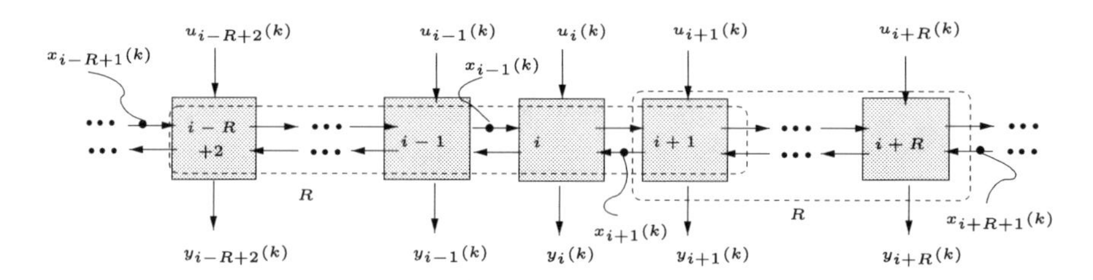

Figure 4.4 The 1D network of R neighbouring systems on the left and right, respectively, of the system at location $i+1$ in the network including the $i+1$th local system. The communication between subsystems as indicated by the state quantities $x_\bullet(k)$ are not observed.

where $x_i(k) \in \mathbb{R}^{n_i}, u_i(k) \in \mathbb{R}^{m_i}, y_i(k) \in \mathbb{R}^{p_i}$ and $e_i(k)$ is a zero-mean white-noise sequence with given covariance matrix.

When lifting the local state, input and output vectors in Equation (4.18) into a global state $x(k)$, input $u(k)$ and output $y(k)$ can be written as:

$$x(k) = \begin{bmatrix} x_1(k) \\ x_2(k) \\ \vdots \\ x_N(k) \end{bmatrix}, \quad u(k) = \begin{bmatrix} u_1(k) \\ u_2(k) \\ \vdots \\ u_N(k) \end{bmatrix}, \quad y(k) = \begin{bmatrix} y_1(k) \\ y_2(k) \\ \vdots \\ y_N(k) \end{bmatrix},$$

and the lifted state-space model is of the form:

$$\begin{cases} x(k+1) = \mathcal{A}x(k) + \mathcal{B}u(k), \\ \quad\;\; y(k) = \mathcal{C}x(k) + e(k), \end{cases} \tag{4.19}$$

where, for this example, only the $\mathcal{A}$ matrix has the block-tridiagonal structure.

A second example discusses the importance of ordering the nodes in a network and stems from finite element modelling, but its use may also apply to network connected systems because node ordering will effect, for example, the adjacency matrix.

Example 4.6 (Reordering network nodes) In finite element modleling nodes are often used. The ordering of these nodes can have great influence on the structure of the matrix that is modelled. For example, consider the graph of a carbon-60 molecule, depicted in Figure 4.5 with 60 nodes. Using the ordering of the nodes as in Figure 4.5 the adjacency matrix of this network has the structure indicated in Figure 4.6(a). Here, the non-zero entries are displayed by a black dot. When using the reverse ordering of the Cuthill–McKee ordering (Cuthill and McKee, 1969) for this network, we get an adjacency matrix as shown in Figure 4.6(b). It is clear that this matrix is more banded and that it can easily be put in the Block Tri-diagonal matrix class as in Equation (4.17).

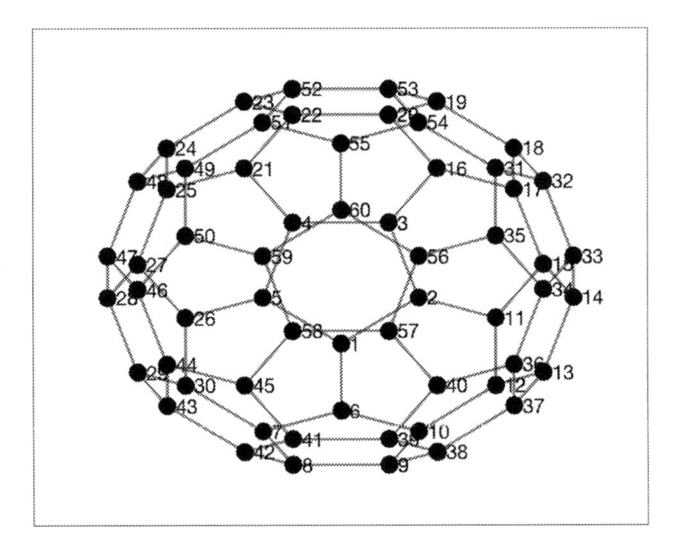

Figure 4.5 The graph of a carbon-60 molecule with 60 nodes for a particular ordering of the nodes.

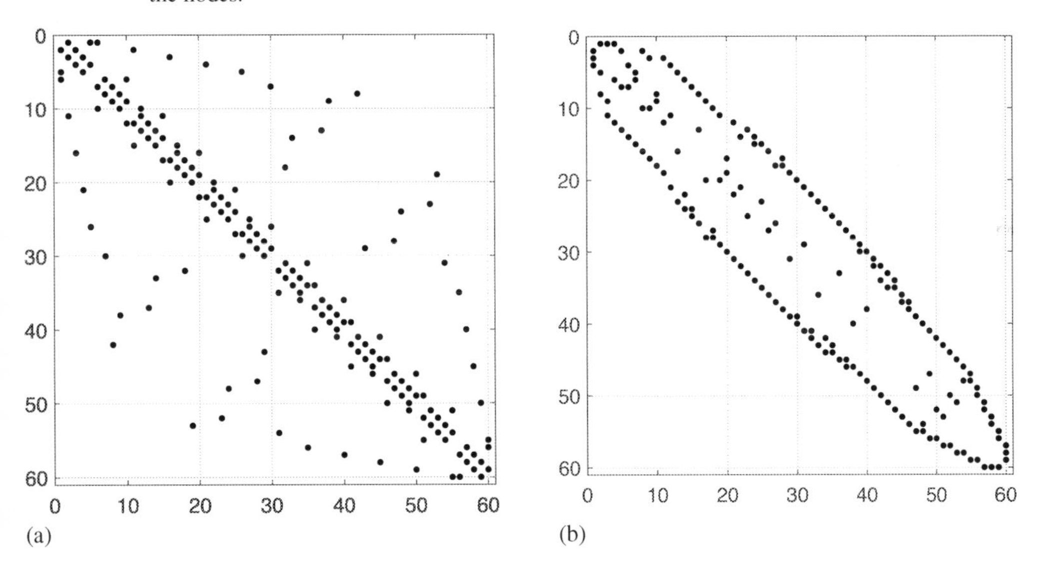

Figure 4.6 Two adjacency matrices corresponding to the nodes and vertex structure in Figure 4.5 for a chosen ordering (a) or the reverse-ordering as calculated by the Cuthill–McKee method (Cuthill and McKee, 1969) (b).

As we are in identification often only interested in the structure (i.e. non-zero element pattern) and as matrices could be considered as weighted adjacency matrices, the methodology illustrated in Example 4.6 could be a motivation for using the state-space model with tri-diagonal system matrices. This is certainly the case when having access to identification methods that only require a partial topology specification.

4.3 Dense Data-Sparse State-Space Models

The state-space model with block-tridiagonal system matrices like the state transition matrix A in Equation (8.11) represents a series connection of dynamic systems (4.18), as depicted in Figure 4.4. It allows for exchange of state information (not directly measured) between neighbouring systems *only*. A drawback of this model class is that information exchange between local systems at a further spatial distance away from the nearest neighbour cannot be considered. The speed at which information is shared within the network is fixed by the pattern of this matrix. The larger the bandwidth, the larger the funnel causality that we depicted with a cone in Figure 1.5 in Chapter 1.

Two multi-dimensional classes of state-space models alleviate this drawback. The first is a generalization of the 1D spatial network of dynamic systems depicted in Figure 4.4 and is described in Section 4.3.1. The second, using multi-linear algebra, is introducted in Section 4.3.2.

4.3.1 Sequentially Semi-separable State-Space Models: 1D Spatial Case

This class of models is based on the work of D'Andrea and Dullerud (2003) that assumes these models to be given for the design of distributed controllers. The class of models introduced in D'Andrea and Dullerud (2003) considers an interconnection of spatially invariant subsystems on an infinite array or on a finite loop. The class was later extended to a finite number of spatially varying (or heterogeneous) subsystems with boundary conditions on a network of systems with a regular interconnection topology in Dullerud and D'Andrea (2004), and on an arbitrary interconnection topology in Langbort et al. (2004). These papers did not consider the identification of these model classes of state-space models. An introduction to formulating these model classes for the purpose of system identification is done in this section.

Here, we follow the modelling approach developed in Rice and Verhaegen (2009) as this methodology enables us to solve distributed control problems for the considered class of models extremely fast, with complexity that is linear in the number of subsystems, and allows us to deal with the heterogeneous case. These models require regular network topologies.

The development of identification methods for these classes of systems fits within the framework of *identification for distributed control* problems that aim to develop efficient identification methods for model classes for which efficient distributed control design methodologies are available or can be developed.

A generalization of the block-tridiagonal state-space model in Equation (4.18) models the ith system Σ_i in the 1D network such as that in Figure 4.4 by the block scheme pictured in Figure 4.7. This subsystem Σ_i has the following (discrete-time[1]) state-space form:

[1] For the continuous-time variant of this model we refer to Rice and Verhaegen (2009).

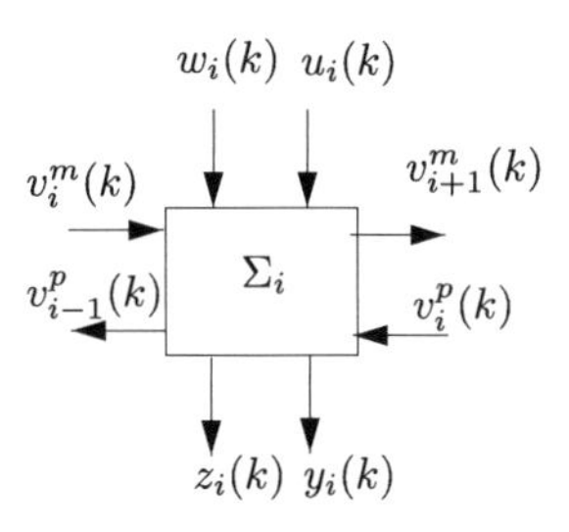

Figure 4.7 Block scheme of the ith subsystem Σ_i in the 1D network like in Figure 4.4 that gives rise to the class of sequentially semi-separable state-space models.

$$\Sigma_i : \quad \begin{bmatrix} x_i(k+1) \\ v_{i-1}^p(k) \\ v_{i+1}^m(k) \\ \hline z_i(k) \\ y_i(k) \end{bmatrix} = \left[\begin{array}{ccc|cc} A_i & B_i^p & B_i^m & B_i^w & B_i^u \\ C_i^p & W_i^p & Z_i^m & L_i^p & V_i^p \\ C_i^m & Z_i^p & W_i^m & L_i^m & V_i^m \\ \hline C_i^z & J_i^p & J_i^m & D_i^{11} & D_i^{12} \\ C_i^y & H_i^p & H_i^m & D_i^{21} & D_i^{22} \end{array}\right] \begin{bmatrix} x_i(k) \\ v_i^p(k) \\ v_i^m(k) \\ \hline w_i(k) \\ u_i(k) \end{bmatrix}. \tag{4.20}$$

Each subsystem Σ_i has spatially mixed causal anti-causal quantities $v_i^p(k)$ and $v_i^m(k)$. This subsystem has unknown interconnections with all its neighbouring subsystems. Each subsystem is defined by a limited set of matrices, which we will call the generators (see Definition 4.5). The overall model models spatial-temporal dynamics.

This model class has also been considered in D'Andrea and Dullerud (2003); Dullerud and D'Andrea (2004), and when neglecting the temporal dynamics, it is also similar to earlier models proposed and analysed in Roesser (1975).

When the terms Z_i^p and Z_i^m in Equation (4.20) are non-zero, they may cause interconnections between subsystems that are not well posed. The topic of the well-posedness of systems like those of Equation (4.20) has been discussed in D'Andrea and Dullerud (2003); Dullerud and D'Andrea (2004); Langbort et al. (2004). In the case of a finite number of subsystems, the well-posedness is guaranteed by making the terms Z_i^p and Z_i^m in Equation (4.20) zero (Rice and Verhaegen, 2009). Therefore, this class of systems is considered here and is given explicitly as:

$$\Sigma_i : \quad \begin{bmatrix} x_i(k+1) \\ v_{i-1}^p(k) \\ v_{i+1}^m(k) \\ \hline z_i(k) \\ y_i(k) \end{bmatrix} = \left[\begin{array}{ccc|cc} A_i & B_i^p & B_i^m & B_i^w & B_i^u \\ C_i^p & W_i^p & 0 & L_i^p & V_i^p \\ C_i^m & 0 & W_i^m & L_i^m & V_i^m \\ \hline C_i^z & J_i^p & J_i^m & D_i^{11} & D_i^{12} \\ C_i^y & H_i^p & H_i^m & D_i^{21} & D_i^{22} \end{array}\right] \begin{bmatrix} x_i(k) \\ v_i^p(k) \\ v_i^m(k) \\ \hline w_i(k) \\ u_i(k) \end{bmatrix}. \tag{4.21}$$

One of the contributions in Rice and Verhaegen (2009) was the realization that lifting the local models in Equation (4.21) into a global state-space model while resolving the interconnection variables yields system matrices that are highly structured, such as

$$\left[\begin{array}{c} x(k+1) \\ \hline z(k) \\ y(k) \end{array}\right] = \left[\begin{array}{c|cc} \mathcal{A} & \mathcal{B}_u & \mathcal{B}_w \\ \hline \mathcal{C}_z & \mathcal{D}_{zu} & \mathcal{D}_{zw} \\ \mathcal{C}_y & \mathcal{D}_{yu} & \mathcal{D}_{yw} \end{array}\right] \left[\begin{array}{c} x(k) \\ \hline u(k) \\ w(k) \end{array}\right]. \tag{4.22}$$

The system matrices are so-called sequentially semi-separable matrices (Dewilde and van der Veen, 1998) and are defined by Definition 4.5. LTI systems modelled by a state space model given by (4.22) with system matrices $\mathcal{A}, \mathcal{B}_u, \ldots, \mathcal{D}_{yw}$ within the class of sequentially semi-separable matrices are called *Sequentially Semi-Separable (SSS) State-Space Models.*

DEFINITION 4.5 (Sequentially semi-separable (SSS) matrices (Dewilde and van der Veen, 1998)) A matrix $\mathcal{A}$ is an SSS-matrix, denoted as:

$$\mathcal{A} = \text{SSS}\left(B_s^m, W_s^m, C_s^m, A_s, B_s^p, W_s^p, C_s^p\right),$$

if these spatial anti-causal part of the 'SSS-generators', i.e. $A_s \in \mathbb{R}^{n_s \times n_s}, B_s^p \in \mathbb{R}^{n_s \times r^p}$, $W_s^p \in \mathbb{R}^{r_p \times r_p}$, $C_s^p \in \mathbb{R}^{r_p \times n_s}$ describe the block diagonal and the (block) upper triangular part of the SSS matrix and the spatial causal part of the SSS-generators, i.e. $B_s^m \in \mathbb{R}^{n_s \times r^m}$, $W_s^m \in \mathbb{R}^{r_m \times r_m}$, $C_s^m \in \mathbb{R}^{r_m \times n_s}$ describe the strickly lower block-triangular part of $\mathcal{A}$ as follows (illustrated for $\mathcal{A} \in \mathbb{R}^{\sum_{s=1}^{5} n_s \times \sum_{s=1}^{5} n_s}$):

$$\mathcal{A} = \left[\begin{array}{ccc|cc} A_{i-3} & B_{i-3}^p C_{i-2}^p & B_{i-3}^p W_{i-2}^p C_{i-1}^p & B_{i-3}^p W_{i-2}^p W_{i-1}^p C_i^p & B_{i-3}^p W_{i-2}^p W_{i-1}^p W_i^p C_{i+1}^p \\ B_{i-2}^m C_{i-3}^m & A_{i-2} & B_{i-2}^p C_{i-1}^p & B_{i-2}^p W_{i-1}^p C_i^p & B_{i-2}^p W_{i-1}^p W_i^p C_{i+1}^p \\ B_{i-1}^m W_{i-2}^m C_{i-3}^m & B_{i-1}^m C_{i-2}^m A_{i-1} & A_{i-1} & B_{i-1}^p C_i^p & B_{i-1}^p W_i^p C_{i+1}^p \\ \hline B_i^m W_{i-1}^m W_{i-2}^m C_{i-3}^m & B_i^m W_{i-1}^m C_{i-2}^m & B_i^m C_{i-1}^m & A_i & B_i^p C_{i+1}^p \\ B_{i+1}^m W_i^m W_{i-1}^m W_{i-2}^m C_{i-3}^m & B_{i+1}^m W_i^m W_{i-1}^m C_{i-2}^m & B_{i+1}^m W_i^m C_{i-1}^m & B_{i+1}^m C_i^m & A_{i+1} \end{array}\right] \tag{4.23}$$

The special structure of these matrices, as illustrated for the $\mathcal{A}$ matrix in Equation (4.23), reveals that the submatrices above (and below) the main block diagonal are block-Hankel matrices that are of low rank. One such upper and lower submatrix is highlighted in Equation (4.23). The low rank property follows from the fact that the block entries in these Hankel matrices can be considered as the Markov parameters of linear time-varying systems.

The set of SSS-matrices defined in Definition 4.5 is closed under addition, multiplication and inversion. This means, for example, that the multiplication of two SSS-matrices is again an SSS-matrix. The main feature of interest to the design of efficient algorithms is that such operations (like multiplication) never have to be executed explicitly but only in terms of the generator matrices of the SSS-matrix. This is the property that has given rise to the design of efficient distributed control methodologies in Rice and Verhaegen (2009). The caveat is that, in general, the algebraic operations increase the rank of the off-diagonal block-Hankel matrices. To reduce this rank, good use can often be made of optimal Hankel norm reduction methods developed for time-varying systems in Dewilde and van der Veen (1998).

REMARK 4.2 (Multi-level SSS) The extension of SSS matrices (and hence systems) to higher than one spatial dimension is, in principle, possible (Qiu, 2015). This has given rise to multi-level SSS matrices in which the block entries are SSS matrices. The computational procedure to operate efficiently with such matrices is, however, more complex and efficiency loss may occur due to the lack of appropriate model reduction via balanced truncation. For that reason we restrict the analysis in this research monograph to SSS systems only.

Example 4.7 (Discretization of PDE: Euler–Bernoulli beam) In Example 4.4, we considered a model of a deformable mirror modelled using thin plate theory. A simplified version of such model is the Euler–Bernoulli beam equation, given for the 1D case (x-direction only):

$$EI\frac{\partial^4 w(x,t)}{\partial x^4} + \rho h\frac{\partial^2 w(x,t)}{\partial t^2} = p(x,t), \tag{4.24}$$

where the quantities have been defined in Example 4.4. Using finite differencing to approximate the partial spatial and temporal derivatives, we can approximate such a beam of finite length by a series of connected mass-spring-damper systems, as shown in Figure 4.8. Applying the central-time central-space finite differencing method to Equation (4.24) (Liu, 2015) yields,

$$\begin{aligned} EI\frac{w(i-2,k) - 4w(i-1,k) + 6w(i,k) - 4w(i+1,k) + w(i+2,k)}{\partial(\Delta X)^4} &+ \\ \rho h\frac{w(i,k+1) - 2w(i,k) + w(i,k-1)}{\partial(\Delta T)^2} &= p(i,t). \end{aligned} \tag{4.25}$$

As in Example 6.1 of Liu (2015), this model can be put into a state-space model of exactly the form as given by Equation (4.21), neglecting the performance output $z_i(k)$ and the external perturbation $w_i(k)$. The output vector of this model is given as:

$$y_i(k) = \begin{bmatrix} y_{i,1}(k) \\ y_{i,2}(k) \end{bmatrix} = \begin{bmatrix} \ddot{w}(i,k) \\ w(i+1,k) - w(i,k) \end{bmatrix}. \tag{4.26}$$

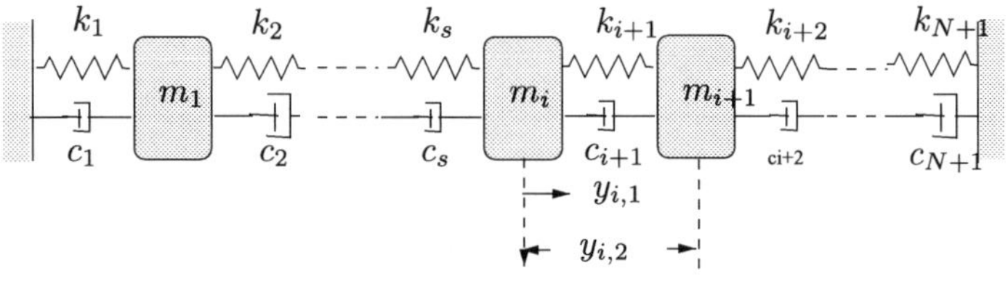

Figure 4.8 Schematic representation of the mass-spring-damper system for approximating an Euler–Bernouilli beam by spatial discretization.

4.3.2 Multi-dimensional State-Space Models

4.3.2.1 Tensor State-Space Models

In line with the tensor VAR models presented in Section 3.6, we present in this section tensor state-space models (TSSM) to model multi-dimensional systems. First, this model class is presented and then a number of key properties relevant for system identification are summarized.

For tensor state-space models, we parametrize the system matrices by Kronecker products of d factor matrices. Though the latter matrices may themselves have additional structure, this will not be further considered as it further complicates the identification problem with additional constraints. Instead, we generalize the class of low Kronecker rank matrices in $\mathcal{K}_{2,r}$ given in Definition 3.7. For that purpose, let d, I, J be three integers. Let $(I_1, \ldots, I_d)$ and $(J_1, \ldots, J_d)$ be two tuples of integers such that $\prod_{j=1}^{d} I_j = I$ and $\prod_{j=1}^{d} J_j = J$. Let $(n_1, \ldots, n_d) \in \mathbb{N}^d$ and $n = \prod_{j=1}^{d} n_j$.

DEFINITION 4.6 (Kronecker rank within $\mathcal{K}_{d,r}$ (Sinquin, 2019)) Let $r \in \mathbb{N}$. The class of matrices $\mathcal{K}_{d,r}$ contains all matrices $X \in \mathbb{R}^{J \times I}$ parametrized as follows:

$$X = \sum_{\ell=1}^{r} X_{d,\ell} \otimes \ldots \otimes X_{1,\ell}, \tag{4.27}$$

for all $(j, \ell) \in \{1, \ldots, d\} \times \{1, \ldots, r\}$ and $X_{j,\ell} \in \mathbb{R}^{J_j \times I_j}$. The number r is the Kronecker-rank and is assumed to be much smaller than $\min(\{J_j, I_j\}_{j=1..d})$.

REMARK 4.3 This class is related to the canonical polyadic decomposition (CPD) expression of a tensor $\mathcal{X}$ (see Definition B.6) of rank r given as,

$$\mathcal{X} = \sum_{\ell=1}^{r} U_{d,\ell} \circ \ldots \circ U_{1,\ell}. \tag{4.28}$$

For the case $d = 2$, a reshuffling operator $\mathcal{R}$ was introduced in Chapter 3 (see Definition 3.6). This operator allows to transform the matrix X into,

$$\mathcal{R}(X) = \sum_{\ell=1}^{r} U_{2,\ell} U_{1,\ell}^{T} = \sum_{\ell=1}^{r} U_{2,\ell} \circ U_{1,\ell}, \tag{4.29}$$

where $U_{j,\ell} = \text{vec}(X_{j,\ell})$ and $\circ$ denotes the outer product (see Appendix B). When the matrix X is as in Equation (4.6), the reshuffled matrix X can be written as,

$$\mathcal{R}(X) = \mathcal{X} = \sum_{\ell=1}^{r} U_{d,\ell} \circ \ldots \circ U_{1,\ell}. \tag{4.30}$$

This expression is a CPD of rank r of the tensor $\mathcal{X}$ (see Appendix B for more details). The reshuffled matrix $\mathcal{R}(X)$ is a tensor in $\mathbf{R}^{J_d I_d \times \cdot \times J_1 I_1}$.

DEFINITION 4.7 (Tensor state-space model (TSSM) (Sinquin, 2019)) An LTI system governed by the state-space model in Equation (4.9) with all system matrices in $\mathcal{K}_{d,1}$ and $v(k) \equiv 0$ is called a tensor state-space model.

For further illustration, the state-space model of the form:

$$\begin{cases} x(k+1) = Ax(k) + Bu(k), \\ y(k) = Cx(k) + e(k), \end{cases} \tag{4.31}$$

with $e(k)$ zero-mean white noise with a covariance matrix that is a multiple of the identity matrix, is a TSSM if and only if we can define for $j \in \{1, \ldots, d\}$ the matrices $A_j \in \mathbb{R}^{n_j \times n_j}$, $B_j \in \mathbb{R}^{n_j \times l_j}$, $C_j \in \mathbb{R}^{J_j \times n_j}$ such that the (A, B, C) have the following expression:

$$\begin{cases} A = A_d \otimes \cdots \otimes A_1, \\ B = B_d \otimes \cdots \otimes B_1, \\ C = C_d \otimes \cdots \otimes C_1. \end{cases} \tag{4.32}$$

An equivalent way to write the state-space model that clearly explains its link with tensor calculus is to use the n-mode matrix tensor product in Definition B.4,

$$\begin{cases} \mathcal{X}(k+1) = \mathcal{X}(k) \times_1 A_1 \times_2 \ldots \times_d A_d + \mathcal{U}(k) \times_1 B_1 \times_2 \ldots \times_d B_d, \\ \mathcal{Y}(k) = \mathcal{X}(k) \times_1 C_1 \times_2 \ldots \times_d C_d + \mathcal{E}(k), \end{cases} \tag{4.33}$$

where $\mathcal{X}(k)$, $\mathcal{U}(k)$, $\mathcal{Y}(k)$, $\mathcal{E}(k)$ are in $\mathbb{R}^{n_1 \times \ldots \times n_d}$, $\mathbb{R}^{m_1 \times \ldots \times m_d}$, $\mathbb{R}^{p_1 \times \ldots \times p_d}$ and $\mathbb{R}^{p_1 \times \ldots \times p_d}$ respectively.

Figure 4.9 depicts the state equation for an autonomous system for the cases $d \in \{1, 2, 3\}$.

The standard LTI equation in vector form is a special case of the tensor state-space Model for $d = 1$. The multi-linear equations are used to represent the dynamics when the state is a dth order tensor for $d > 1$. In that case, the nth mode fiber (see

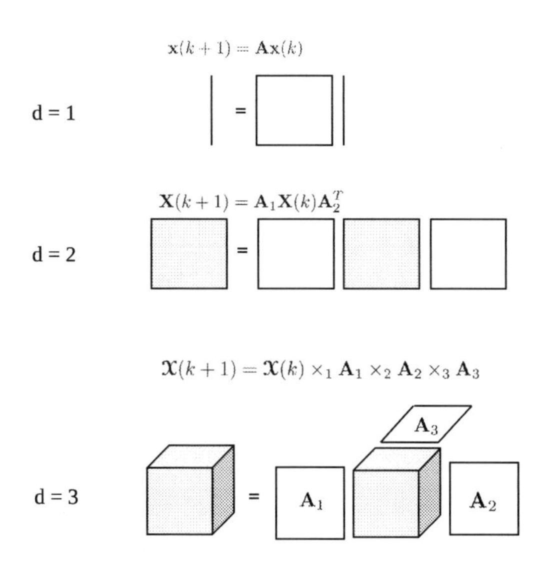

Figure 4.9 Schematic representation of the state-update equation (autonomous case only) when (Top) $d = 1$ (Middle) $d = 2$ and (Bottom) $d = 3$. Reprinted/adapted from Sinquin (2019) with permission from TU Delft

Definition B.3) of the state at instant k is multiplied with rows of the matrix A_n. For example, for the case $d = 2$ (see Figure 4.9), the rows of the state matrix $X(k)$ are multiplied with the rows of the matrix A_2 and subsequently the columns of that result are multiplied with the rows of the matrix A_1. The special case $d = 2$ can be generalized as in Equation (3.19) to represent the coefficient matrices of the Quarks model. This would then give rise to a representation of the (A, B, C) matrices in Equation (4.31) given as:

$$\begin{cases} A = \sum_{\ell=1}^{r_A} A_{\ell,2} \otimes A_{\ell,1}, \\ B = \sum_{\ell=1}^{r_B} B_{\ell,2} \otimes B_{\ell,1}, \\ C = \sum_{\ell=1}^{r_C} C_{\ell,2} \otimes C_{\ell,1}. \end{cases} \tag{4.34}$$

Increasing the Kronecker ranks r_A, r_B, r_C within $\mathcal{K}_{2,r}$ would enable us to represent any matrix-valued linear operator (Bamieh and Filo, 2018). However, as done for the parametrization of the coefficient matrices of tensor VAR models in Equation (3.26), both representations in Equations (4.32) and (4.34) could be combined to yield:

$$\begin{cases} A = \sum_{\ell=1}^{r_A} A_{\ell,d} \otimes \cdots \otimes A_{\ell,1}, \\ B = \sum_{\ell=1}^{r_B} B_{\ell,d} \otimes \cdots \otimes B_{\ell,1}, \\ C = \sum_{\ell=1}^{r_C} C_{\ell,d} \otimes \cdots \otimes C_{\ell,1}. \end{cases} \tag{4.35}$$

The analysis of multi-dimensional state space models (MSSM) with Kronecker ranks larger than one differs especially regarding the meaning of r_A that would imply Markov parameters with exponentially increasing Kronecker ranks.

4.3.2.2 Observability, Controllability and Equivalent TSSM

We restrict the analysis of MSSM to the system matrices as given in Equation (4.32) as a pioneer for later contributions for more general system matrix structures. For this class of systems, a number of properties are recalled.

First, the state-space matrices given in Equation (4.32) are unique up to the trivial indeterminacies (scaling and permutation). Second, the stability of the A-matrix is related to that of its factor matrices as stated by the following Lemma.

LEMMA 4.1 (Lemma 4.1 of Sinquin [2019]) *Let $d \in \mathbb{N}$. If $\lambda_{max}(A_j) < 1$ for all $j \in \{1, \cdots, d\}$, then $\lambda_{max}(A_d \otimes \cdots \otimes A_1) < 1$. The reverse is not true in general.*

Proof By induction. For $d = 1$, the lemma is true. Let us assume that the lemma holds for some d, we then prove it holds for $d + 1$. Then let $\tilde{A} := A_d \otimes \cdots \otimes A_1$ satisfy, $\lambda_{\max}(\tilde{A}) < 1$. Now consider $A_{d+1} \otimes \tilde{A}$, then by lemma 3.1. of Sinquin (2019) the eigenvalues of $A_{d+1} \otimes \tilde{A}$ are the product of the eigenvalues of A_{d+1} and $\tilde{A}$ and hence the condition holds for $d + 1$ and the proof is completed. □

The following two Lemmas relate the observability and controllability of the factored systems governed by the factor matrices (A_i, B_i, C_i) to that of the global system governed by (A, B, C) when both are related as in Equation (4.32).

LEMMA 4.2 (Lemma 4.2 of Sinquin [2019]) *Let $d \in \mathbb{N}$ and $j \in \{1,\dots,d\}$. If the pair $(A_d \otimes \cdots \otimes A_1, C_d \otimes \cdots \otimes C_1)$ is observable, then each of the pairs (A_j, C_j) is observable. The reverse is not true in general.*

Proof By induction and using lemma 3.2 of Sinquin (2019). □

The dual result holds for controllability.

LEMMA 4.3 (Lemma 4.3 of Sinquin [2019]) *Let $d \in \mathbb{N}$ and $j \in \{1,\dots,d\}$. If the pair $(A_d \otimes \cdots \otimes A_1, B_d \otimes \cdots \otimes B_1)$ is controllable, then each of the pairs (A_j, B_j) is controllable. The reverse is not true in general.*

Proof By induction and using the dual of lemma 3.2 of Sinquin (2019) dealing with the controllability matrix. □

LEMMA 4.4 (Lemma 4.4 of Sinquin [2019]) *Two TSSM systems of the form as in Equation* (4.31) *for two different sets of generator matrices:*

$$\Big((A_d^{(1)},\dots,A_1^{(1)}),(B_d^{(1)},\dots,B_1^{(1)}),(C_d^{(1)},\dots,C_1^{(1)})\Big)$$

and

$$\Big((A_d^{(2)},\dots,A_1^{(2)}big),(B_d^{(2)},\dots,B_1^{(2)}),(C_d^{(2)},\dots,C_1^{(2)})\Big),$$

are equivalent (i.e. yield the same input-output response) if there exists a similarity transformation $T \in \mathbb{R}^{n\times n}$ such that $T = T_d \otimes \cdots \otimes T_1$ and:

$$\begin{cases} A_d^{(1)} \otimes \cdots \otimes A_1^{(1)} = T^{-1}(A_d^{(2)} \otimes \cdots \otimes A_1^{(2)})T, \\ B_d^{(1)} \otimes \cdots \otimes B_1^{(1)} = T^{-1}(B_d^{(2)} \otimes \cdots \otimes B_1^{(2)}), \\ C_d^{(1)} \otimes \cdots \otimes C_1^{(1)} = (C_d^{(2)} \otimes \cdots \otimes C_1^{(2)})T. \end{cases} \tag{4.36}$$

Proof The proof follows readily by writing T^{-1} as (see Appendix B):

$$T^{-1} = T_d^{-1} \otimes \cdots \otimes T_1^{-1}.$$

Let us write down a first TSSM:

$$\begin{aligned} x^{(1)}(k+1) &= A_d^{(1)} \otimes \cdots \otimes A_1^{(1)} x^{(1)}(k) + B_d^{(1)} \otimes \cdots \otimes B_1^{(1)} u(k), \\ y(k) &= C_d^{(1)} \otimes \cdots \otimes C_1^{(1)} x^{(1)}(k). \end{aligned}$$

Then, substituting the equalities in Equation (4.36), and using the given expression for T^{-1} directly shows that the TSSM:

$$\begin{aligned} x^{(2)}(k+1) &= A_d^{(2)} \otimes \cdots \otimes A_1^{(2)} x^{(2)}(k) + B_d^{(2)} \otimes \cdots \otimes B_1^{(2)} u(k), \\ y(k) &= C_d^{(2)} \otimes \cdots \otimes C_1^{(2)} x^{(2)}(k), \end{aligned}$$

gives the same input-output behaviour for $x^{(2)}(k) = Tx^{(1)}(k)$. If the similarity transformation T belongs to $\mathcal{K}_{d,1}$ and Equation (4.36) holds, the input-output behaviour for the two sets is the same. □

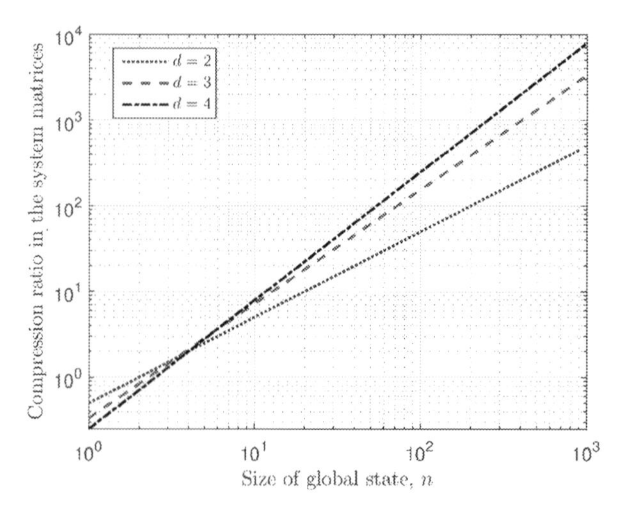

Figure 4.10 Compression ratio $\frac{\prod_{j=1}^{d} J_j \prod_{j=1}^{d} I_j}{rd\sum_{j=1}^{d} J_j I_j}$ for different tensor order d and sizes of the state vector n.

4.3.2.3 Compression Capabilities of TSSM

To represent a tensor in terms of its CPD factors, as in Equation (4.27), we need $rd\sum_{j=1}^{d} J_j I_j$ parameters, while for the unstructured case we need $\prod_{j=1}^{d} J_j \prod_{j=1}^{d} I_j$ parameters. The compression ratio between both is illustrated in Figure 4.10. The complexity of calculating the state-update is linear in the size of the state-vector. We refer to Appendix B for a detailed count.

4.3.2.4 Similarities between SSS and Low-Kronecker Rank Frameworks

The structure-preserving properties of SSS matrices are remarkable and have shown applications for solving the discrete algebraic Riccati equation (DARE). If the set of SSS matrices is closed under addition, multiplication and inversion, the set of sums-of-Kronecker matrices shares some of these properties. More particularly, the system order of the underlying LTI representation in SSS matrices plays the same role as the Kronecker rank. Adding two SSS matrices results in a matrix with SSS order equal to the sum of the orders. Multiplying them is such that the orders are multiplied as well. Adding two matrices within $\mathcal{K}_{d,r}$ results in a matrix within $\mathcal{K}_{d,2r}$. Multiplying them multiplies the Kronecker rank of the product. The Kronecker rank of the inverse of a low Kronecker rank matrix may however be very large: this is a difference of prime importance with the SSS, preventing the extension of a large span of results for the design of controllers to the class of low Kronecker rank matrices.

The system order in SSS matrices relates to the amount of information shared with neighbouring subsystems. The Kronecker rank on the other hand relates to the index of separability of the underlying multi-dimensional function. This can be interpreted as the amount of decoupling between the dimensions. The notion of localizability does not exist in the Kronecker models unless the factors are further parametrized, e.g. with SSS matrices. This has practical consequences: it is only possible to think of distributed controllers embedded in each subsystem when using the SSS structure whereas controllers built upon the framework of sums-of-Kronecker matrices require a centralized computing unit.

It is possible to think of strings of interconnected systems also with MSSM (or TSSM) by parametrizing the factors with SSS matrices. To illustrate this point, consider an autonomous system,

$$x(k+1) = \left(\bar{A}^h \otimes \bar{A}^v\right)x(k). \tag{4.37}$$

Then for $x(k) = x^h(k) \otimes x^v(k)$, Equation (4.37) decouples into two independent *small-scale* autonomous systems given as in Equation (4.38). The first autonomous system may be viewed as a spatially horizontal model, while the second is modelling the vertical dimension,

$$\begin{cases} x^h(k+1) = \bar{A}^h x^h(k), \\ x^v(k+1) = \bar{A}^v x^v(k). \end{cases} \tag{4.38}$$

When considering $x^h(k) = \left[\ldots\ x^T_{h,i-1}(k)\ x^T_{h,i}(k)\ \ldots\right]^T$ and $x^v(k) = \left[\ldots\ x^T_{v,i-1}(k)\ x^T_{v,i}(k)\ \ldots\right]^T$, we can depict the operation in Equation (4.37) as in Figure 4.11. This respresentation highlights the similarity with the multi-level SSS matrix represenation.

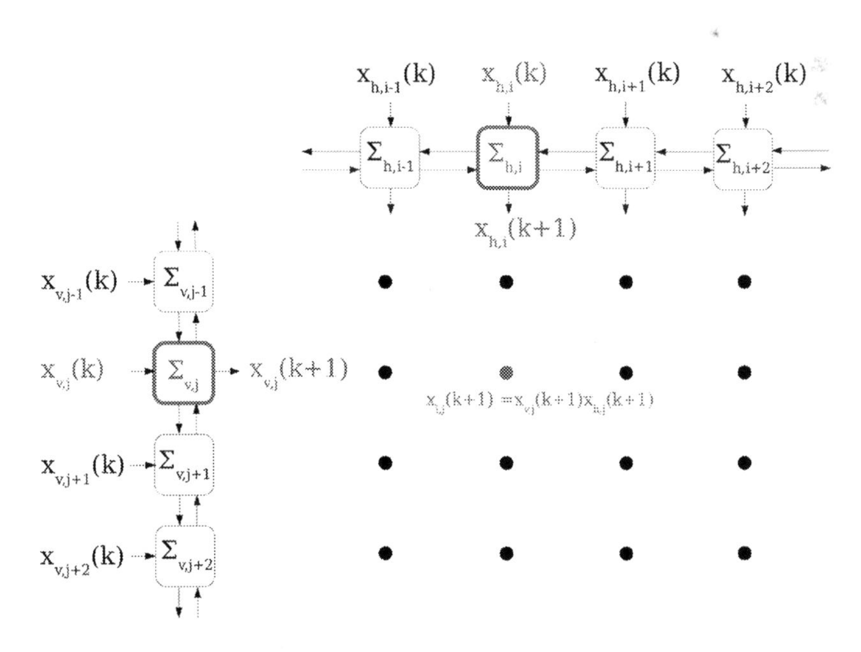

Figure 4.11 Schematic illustrating the autonomous system in Equation (4.37) as a 2D network of interconnected systems with a 'vertical' and 'horizontal' component. Only the case $r = 1$ is depicted. The state at each position in the grid is obtained by multiplying the horizontal and vertical (virtual) states obtained with a forward and backward pass on the SSS string. Reprinted/adapted from Sinquin (2019) with permission from TU Delft

4.4 Conclusions

State-space models have been presented for modelling the dynamics of large-scale networks. A great challenge is to limit the number of parameters to parametrize the system matrices in these models. Knowledge of the network topology might be helpful in determining the topology (or structure) of these matrices. The challenge remaining is to exploit these data-sparse parametrizations in design synthesis or identification algorithms.

A distinction is made between a sparse and dense, data-sparse matrix representations. The sparse matrix representations include decomposable systems for interconnections between homogeneous systems and interconnected strings of subsystems with a limited spatial communication resulting in a block tri-diagonal transition matrix. The dense, data-sparse matrix representations include the sequentially semi-separable matrix structures and tensor state-space models. Interestingly, we can relate the different model structures representing a 1D heterogeneous NDS as a special case of the SSS model. Additionally, the low Kronecker rank matrices may be considered as an extension of the decomposable systems in the case when the topology is unknown and for heterogeneous systems.

5 Classification of Models of Networks of LTI Systems

5.1 Introduction

In this chapter, an attempt is made to classify the model defined in the previous chapters. This is commonly done for models used for system identification; see for example Ljung (1999). Apart from the generic division for LTI systems in transfer function models and state-space models, the main classification in lumped parameter models is how the external noise transfer is modelled. This gives rise to models like AR(MA)X, Box-Jenkins, output-error models, etc.

Although such model classes can, in principle, also be defined for networks of dynamical systems, the structures that we have considered so far have an additional layer, namely the parametrization of the model coefficient or model system matrices.

The goal of this chapter is twofold. We start by structuring the different models in Section 5.2. Then, in Section 5.3, we consider the effects of transforming these models with first, the discretization of continuous time models with coefficient or system matrices belonging to one of the model classes defined (for discrete-time systems); and second, the transformation from one model class to another.

5.2 A Tree of Models for the Identification of NDS

The overview of the different model classes for networks of dynamic systems (NDS) given in the previous two chapters have been and will be investigated within a system identification context. Other model classes may be added in the future. Before presenting a number of identification methods for identifying some (variants) of these model classes in the three next chapters, an overview is given of the prior knowledge about the network structure or network topology required from the user. Before starting this overview, we state the following definition of network topology.

DEFINITION 5.1 (Network topology) The network topology is the interconnection pattern of local systems operating in the network. This interconnection may be physical due to the physical constraints these local systems have on one another, or logical due to the exchange of data between them.

This definition was specialized earlier on for the pattern of non-zero elements in structured systems matrices of models of NDS in Definitions 4.1 and 4.2.

These notions are not identical but may be related. For example, the pattern of a NDS may give rise to the pattern matrix to describe decomposable (state-space) models (defined in Section 4.2.2). As such, the network topology may partially define the parametrization of the system matrices of a state-space model.

Prior knowledge of the network topology information can be classified on two different levels, as outlined in the following subsections. The first level is the (ir-)regularity of the network topology. The second level refers to the homogeneity or heterogeneity of the systems in the network.

For system identification a further classification is relevant. This has to do whether prior knowledge of the network topology or the topology (or structure) of the coefficient or system matrices is available. This classification will be used to organize the different models that have been presented in the previous two chapters. However, as will be seen, the organization will not be crisp as some models can be used to represent NDS with known or unknown topology.

In the following two sections we classify, respectively, the models that can be used when the network topology is known followed by the case when no prior knowledge is available about the (sensor-actuator) network topology.

5.2.1 Known Network Topology

When the network topology is known, the systems in the network may be (physically or logically) connected, as illustrated in Figure 5.1. The model classes that can use such information or, more usefully, may benefit from such information are

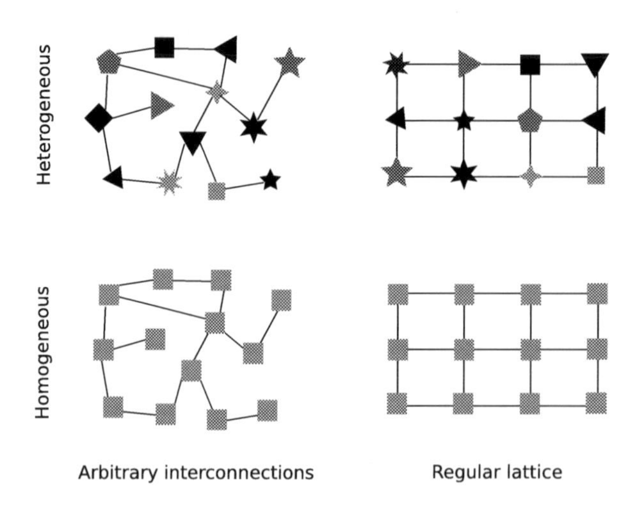

Figure 5.1 Classification of topologies of networks of dynamical systems (NDS). Here, the edges indicate the physical or logical connection between the systems. The latter are indicated by a node (box, star, etc). Nodes of the same shape indicate identical systems, while different shapes indicate different systems. Reprinted/adapted from Massioni (2015) with permission TU Delft

Table 5.1 Model classes that may benefit from knowledge of the topology of NDS.

	Arbitrary interconnections	Regular Lattice
Heterogeneous	• A priori parametrized state-space models (Section 4.2.1) • Clusters of small NDS • α-decomposable systems (Massioni, 2014) • ...	• A priori parametrized state-space models (Section 4.2.1) • State-space Models (1D) (Section 4.3) • Block tri-diagonal state-space models (Section 4.2.3) • ...
Homogeneous	• A priori parametrized state-space models (Section 4.2.1) • decomposable systems (Section 4.2.2) • ...	• A Priori parametrized state-space models (Section 4.2.1) • State-space models (1D) (Section 4.3) • Block tri-diagonal state-space models (Section 4.2.3) • ...

summarized in Table 5.1. Such knowledge on arbitrary interconnections of either homogeneous or heterogenous systems in the network at a systems level is available in a number of applications. See, for example, Borelli and Keviczky (2008); Fax and Murray (2004); Langbort et al. (2004); Motee and Jadbabaie (2008). For regular networks (or lattices), examples can be found in D'Andrea and Dullerud (2003); Jovanović and Bamieh (2005); Recht and D'Andrea (2009); Rice and Verhaegen (2009); Shamma and Arslan (2006). Such lattice structures have been used, see e.g. Recht and D'Andrea (2009), to use the formalism of spatial group theory to design distributed controllers. From an identification perspective, the knowledge on the network topology may be used to specify a partial topology of the DNS state-space model; see Definition 4.2. For example, for decomposable systems knowledge of the topology might translate into knowledge of the pattern matrix $\mathcal{P}$ in the definition of the system matrices of a global state-space model as given by Equation (4.9). The system identification challenge should then be to use this partial topology to estimate the size of the parametrized matrix blocks as well as their entries. This includes assuming that each system matrix of the global state-space model (4.9) is in the class of decomposable matrices (defined in Definition 4.3) and estimating the size of the pairs (M_a, M_b) for each system matrix as well as its parameter values.

REMARK 5.1 (Definition of the pattern matrix $\mathcal{P}$ for decomposable systems) As illustrated in Example 4.4, the pattern matrix $\mathcal{P}$ may also follow from the spatial discretization of the underlying partial differential equation describing the system at hand. Such a pattern matrix may then be considered as a finite-dimensional approximation of the 'physical' connection between subsystems.

In dealing with large scale DNS the known network topology may additionally be useful for system identification in attempts to identify local systems in the network using local data *only*. This will be a key problem in the development of identification methods for the class of (a) block tri-diagonal state-space models (defined in Section 4.2.3) and (b) 1D state space models (defined in Section 4.3).

5.2.2 Unknown Network Topology Using Regular or Irregular Sensor (Actuator) Arrays

When use is made of (large) arrays of sensors such as those in geographic information systems or Example 3.5, it might not be straightforward to a priori specify a pattern between the dynamic systems, i.e. to specify the DNS topology.

For this reason, model classes are proposed that start from the signals in the network and try to use these to estimate a DNS topology from the sensor-actuator arrays in the network. Here, for *regular data arrays* the models listed in the left column of Table 5.2 can be used. The same model classes can also be used when the data arrays are *irregular*. However, in this case, then much more care needs to be taken in organizing the data.

This table gives rise to a number of further remarks:

1. From this table, it is clear that the model classes listed can both work with regular and irregular data arrays. The distinctive feature is in the specification of the data structures of the data involved in these models. Example 3.6 showed how a 2D regular sensor array may help the tensorization of the data. On irregular data arrays, such information may be more difficult to retrieve. Then, bringing sparsity, or defining the tensor from data, might be integrated in the algorithmic solutions.

Table 5.2 Model classes for a priori unknown NDS topology with regular data collection grids (left) and irregular data collection grids (right).

	Regular Data Arrays	Irregular Data Arrays
Full Model	• Full DNS/DSF models (Section 2.2.2)	• Full DNS/DSF models (Section 2.2.2)
Sparse Model	• Local or sparse DNS/DSF models (Section 3.4) • Sparse VAR models (Section 3.3)	• Local or sparse DNS/DSF models (Section 3.4) • Sparse VAR models (Section 3.3)
Data-Sparse Model	• Kronecker-based VAR models (Quarks) (Section 3.5) • Tensor VAR models (Section 3.6) • Tensor state-space models (Section 4.3.2)	• Kronecker-based VAR models (Quarks) (Section 3.5) • Tensor VAR models (Section 3.6) • Tensor state-space models (Section 4.3.2)

For that, we refer to Chapter 6 where, for example, use is made of regularization to induce sparsity in the models.

2. For small-size data arrays, full models (like those on the top row of Table 5.2) can be estimated first and then subsequently try to decide which local 'transfer functions' can be neglected (statistically).
3. In order to be able to deal with large data arrays when solving identification problems with the model class in the top row of Table 5.2, the identification of local transfer functions only using local(ly processed) data becomes of the utmost importance.
4. When networks are not regular, interpolation techniques may be used to introduce regularity.

5.3 Transforming NDS Models

The model classes presented in the previous two chapters and classified in the previous two sections are (nearly) all presented in discrete time and may result from a temporal-spatial discretization of distributed parameter models like partial differential equations.

As prior information about the structure (or topology) of (the matrices in) these models is often based on the analysis of such continuous time (distributed parameter) models (see e.g. Yeung et al. [2010]), we briefly investigate two related issues. First, to what extent temporal discretization impacts continuous analogues of some of the presented models. This is done in Section 5.3.1. Second, we study the change in structure when attempting to transfer one model class into another. Here, we restrict ourselves to the transformation of state-space model classes to some other input-output type of models as the inverse falls under the topic of *system realization* and this problem is treated within the development of identification methods. This second topic is discussed in Section 5.3.2.

5.3.1 Effect of Temporal Discretization

It is assumed that the spatial dimension of the network has been discretized. An example is given in the spatial discretization of the PDE considered in Example 4.4. This assumption is often made in the analysis of distributed systems; see e.g. D'Andrea (1999); D'Andrea and Dullerud (2003). Further, in a number of applications, such as formation flying, the systems are distributed in space as distinct lumped systems. Therefore, the focus is on the effect of temporal discretization on the structure of (some of) the continuous-time analogues of the model classes considered. We also restrict the discussion on the temporal discretization of continuous-time state-space models. The following Lemma considers two standard approximate integration rules.

Table 5.3 Transformation of the system matrices of the continuous-time state-space model (5.1) into the approximate discrete-time model (5.2) for two different integration rules indicated in the first column. Here, s is the Laplace variable and z the variable of the z-transform. Zero initial conditions are assumed.

Approximation of $\frac{1}{s}$	Discrete-time quadruple $\begin{bmatrix} A & B \\ C & D \end{bmatrix}$ in Equation (5.2)
Euler Forward $\frac{\Delta T}{z-1}$	$\begin{bmatrix} I + A_c \Delta T & B_c \Delta T \\ C_c & D_c \end{bmatrix}$
Trapezoidal $\frac{\Delta T}{2}\frac{z+1}{z-1}$	$\begin{bmatrix} \left(I - A_c\frac{\Delta T}{2}\right)^{-1}\left(I + A_c\frac{\Delta T}{2}\right) & \left(I - A_c\frac{\Delta T}{2}\right)^{-1} B_c \Delta T \\ C_c\left(I - A_c\frac{\Delta T}{2}\right)^{-1} & D_c + C_c\left(I - A_c\frac{\Delta T}{2}\right)^{-1} B_c\frac{\Delta T}{2} \end{bmatrix}$

LEMMA 5.1 (Approximate integration of continuous-time models) *Consider the continuous-time LTI, finite-dimensional state-space model:*

$$\begin{aligned} \dot{x}_c(t) &= A_c x_c(t) + B_c u(t) \quad x_c(t) \in \mathbb{R}^n,\ u(t) \in \mathbb{R}^m, \\ y_c(t) &= C_c x_c(t) + D_c u(t) \quad y_c(t) \in \mathbb{R}^p, \end{aligned} \tag{5.1}$$

with A_c asymptotically stable (having all its eigenvalues in the left half-plane).

This model is discretized using a constant sampling period ΔT and it is assumed that the inverse $\left(I - A_c\frac{\Delta T}{2}\right)$ exists. The two approximate integration rules listed in the first column of Table 5.3 are considered. Assuming zero initial conditions, the following approximate discrete-time state-space model is obtained:

$$\begin{aligned} x(k+1) &= Ax(k) + Bu(k), \\ y(k) &= Cx(k) + Du(k), \end{aligned} \tag{5.2}$$

with the system matrices for the two integration rules given in the second column of Table 5.3.

Proof The proof is restricted to the trapezoidal or Tustin integration rule. For zero initial conditions, the Laplace transform of Equation (5.1) equals:

$$\begin{aligned} sX_c(s) &= A_c X_c(s) + B_c U(s), \\ Y_c(s) &= C_c X_c(s) + D_c U(s). \end{aligned} \tag{5.3}$$

For the given approximate trapezoidal integration rule in Table 5.3, the (z-transform) of the discrete time model is given as:

$$\begin{aligned} \frac{1}{\Delta T/2}\frac{z-1}{z+1} X(z) &= A_c X(z) + B_c U(z), \\ Y(z) &= C_c X(z) + D_c U(z). \end{aligned}$$

After some algebraic manipulation, this becomes:

$$zX(z) = \underbrace{\left(I - \frac{\Delta T}{2}A_c\right)^{-1}\left(I + \frac{\Delta}{2}A_c\right)}_{A} X(z) + \underbrace{\left(I - \frac{\Delta T}{2}A_c\right)^{-1} B_c \frac{\Delta T}{2}}_{B'}(z+1)U(z),$$

$$Y(z) = C_c X(z) + D_c U(z).$$

From this, it follows that (a) the spectral radius of A is smaller than 1 and (2) $Y(z)$ equals:

$$\begin{aligned}
Y(z) &= \left(D_c + (z+1)C_c(zI - A)^{-1}B'\right)U(z) \\
&= \left(D_c + C_c(zI - A)^{-1}B' + zC_c(zI - A)^{-1}B'\right)U(z) \\
&= \left(D_c + C_c(zI - A)^{-1}B' + C_c(I - z^{-1}A)^{-1}B'\right)U(z) \\
&= \left(D_c + C_c(zI - A)^{-1}B' + C_cB' + C_cA(zI - A)^{-1}B'\right)U(z) \\
&= \left(D_c + C_cB' + C_c(I + A)(zI - A)^{-1}B'\right)U(z) \\
&= \left(D_c + C_cB' + C_c(zI - A)^{-1}2B'\right)U(z).
\end{aligned}$$

From which, the result in the bottom line on the right in Table 5.3 follows. □

REMARK 5.2 (Zero-order hold) This is another widely used discretization method that is exact when making use of a zero-order hold filter that keeps the input constant over the sample. Then, for the autonomous case:

$$\dot{x}_c(t) = A_c x_c(t).$$

The exact discretization to represent the state at the sample times $t = k\Delta$ is given by the difference equation:

$$x_c\big((k+1)\Delta T\big) = Ax_c\big(k\Delta T\big), \quad A = e^{A_c\Delta T}. \tag{5.4}$$

5.3.1.1 Discretizing Continuous-Time Decomposable Systems

As illustrated in Example 4.4, spatial discretization of a first principles (partial differential equation) model of a flexible membrane yields a continuous-time decomposable system governed by Equations (4.16). Let us, for the sake of brevity, focus on the discretization of the autonomous part, now denoted as:

$$\dot{x}_c(t) = (I \otimes A_{a,c} + P \otimes A_{b,c})x_c(t). \tag{5.5}$$

If the spectral radius of $A_c = (I_N \otimes A_{a,c} + P_N \otimes A_{b,c})$ is smaller than one, we will restrict the analysis to aproximating the matrix exponential in Equation (5.4) by the following series expansion:

$$e^{A_c\Delta T} \approx I + A_c\Delta T + A_c^2\frac{\Delta T^2}{2}.$$

When this is applied, due to A_c being a decomposable matrix (see Definition 4.3), we obtain the following difference equation that approximates the state $x_c(t)$ by $x(t)$ for $t = k\Delta T$:

$$x\big((k+1)\Delta T\big) = Ax\big(\Delta T\big),$$

for

$$\begin{aligned} A &= I + (I \otimes A_{a,c} + P \otimes A_{b,c})\Delta T + (I \otimes A_{a,c} + P \otimes A_{b,c})^2 \frac{\Delta T^2}{2} \\ &= I \otimes \left(I + A_{a,c}\Delta T + A_{a,c}\frac{\Delta T^2}{2} \right) \\ &\quad + P \otimes \left(A_{b,c}\Delta T + \left(A_{a,c}A_{b,c} + A_{b,c}A_{a,c}\right)\frac{\Delta T^2}{2} \right) + P^2 \otimes A_{b,c}^2 \frac{\Delta T^2}{2}. \end{aligned}$$

Two conclusions can be drawn from this brief analysis:

1. For the case that either ΔT is small such that $\frac{\Delta T^2}{2}$ and/or $A_{b,c}$ is nilpotent such that $A_{b,c}^2$ is zero, the discrete-time system matrix A remains decomposable.
2. When the above does not hold, A becomes a *special case* of a multi-linear system with system matrices defined in Equation (4.34), where some of the factor matrices are known (in terms of the pattern matrix P).

This analysis demonstrates, on one hand, that it is still useful to acquire information about the structure from first principles (continuous-time) modelling. On the other hand, it also shows that a variety of model classes is necessary as phenomena like discretization may transform (approximately) one model class into another.

5.3.1.2 Discretizing Continuous-Time State-Space Models

The class of block tridiagonal systems is a special case of the class of 1D state-space systems. For this reason they are treated within the scope of state-space systems.

Consider the (deterministic) continuous-time state-space model in Equation (5.1) with the system matrices A_c, B_c, C_c and D_c within the class of state-space system matrices. Since this class of matrices is closed under addition, multiplication and inversion, it follows that the discretized state-space model as given in Equation (5.2) also is an state-space system when using the integration rules in Lemma 5.1.

This same preservation holds when the matrix exponential $e^{A_c\Delta T}$ is approximated by only a finite number of terms.

5.3.1.3 Discretizing Continuous-Time Multi-Dimensional State-Space Models

Consider the two-dimensional, autonomous state-space model in continuous-time:

$$\dot{x}_c(t) = (A_{c,1} \otimes A_{c,2})x_c(t). \tag{5.6}$$

A two-term approximation of the matrix exponential yields a discrete-time transition matrix in the following form:

$$A = I + (A_{c,1} \otimes A_{c,2})\Delta T + \left(A_{c,1}^2 \otimes A_{c,2}^2\right)\frac{\Delta T^2}{2}. \tag{5.7}$$

This shows that the discrete-time approximation is still a multi-dimensional state model, although no longer with a system matrix of Kronecker rank one within $\mathcal{K}_{2,r}$.

Even when simplifying to a first-order approximation of the exponential, a different model class arises that contains an affine term. To see this, consider the discrete-time first-order approximation (autonomous case only) in terms of the 2D state matrix $X(k)$ (with $\text{vec}(X(k)) = x(k)$):

$$X(k+1) = X(k) + A_{c,1} X(k) A_{c,2}^T \Delta T. \tag{5.8}$$

5.3.2 Transforming Structured State-Space Models to Input-Output Models

The 'state-space model characterizes the complete computational structure of actual processes' (Yeung et al., 2010), which is also often the case for complex networks of dynamical systems. For that reason, we discuss in what way structural information is preserved (or not) when transforming some classes of structured state-space models to the input-output type of models.

Only the following three cases are considered:

1. Transforming block tridiagonal state-space models of Section 4.2.3 to the DNF/DSF of Section 3.4.
2. Transforming approximations of block tridiagonal state-space models given in innovation form into sparse VARMAX models. The latter are an extension of the model class considered in Section 3.3.
3. Transforming a two-dimensional state-space model to a FIR model with coefficient matrices in $\mathcal{K}_{2,r}$.

When a systematic treatment is not possible or still missing, examples are used to illustrate possible effects of the transformation.

5.3.2.1 Transforming Block Tridiagonal State-Space Models to DNF/DSF

Consider the class of block tridiagonal state-space models given by Equation (4.18). Such a model may, for example, arise from the block scheme of NDS in Figure 4.4. In the Example 5.1, we illustrate that the abundant sparsity in the state-space model (almost) gets lost when transforming it to either DNF or DSF. Prior to this example, we summarize the computational procedure to make such transformation for general state-space models. We restrict the discussion here to the deterministic transfer from input to output only.

Consider the following state-space model:

$$\begin{aligned} \tilde{x}(k+1) &= \tilde{A}\tilde{x}(k) + \tilde{B}u(k), \quad \tilde{x}(k) \in \mathbb{R}^n, \;\; u(k) \in \mathbb{R}^m, \\ y(k) &= \tilde{C}\tilde{x}(k) \quad y(k) \in \mathbb{R}^p. \end{aligned} \tag{5.9}$$

Then the first step is to find the invertible transformation matrix $T \in \mathbb{R}^{n\times n}$ such that,

$$\tilde{C}T^{-1} = C = \begin{bmatrix} I_p & 0 \end{bmatrix}.$$

Such a transformation matrix can always be found when $\tilde{C}$ has full row rank, an assumption that is always made in such transformations (Woodbury, 2019). Applying this transformation to the state equation defines the following matrices and a new state vector:

$$T\tilde{A}T^{-1} = A, \quad T\tilde{B} = B, \quad x(k) = T\tilde{x}(k).$$

Next, partition this transformed state equation according to the partitioning of the C-matrix as follows:

$$\begin{aligned} \begin{bmatrix} x_1(k+1) \\ x_2(k+1) \end{bmatrix} &= \begin{bmatrix} A_{11} & A_{12} \\ A_{21} & A_{22} \end{bmatrix} \begin{bmatrix} x_1(k) \\ x_2(k) \end{bmatrix} + \begin{bmatrix} B_1 \\ B_2 \end{bmatrix} u(k), \\ y(k) &= x_1(k). \end{aligned} \tag{5.10}$$

Now assume zero initial conditions and consider the bottom row of the above state equation, and assuming all z-transforms exist, we can write $X_2(z)$ as:

$$X_2(z) = (zI - A_{22})^{-1} A_{21} X_1(z) + (zI - A_{22})^{-1} B_2 U(z),$$

and $X_1(z)$ becomes:

$$X_1(z) = (A_{11} + A_{12}(zI - A_{22})^{-1} A_{21}) X_1(z) + (B_1 + A_{12}(zI - A_{22})^{-1} B_2) U(z).$$

Since $y(k) = x_1(k)$, we therefore have the following DNF formulation:

$$Y(z) = W(z)Y(z) + V(z)U(z), \tag{5.11}$$

where $W(z) = (A_{11} + A_{12}(zI - A_{22})^{-1} A_{21})$ and $V(z) = (B_1 + A_{12}(zI - A_{22})^{-1} B_2)$.

To find the corresponding DSF, the principle of hollow abstraction is mentioned before Equation (2.12), with its result in Equation (2.13). This result applied to Equation (5.11) yields:

$$\begin{aligned} Y(z) &= (I_p - \operatorname{diag}(W(z))^{-1}(W(z) - \operatorname{diag}(W(z))Y(z) \\ &\quad + (I_p - \operatorname{diag}(W(z))^{-1} V(z)U(z) \\ &= Q(z)Y(z) + P(z)U(z). \end{aligned} \tag{5.12}$$

Example 5.1 (Transformation triadiagonal SSM to DNF/DSF) Consider the series of systems in Equation (4.18) for $n_i = 2$ and $i = 1:4$. Each $u_i(k)$ and $y_i(k)$ are scalar, and the lifted triplet (A, B, C) of system matrices is:

$$A = \begin{bmatrix} 0.2858 & 0.1622 & 0 & 0 & 0 & 0 & 0 & 0 \\ 0.7792 & 0.7572 & 0.7943 & 0 & 0 & 0 & 0 & 0 \\ 0 & 0.9340 & 0.7537 & 0.3112 & 0 & 0 & 0 & 0 \\ 0 & 0 & 0.1299 & 0.3804 & 0.5285 & 0 & 0 & 0 \\ 0 & 0 & 0 & 0.5688 & 0.5678 & 0.1656 & 0 & 0 \\ 0 & 0 & 0 & 0 & 0.4694 & 0.0759 & 0.6020 & 0 \\ 0 & 0 & 0 & 0 & 0 & 0.0119 & 0.0540 & 0.2630 \\ 0 & 0 & 0 & 0 & 0 & 0 & 0.3371 & 0.5308 \end{bmatrix},$$

$$B = \begin{bmatrix} 1.3790 & 0 & 0 & 0 \\ -1.0582 & 0 & 0 & 0 \\ 0 & -0.4686 & 0 & 0 \\ 0 & -0.2725 & 0 & 0 \\ 0 & 0 & 1.0984 & 0 \\ 0 & 0 & -0.2779 & 0 \\ 0 & 0 & 0 & 0.7015 \\ 0 & 0 & 0 & -2.0518 \end{bmatrix},$$

$$C = \begin{bmatrix} -2.1384 & -0.8396 & 0 & 0 & 0 & 0 & 0 & 0 \\ 0 & 0 & 1.3546 & -1.0722 & 0 & 0 & 0 & 0 \\ 0 & 0 & 0 & 0 & 0.9610 & 0.1240 & 0 & 0 \\ 0 & 0 & 0 & 0 & 0 & 0 & 1.4367 & -1.9609 \end{bmatrix}.$$

This triplet is minimal. Figures 5.2 and 5.3 show the Bode plots of the transfer functions $W(z)$ and $V(z)$ of the DNF given by Equation (5.11). The Bode plot of the transfer function $P(z)$, derived in Equation (5.12), is shown in Figure 5.4.

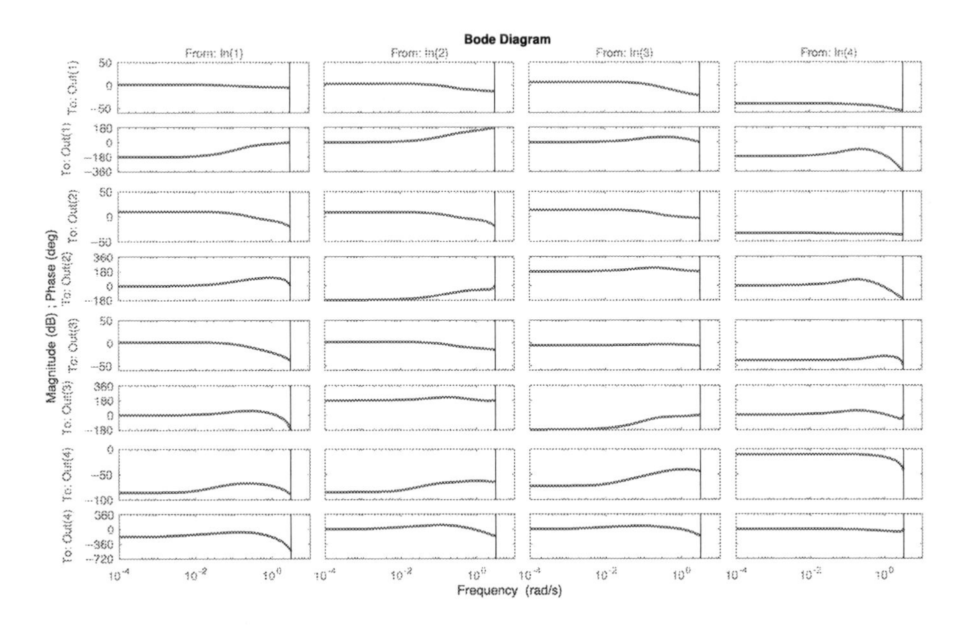

Figure 5.2 Bode plot of the 4×4 transfer function $W(z)$ in Equation (5.11) of the DNF derived from the state-space model in block tridiagonal form given by the matrices (A, B, C).

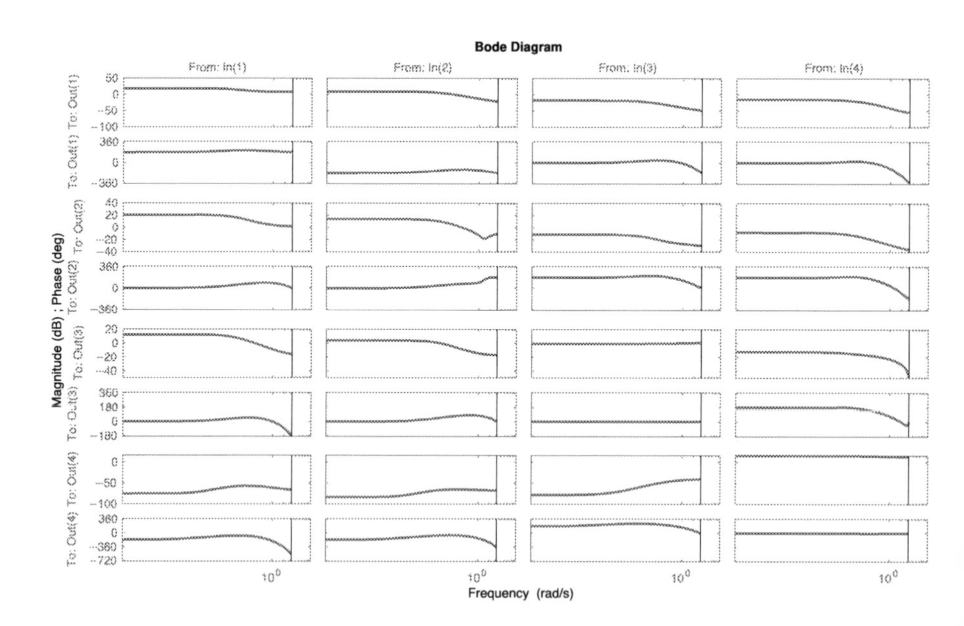

Figure 5.3 Bode plot of the 4×4 transfer function $V(z)$ in Equation (5.11) of the DNF derived from the state-space model in block tridiagonal form given by the matrices (A, B, C).

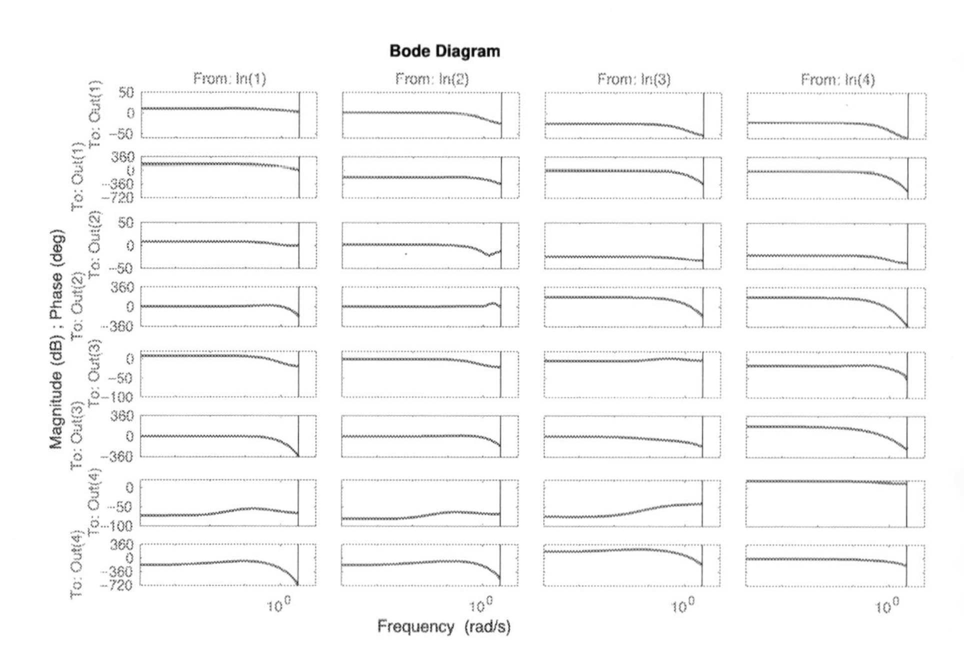

Figure 5.4 Bode plot of the 4×4 transfer function $P(z)$ in Equation (5.12) of the DSF derived from the state-space model in block tridiagonal form given by the matrices (A, B, C).

From the example, the following observations can be made:

1. *The loss of sparsity*: Though the system matrices (A, B, C) are very sparse, little of that sparsity can be seen in either the DNF models $W(z), V(z)$ or in the DSF indicated by $P(z)$.
2. *Identifiability*: The transfer function $P(z)$ turns out to be non-diagonal. For uniqueness or identifiability of a DSF for this deterministic case with square $P(z)$, it is often indicated to take $P(z)$ diagonal (Yue et al., 2018c). The diagonal property is not preserved when going from the given state-space model to a DSF.

REMARK 5.3 There have been attempts to parametrize state-space models for specific cases of square systems and special noise conditions in order to guarantee identifiability of the DSF (Yue et al., 2018c). In this paper it is also stated that sparsity-inducing techniques, discussed in the next chapter, bringing sparsity into the system matrices of the state-space model might lead to sparse DSF. However Example 5.1 shows that this is not always the case. The reverse also does not hold. That is 'non-sparse A matrices of a state-space model may lead to sparse $Q(z)$ of a DSF' (Yue et al., 2018c).

5.3.2.2 Transforming Block Tridiagonal Innovation State-Space Models to Sparse VARX Models

To discuss this transformation, we first discuss a standard least-squares method to estimate the Kalman gain in an innovation state-space model from input-output data without knowledge of the noise covariance matrices. This is often lacking in practical case studies. The method is based on Juang et al. (1993).

We start with a general formulation and then return to the topic of this subsection. Consider the following innovation from Verhaegen and Verdult (2007):

$$\hat{x}(k+1) = \underbrace{(A - KC)}_{\Phi}\hat{x}(k) + Ky(k),$$
$$y(k) = C\hat{x}(k) + e(k), \tag{5.13}$$

where $e(k)$ is a zero-mean white-noise sequence. As with many practical applications, the covariance matrix of the noise sequence is unknown, and hence one considers the estimation of the Kalman gain K from data.

The output of Equation (5.13) can be written as:

$$y(k) = \left(\sum_{i=1}^{k} C\Phi^{i-1}Ky(k-i)\right) + e(k) + C\Phi^k\hat{x}(0). \tag{5.14}$$

As the spectral radius of the matrix ϕ may be assumed to be strictly smaller than one, taking $k \geq s$ for s large enough it will hold that $\Phi^k\hat{x}(0)$ can be neglected. Therefore, defining $\mathcal{L}$ as:

$$\mathcal{L} = \begin{bmatrix} CK & C\Phi K & \cdots & C\Phi^{s-1}K \end{bmatrix}, \tag{5.15}$$

we can define the following least-squares problem to estimate $\mathcal{L}$:

$$\min_{\mathcal{L}} \left\| \begin{bmatrix} y(s+1) & \cdots & y(N_t) \end{bmatrix} - \mathcal{L} \begin{bmatrix} y(s) & \cdots & y(N_t-1) \\ y(s-1) & \cdots & y(N_t-2) \\ \vdots & \ddots & \vdots \\ y(1) & \cdots & y(N_t-s) \end{bmatrix} \right\|_F^2 . \tag{5.16}$$

This is precisely the equation used to estimate the parameters of the VAR model. For retrieving a sparse VAR model, we refer to Chapter 6.

However, in case partial information about this model in terms of knowledge of the pair (A, C) is known, we can, as outlined in Juang et al. (1993), define a dedicated least-squares problem for estimating K only. For this purpose, we define the following two pairs of quantities:

$$\alpha_i = C\Phi^{i-1}K,$$
$$\beta_i = CA^{i-1}K,$$

and both quantities are related recursively as:

$$\beta_i = \alpha_i + \sum_{j=1}^{i-1} \alpha_{i-j}\beta_j .$$

From the estimates of α_i, obtained by solving the least squares problem in Equation (5.16), the above equation allows us to recursively find estimates of β_i. Denote these estimates as $\hat{\beta}_i$, then the definition of β_i allows to formulate the following least-squares problem in K:

$$\min_{K} \left\| \begin{bmatrix} C \\ CA \\ \vdots \\ CA^{s-1} \end{bmatrix} K - \begin{bmatrix} \hat{\beta}_1 \\ \hat{\beta}_2 \\ \vdots \\ \hat{\beta}_s \end{bmatrix} \right\|_F^2 . \tag{5.17}$$

When the pair (A, C) has the block triangular form, this form can be fully exploited to estimate K efficiently while (possibly) enforcing the latter gain matrix to be sparse (block triangular) as well.

It should be remarked that when the Kalman gain matrix K is block tridiagonal, the coefficient matrices in $\mathcal{L}$ in Equation (5.15) also have that structure, although their bandwith might increase depending on the multiplication in the definition of these coefficient matrices. The consequence is that in estimating the coefficient matrices of a VAR model, the latter might then also be assumed to be block triangular.

5.3.2.3 Transforming Two-Dimensional State-Space Models to Tensor FIR Models

When considering the two-dimensional state-space model in Equation (4.31) with the system matrices (A, B, C) given as in Equation (4.32), we have the following TSSM:

$$
\begin{aligned}
X(k+1) &= A_1 X(k) A_2^T + B_1 U(k) B_2^T, \\
Y(k) &= C_1 X(k) C_2^T + E(k).
\end{aligned} \tag{5.18}
$$

When assuming the data to be recorded or stored in a square grid, the data matrices have the following form:

$$
U(k) = \begin{bmatrix} u_{1,1}(k) & \cdots & u_{1,N}(k) \\ \vdots & \ddots & \vdots \\ u_{N,1}(k) & \cdots & u_{N,N}(k) \end{bmatrix},
$$

where each $u_{i,j}(k) \in \mathbb{R}^m$. The output matrix $Y(k)$ is defined similarly from the local signal $y_{i,j}(k)$. In a compatible way the state matrix $X(k)$ is defined. The correspondence of the TSSM to input-output models is briefly investigated by simulating these type of models. That is, for a given input (matrix) sequence we then question how the output (matrix) evolves. Given Equation (5.18), the input-output relationship in matrix form reads,

$$
Y(k) = C_1 A_1^{k-1} X(1) \big(C_2 A_2^{k-1}\big)^T + \sum_{i=1}^{k-1} C_1 A_1^{k-i-1} B_1 U(i) \big(C_2 A_2^{k-i-1} B_1\big)^T + E(k). \tag{5.19}
$$

When both local system matrices A_i for $i = 1, 2$ have spectral radius smaller than one, for 'large' k the effect of the initial condition $X(1)$ decays out. Therefore, when these assumptions hold the output may be approximated by the following FIR model:

$$
\begin{aligned}
Y(k) &\approx \sum_{i=1}^{s} C_1 A_1^{i-1} B_1 U(k-i) \big(C_2 A_2^{i-1} B_1\big)^T + E(k), \\
y(k) &\approx \sum_{i=1}^{s} \Big(C_2 A_2^{i-1} B_2 \otimes C_1 A_1^{i-1} B_1\Big) u(k-i) + e(k),
\end{aligned} \tag{5.20}
$$

where it is assumed that $y(k) = \text{vec}(Y(k))$ and similarly for the input.

The FIR model on the right-hand side of Equation (5.20) has the form of the Quarks model (3.23) with the factor matrices of Kronecker rank one in $\mathcal{K}_{2,r}$ (see Definition 3.7).

5.4 Conclusions

The model classes defined in Chapters 3 and 4 can easily be transformed into another class for the case of lumped parameter systems where we have dense coefficient or system matrices. However, this picture completely changes for the case of large DNS.

Taking the sparsity of the coefficient/system matrices or of the data into account, the transformation from one model class to another can completely destroy this model sparsity.

This may also happen when discretizing continuous-time models. This is unfortunate as most structural information about sparsity patterns in matrices in dynamic models comes from continous-time models. What the best model structure is for a particular application now becomes (much) more difficult due to these additional problems.

Part II

The Identification Methods

6 Identification of Signal Flow Dynamic Networks

In this chapter, we focus on identification methods for signal flow dynamic networks. The identification methods presented belonging to the following three classes of model:

1. Sparse VARX models (Section 3.3). A brief overview of the methods for inducing sparsity in the coefficient matrices of autoregressive models such as the use of non-differentiable priors like the ℓ_1 norm is presented. If the pattern of the non-zero entries in the coefficient matrices is known, e.g. coefficient matrices are banded, then the computations can be carried out in parallel using only local data.
2. Dense data-sparse VARX models including Quarks (Section 3.5) and tensor VARX models (Section 3.6). This section is particularly targeted at multi-dimensional systems, e.g. where the data stems from a sensor array. If the spatial dynamics are separable in the main spatial directions of the system, then the coefficient matrices are compactly written with a sum of few Kronecker products. In this section, the way to exploit this structure for solving large least squares in a scalable manner and requiring less temporal samples is reviewed. It is first presented for 2D systems, which feature matrices of sensor measurements instead of the usual vectors, and is then extended to the more general case when sensor measurements are re-casted into tensors.
3. Dynamic network function (DNF) and dynamic structure function (DSF) models. Here, the focus is on two global identification methods for such networks. The first treats the estimation of the rational transfer functions in the model as a series of multi-input single-output (MISO) prediction error problems. The second method directly develops a one-step ahead predictor for DSF models. We end the chapter by discussing ways developed in the context of DSF to estimate single transfer functions in the model.

This chapter is the dual from Chapter 3 that introduced these model classes.

6.1 Sparse VAR(X) Models

PROBLEM 6.1 The generalized VARX model is:

$$A_0 y(k) = \sum_{i=1}^{n_a} A_i y(k-i) + \sum_{i=0}^{n_b} B_i u(k-i) + w(k), \tag{6.1}$$

where $y(k) \in \mathbb{R}^{Np}$ and $u(k) \in \mathbb{R}^{Nm}$, and such that $w(k)$ is a zero-mean white noise with diagonal covariance matrix, and A_0 is a lower triangular matrix with ones on the diagonal. Estimate sparse coefficient matrices A_i, B_i thereby establishing a trade-off between compactness and accuracy of the model representation.

This problem does not focus on alleviating the computational load, but rather on finding sparse estimates of coefficient matrices, which is subsequently useful for controller design. A clear distinction in the possible strategies is whether the position of the non-zero entries is known.

The triangular matrix A_0 is the square root (up to a diagonal matrix scaling) of the inverse of a possibly full covariance matrix. The uncorrelatedness between stochastic variables, and hence the topology of the network, is reflected by zero entries in the inverse of the covariance matrix of these stochastic variables (Dahlhaus and Eichler, 2003). The combination of the stochastic properties of $w(k)$ and the triangular structure of the coefficient matrix A_0 allow us to decompose the estimation of the coefficient matrices of the VARX model block-row by block-row. When just assuming the data vectors to be large vectors with no further partitioning, we can then formulate the following series of independent least-squares problems, for all j in the set $\{1, \ldots, Np\}$:

$$\min_{X_j} \sum_{k=max(n_a,n_b)+1}^{N_t} \Bigg\| y_j(k) + A_0(j, 1:j-1)y_{1:j}(k) - \sum_{i=1}^{n_a} A_i(j,:)y(k-i) - \sum_{i=1}^{n_b} B_i(j,:)u(k-i) \Bigg\|_F^2 + \gamma g(X_j), \quad (6.2)$$

where $y_j(k)$ indicates the jth entry in the vector $y(k)$ and $y_{r:s}$ denotes the partitioned part of $y(k)$ from the rth row until the sth one. The convention is that a vector vanishes if one of its arguments becomes smaller or equal to zero. The vector X_j (with growing dimension with increasing index j) contains the unknowns of the least-squares problem and is defined explicitly as,

$$X_j = \begin{bmatrix} A_0(j, 1:j-1) & \ldots & A_i(j,:) & \ldots & B_i(j,:) & \ldots \end{bmatrix}.$$

The cost function $g(X_j)$ strives to introduce sparsity in the coefficient matrices, for example when minimizing the number of non-zero entries using the *card* operator,

$$g(X_j) = \text{card}(X_j). \quad (6.3)$$

This operator makes the regularized least-squares Problem (6.2) non-convex. Proxies have been proposed to retain convexity in e.g. Efron et al. (2004); Tibshirani (1996); Wen et al. (2018). These include the following:

$$\begin{aligned} g_1(X_j) &= \|\text{vec}(X_j)\|_1, \\ g_2(X_j) &= \sum_{i=1}^{p-1} \|A_0(j, 1:j-1)\|_F^2 + \sum_{i=1}^{n_a} \|A_i(j,:)\|_F^2 + \sum_{i=1}^{n_b} \|B_i(j,:)\|_F^2. \end{aligned} \quad (6.4)$$

Regularization to sparsify the coefficient matrices of autoregressive models is proposed in Chiuso and Pillonetto (2012) by weighting the system's spatial and temporal impulse responses. The temporal stability can be influenced by constraining the impulse response by a Gaussian process with mean zero and chosen covariance matrix. Such a constraint ensures that the parameters of the impulse response decay in magnitude. In case of high spatial coupling between the nodes in the network, this allows us to tune the strong or weak influence between neighbouring systems by tuning the coefficients of such Gaussian processes. This constraint is taken as a regularization to the cost function for estimating the model parameters of the network. This is an alternative to ℓ_1-norm regularization that approximates the attempt to minimize the number of non-zero elements in the parameter vector. Such regularization does not, however, reduce the computational complexity for estimating the coefficient matrices. Further prior knowledge can be added, when available, such as the position of the non-zero entries.

Graphs or regular networks that result from (spatial) discretization of PDEs are such that each node is connected to only a small number of other nodes, which are small compared to the size of the network. When only a small number of nodes are connected to one another, the adjacency (or coefficient) matrix will become a multi-banded structure (see Example 4.6). As an example of an application in modelling for control of adaptive optics systems, we refer to Piscaer (2016) where constrained regions on the indices of the coefficient matrices are proposed in which the coefficients are zero based on physical insight. The coefficient matrices have very few non-zero entries, and the position of these non-zero entries is known, thereby considerably reducing the computational cost for the estimation.

6.2 Quarks Models

In this section, which is based on Sinquin and Verhaegen (2019b), we restrict ourselves to the class of VARX models without external input. Much of the machinery explained in this section can be extended to other model classes such as FIR models and possibly even further ARMAX, Box Jenkins, etc. models.

6.2.1 The Model Class

The Quarks model was defined in Equation (3.19), and is repeated here,

$$y(k) = \sum_{i=1}^{n} \left(\sum_{j=1}^{r_j} M_{i,j,2} \otimes M_{i,j,1} \right) y(k-i) + w(k), \tag{6.5}$$

with $y(k) = \text{vec}(Y(k))$ for $Y(k) \in \mathbb{R}^{pN \times N}$ as,

$$Y(k) = \begin{bmatrix} y_{1,1}(k) & y_{1,2}(k) & \cdots & y_{1,N}(k) \\ y_{2,1}(k) & y_{2,2}(k) & & y_{2,N}(k) \\ \vdots & \vdots & \ddots & \vdots \\ y_{N,1}(k) & y_{N,2}(k) & \cdots & y_{N,N}(k) \end{bmatrix} \tag{6.6}$$

and the coefficient matrices $(M_{i,j,1}, M_{i,j,2}) \in \mathbb{R}^{pN \times pN} \times \mathbb{R}^{N \times N}$. This model can also be written in matrix form; this is more commonly used when deriving scalable algorithms such as Equation (3.20):

$$Y(k) = \sum_{i=1}^{n} \sum_{j=1}^{r_i} \left(M_{i,j,1} Y(k-i) M_{i,j,2}^T \right) + W(k). \tag{6.7}$$

6.2.2 The Identification Problem

PROBLEM 6.2 Given the structure of the Quarks model Equation (6.7), with the output data stored in matrix format, the problem of identifying this model from measurement sequences $\{Y(k)\}_{k=1\ldots N_t}$ requires finding the following four sets of data:

1. The temporal order index n.
2. The spatial order index r_i for each coefficient matrix.
3. The parameter vectors $a_i^{(j)}$ and $b_i^{(j)}$ that parametrize the matrices $M\left(a_i^{(j)}\right)$ and $M\left(b_i^{(j)}\right)$. Examples of such parametrizations can be induced by selecting the coefficient matrices to be Toeplitz or banded.
4. The estimation of the parameter vectors $a_i^{(j)}$, $b_i^{(j)}$ up to an ambiguity in the model Equation (6.7) (that allows us to scale the matrices $M_{i,j,1}$ by a scalar and $M_{i,j,2}$ by the inverse scaling without changing the input-output behaviour), via the minimization of the cost function for data batches with N_t temporal points:

$$\min_{a_i^{(j)}, b_i^{(j)}} \sum_{k=n+1}^{N_t} \left\| Y(k) - \sum_{i=1}^{n} \left(\sum_{j=1}^{r_i} M\left(a_i^{(j)}\right) Y(k-i) M\left(b_i^{(j)}\right) \right) \right\|_F^2 \tag{6.8}$$

It is assumed that the vectors $a_i^{(j)}$ and $b_i^{(j)}$ parametrize the coefficient matrices $M\left(a_i^{(j)}\right)$ and $M\left(b_i^{(j)}\right)$ in an affine manner.

By the selection of the parameter n and the particular choices of the parametrization in step 3 above, various special cases of restricting the coefficient matrices A_i to particular sets such as $\mathcal{K}_{2,r_i}$ can be considered. Additional constraints to the above least-squares cost function in Equation (6.8) might be added to introduce sparsity (or small values) in the parameter vectors $a_i^{(j)}$ and $b_i^{(j)}$.

An important challenge in solving the minimization Problem (6.8) is restricting as much as possible the *computational cost* and this especially the case when the size of the array N is large.

The cost function Equation (6.8) does not have a unique solution by the highlighted ambiguity in part 4 of Problem 6.2. However, as highlighted in the solution of this problem, this is not an issue when using the model for predicting the output.

One way to solve this optimization problem is by writing it as:

$$\begin{aligned} \min_{A_i, a_i^{(j)}, b_i^{(j)}} \quad & \sum_{k=n+1}^{N_t} \left\| y(k) - \sum_{i=1}^{n} A_i y(k-i) \right\|_2^2 \\ \text{such that} \quad & A_i = \sum_{j=1}^{r_i} M\left(b_i^{(j)}\right)^T \otimes M\left(a_i^{(j)}\right). \end{aligned} \tag{6.9}$$

As the re-shuffling operator $\mathcal{R}$ defined (for fixed dimensions) in Example B.1 is bijective in $\mathbb{R}^{pN^2 \times pN^2}$, this minimization problem is rewritten with,

$$\begin{aligned} \min_{A_i, U_i, V_i} \quad & \sum_{k=n+1}^{N_t} \left\| y(k) - \sum_{i=1}^{n} A_i y(k-i) \right\|_2^2 \\ \text{such that} \quad & \mathcal{R}(A_i) = U_i V_i^T, \end{aligned} \tag{6.10}$$

where:

$$\begin{aligned} U_i &= \left[\text{vec}\left(M(a_i^{(1)})\right) \quad \dots \quad \text{vec}\left(M(a_i^{(r_i)})\right), \right] \\ V_i &= \left[\text{vec}\left(M(b_i^{(1)})\right) \quad \dots \quad \text{vec}\left(M(b_i^{(r_i)})\right). \right] \end{aligned}$$

For a non-singular transformation $T_i \in \mathbb{R}^{r_i \times r_i}$, the constraint in Equation (6.10) can be equivalently written as:

$$\mathcal{R}(A_i) = \widetilde{U}_i \widetilde{V}_i^T, \tag{6.11}$$

where $\widetilde{U}_i = U_i T_i$ and $\widetilde{V}_i = V_i T_i^{-T}$. The solution to problem in Equation (6.9) remains non-unique as the factor matrices can be scaled by reciprocal values. This is not an issue for practical use of Quarks models when using these models for prediction.

6.2.2.1 General Algorithmic Strategy

The general idea of the algorithm is first, to add regularization to enforce less influence from the sensor measurements with increasing spatial or temporal distance; second, to estimate the coefficient matrices without forming the Kronecker products as this would kill all computational benefits. Instead of solving a large convex optimization problem, the coefficient matrices are parametrized with a sum of few Kronecker products which allows us to write another optimization problem that is bilinear in its unknowns although with a reduced set of unknowns.

6.2.3 Regularization Inducing Stability and Sparsity

Two additional regularizations are added to the cost function Equation (6.9) that do not alter the convergence properties of the iterative solution developed.

We assume that the Kronecker rank is independent of i, and introduce the following notation for gathering all left factor matrices:

$$M_{a_i} = \left[M\left(a_i^{(1)}\right) \quad \cdots \quad M\left(a_i^{(r)}\right) \right], \quad M_a = \left[M_{a_1} \quad \cdots \quad M_{a_n} \right].$$

The matrix M_b is defined similarly. Here, we simplify the notation so that all r_is are taken equal to be r.

6.2.3.1 Stability of VAR Models

A diagonal-correlated kernel (Chen et al., 2012) is used to regularize Equation (6.8) and induce the identification of bounded-input bounded-output (BIBO) stable VAR models. The associated positive definite matrix P_t is defined as,

$$P_t(i,j) = \xi^{\frac{i+j}{2}} \eta^{|i-j|}, \tag{6.12}$$

for $i,j \in \{1,\ldots,n\}$, and where the hyperparameters η, ξ are such that $-1 \leq \eta \leq 1$ and $0 \leq \xi < 1$. These allow us to influence both the decay rate of the impulse response and its smoothness. Let W_t be a square root of the matrix P_t^{-1}, let $Q_t = W_t \otimes I_{N^2}$ and further assume that there is no prior information to distinguish between the different factor matrices, then restricting the decay rate of the impulse response can be influenced by adding the following regularization term to the cost function in Equation (6.10):

$$r_t(M_a, M_b) = \sum_{j=1}^{r} \left\| Q_t \begin{bmatrix} U_1(:,j)V_1(:,j)^T \\ \vdots \\ U_n(:,j)V_n(:,j)^T \end{bmatrix} \right\|_F^2 . \tag{6.13}$$

This regularization term r_t is bilinear in its unknowns.

6.2.3.2 Spatial Sparsity

For systems that stem from a discretized PDE, the influence of the neighbours decays with distance. The entries of the factor matrices then decay away from the main diagonal (in absolute value). To favour the identification of such patterns, regularization is proposed with an appropriate weighting. The following notation is helpful.

NOTATION 6.1 (Diagonal operator) Let $i \in \{0,\ldots,N-1\}$. For a matrix $X \in \mathbb{R}^{N\times N}$, the notation $\text{diag}(X,i)$ is used to indicate the ith diagonal above the main diagonal, while $\text{diag}(X,-i)$ denotes the ith diagonal below the main diagonal. Both vectors are concatenated into a vector d_i as follows:

$$d_i = \left[\text{diag}(X,i)^T \quad \text{diag}(X,-i)^T \right].$$

With this definition, we can reshape the entries of this square matrix X diagonal-wise as defined in the following operator $\mathcal{D}(X)$:

$$\mathcal{D}(X) = \left[\text{diag}(X,0)^T \quad d_1^T \quad \ldots \quad d_{N-1}^T \right]^T .$$

This operation defines a vector in $\mathbb{R}^{N^2}$.

A diagonal matrix $K_s \in \mathbb{R}^{N^2 \times N^2}$ that is used to weight the entries in the factor matrices is introduced:

$$K_s = \begin{bmatrix} I_N k_1 & 0 & \dots & 0 \\ 0 & I_{2(N-1)} k_2 & \ddots & \vdots \\ \vdots & \ddots & \ddots & 0 \\ 0 & \dots & 0 & I_2 k_N \end{bmatrix}, \tag{6.14}$$

where the scalars k_i are such that $0 < k_i < k_{i+1}$. A possible choice of these scalars is $k_i = e^{\zeta i}$ for $\zeta > 0$. The prior information in order for the entries above and below the diagonal of the factor matrices $M(a_i^{(j)}), M(b_i^{(j)})$ to decay away from the main diagonal is then induced by adding the following constraint to Equation (6.10):

$$r_s(M_a, M_b) = \sum_{i=1}^{n} \sum_{j=1}^{r} \left\| K_s \mathcal{D}\left(M(a_i^{(j)})\right) \mathcal{D}\left(M(b_i^{(j)})\right)^T K_s^T \right\|_F^2$$

6.2.3.3 The Regularized Cost Function for Quarks Identification

The cost function for the identification of sparse stable Quarks models reads:

$$\min_{a_i^{(j)}, b_i^{(j)}} \quad \sum_{k=n+1}^{N_t} \left\| Y(k) - \sum_{i=1}^{n} \left(\sum_{j=1}^{r_i} M\left(a_i^{(j)}\right) Y(k-i) M\left(b_i^{(j)}\right) \right) \right\|_F^2$$
$$+ \mu r_t(M_a, M_b) + \lambda r_s(M_a, M_b), \tag{6.15}$$

where μ, λ are two regularization coefficients. The optimization problem (6.15) is a multi-convex optimization problem. That is when we fix but one of its decision vectors, that is, either $a_i^{(j)}$ or $b_i^{(j)}$, the remaining optimization problem is convex. The regularization terms r_t and r_s aim to make the estimation problem more resilient to noise and the shortness of the data batches. These are contrary to the regularization in Li et al. (2013) and not meant for speeding up the convergence.

REMARK 6.1 The regularization in optimization problem (6.15) is *bilinear*, contrary to the one analyzed in Baldi and Hornik (1989) and Udell et al. (2016) within the framework of principal component analysis (PCA). Based on the latter, a regularization for minimization problem (6.10) would minimize a (weighted) sum of the Frobenius norm of the factor matrices.

6.2.4 An Alternating Least-Squares Approach

The regularized least-squares cost function (6.15) is bilinear in its unknowns. The latter are compact factor matrices. This has the advantage that additional structure can be more compactly imposed on the factor matrices compared to imposing constraints on the large-scale reshuffling of the coefficient matrices in optimization problem (6.10) via a low-rank constraint.

The optimization problem (6.9) can be done in a couple of manners. One way is using the non-linear optimization framework based on separable least-squares as outlined in Bruls et al. (1999). A second way is via iterative hierarchical algorithms that have been derived as a generalization of the linear Gauss–Seidel iterations for solving coupled Sylvester matrix equations in Ding and Chen (2005). A third way, followed in this book, is based on the work of Hoff (2015), and uses alternating least squares (ALS). The latter is a special case of the block *non-linear* Gauss–Seidel method highlighted in Li et al. (2013).

The ALS algorithm is summarized in Algorithm 6.1 with the help of the additional notation,

$$M_{a_i} \lozenge Y(k) = \left[M\left(a_i^{(1)}\right) Y(k) \quad \ldots \quad M\left(a_i^{(r)}\right) Y(k) \right],$$
$$M_{b_i} \lozenge Y(k) = \left[M\left(b_i^{(1)}\right) Y(k) \quad \ldots \quad M\left(b_i^{(r)}\right) Y(k) \right]. \tag{6.16}$$

The least-squares problems in lines 4 and 5 of Algorithm 6.1 are alternately solved in an iterative manner. This iteration starts with some random initial guess for M_a. The end of the iteration cycle is controlled by the stopping rule in line 3 of the Algorithm.

REMARK 6.2 (Convergence of Algorithm 6.1) The convergence of Algorithm 6.1 has been studied in Sinquin and Verhaegen (2019b) for special conditions. The convergence analysis is based on an extension of the work in Li et al. (2015) to consider matrices as unknown rather than only vectors. At each iteration, it rescales the solution $M_b^{(\kappa)}$ such that the norm of each column coincides with the value of the corresponding column of the true M_b. Such knowledge is, however, lacking in general. However, without such a normalization, (monotonic) convergence has been observed in practical simulation studies.

The computational complexity of Algorithm 6.1 is dominated mainly by the cost of solving the least-squares problems in lines 4 and 5. When solving a least-squares problem of the form,

$$\min_x \|y - Ax\|_2, \quad y \in \mathbb{R}^m \quad x \in \mathbb{R}^n,$$

with the RQ factorization, the computational complexity is $\mathcal{O}(2mn^2)$ (Golub and Van Loan, 1996). Therefore, the complexity is as summarized in Table 6.1.

6.2.5 Numerical Example

Two examples will be shown. The first example in this section is a randomly generated Quarks model that we use to demonstrate the convergence properties of Algorithm 6.1, and the second that is presented in Chapter 10 is an application to adaptive optics.

6.2.5.1 Illustrating Convergence

In this example we consider a Quarks model whose temporal order is known and the coefficient matrices are randomly generated and of known Kronecker rank. A Quarks model of the following (matrix) finite impulse response (FIR) form is considered,

Algorithm 6.1: ALS for QUARKS identification

Input : $\{Y(k)\}_{k=1\ldots N_t}, r, n$
Output: $\{\widehat{M}_{a_j}, \widehat{M}_{b_j}\}_{j=1\ldots n}$

```
/* Default values                                                */
```
1 $\kappa = 1, \kappa_{max} = 50, \epsilon = \infty, \epsilon_{min} = 10^{-3}$
```
/* Initial guesses                                               */
```
2 $M_a =$ `rand(N,nrN)` $\quad c^{(0)} = 0$
```
/* Start ALS                                                     */
```
3 **while** $\kappa < \kappa_{max}$ *and* $\epsilon > \epsilon_{min}$ **do**

/* Least Squares Problem over M_b */

4 $\quad \min_{b_i^{(j)}} \sum_{k=n+1}^{N_t} \left\| Y(k) - \left[M_{a_1}^{(\kappa-1)} \Diamond Y(k-1) \quad \ldots \quad M_{a_n}^{(\kappa-1)} \Diamond Y(k-n) \right] \begin{bmatrix} M_{b_1}^T \\ \vdots \\ M_{b_n}^T \end{bmatrix} \right\|_F^2$

/* Least Squares Problem over M_a */

5 $\quad \min_{a_i^{(j)}} \sum_{k=n+1}^{N_t} \left\| Y(k)^T - \left[M_{b_1}^{(\kappa)} \Diamond Y(k-1)^T \quad \ldots \quad M_{b_n}^{(\kappa)} \Diamond Y(k-n)^T \right] \begin{bmatrix} M_{a_1}^T \\ \vdots \\ M_{a_n}^T \end{bmatrix} \right\|_F^2$

/* Check stopping criterion */

6 $\quad c^{(\kappa)} \leftarrow \sum_{k=n+1}^{N_t} \left\| Y(k)^T - \left[M_{b_1}^{(\kappa)} \Diamond Y(k-1)^T \quad \ldots \quad M_{b_n}^{(\kappa)} \Diamond Y(k-n)^T \right] \begin{bmatrix} (M_{a_1}^{(\kappa)})^T \\ \vdots \\ (M_{a_n}^{(\kappa)})^T \end{bmatrix} \right\|_F^2$

7 $\quad \epsilon \leftarrow |c^{(\kappa)} - c^{(\kappa-1)}|$

8 $\quad \kappa \leftarrow \kappa + 1$

9 **end**

$$Y(k) = \sum_{i=1}^{n} \sum_{j=1}^{r} M\left(a_i^{(j)}\right) U(k-i) M\left(b_i^{(j)}\right), \tag{6.17}$$

for a sensor array with $N = 10$, and a single sensor measurement at each node, $p = 1$. The random distribution of the generated factor matrices $M(a_i^{(j)})$ and $M(b_i^{(j)})$ is uniform, and the entries of the input matrix $U(k)$ are independent random variables with a gaussian distribution with mean zero and unit variance. The number of temporal samples N_t is set to 1 000. Both regularization parameters λ, μ in the Quarks cost function (6.15), are straightforwardly adapted here to the FIR model, are set to zero. Figure 6.1 illustrates the global convergence of the least squares cost function after a number of iterations. This is obtained whatever the random initial guess.

Table 6.1 Computational complexity of Algorithm 6.1 compared with the unstructured VAR case.

Operation	Flops
Solving (6.10)	
without structural constraints	$\mathcal{O}\big(n^2(N_\mathrm{t}-n)N^6\big)$
QUARKS estimation (for each iteration)	
Line 4 (constructing the 'A'-matrix)	$\mathcal{O}\big(nr(N_\mathrm{t}-n)N^2\big)$
Line 4 (solving the least squares problem)	$\mathcal{O}\big(n^2r^2(N_\mathrm{t}-n)N^3\big)$
Lines 5 (constructing the 'A'-matrix)	$\mathcal{O}\big(nr(N_\mathrm{t}-n)N^2\big)$
Line 5 (solving the least squares problem)	$\mathcal{O}\big(n^2r^2(N_\mathrm{t}-n)N^3\big)$
Total	$\mathcal{O}\big(\#\text{iterations } n^2r^2(N_\mathrm{t}-n)N^3\big)$

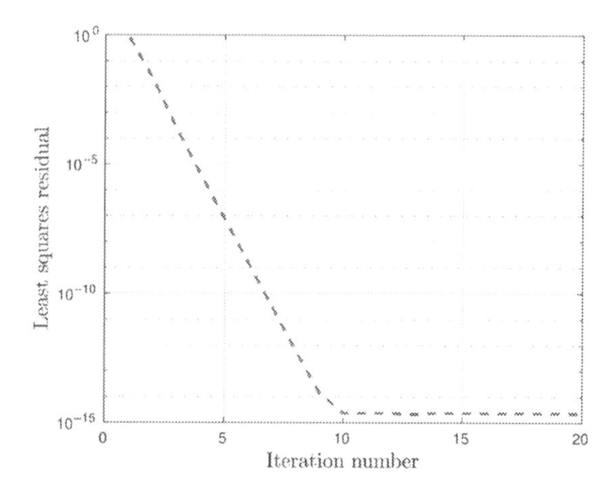

Figure 6.1 Convergence of the alternating least squares used for identifying a Quarks model.

6.2.5.2 Scalability

The main advantage of using Quarks models is the reduction of the computational complexity in estimating large-scale VARX models, i.e. Algorithm 6.1 scales with $\mathcal{O}(N^3N_t)$ compared to the theoretical $\mathcal{O}(N^6+N^4N_t)$ with unstructured coefficient matrices. The number of time samples for Quarks identification is tuned such that $N_t = 10nrN$ and $N_t = 50N^2$ in the unstructured case. These values were fixed such that the prediction error is comparable in both the Quarks and unstructured case. In Figure 6.2 we fit a linear model $log10(\text{time}) = a \times log10(N) + b$ to the time measurements and this shows especially that the regression a significantly decreases using the Quarks model, from 4.56 to 2.52, while the relative root mean square error remains below 10^{-3}.

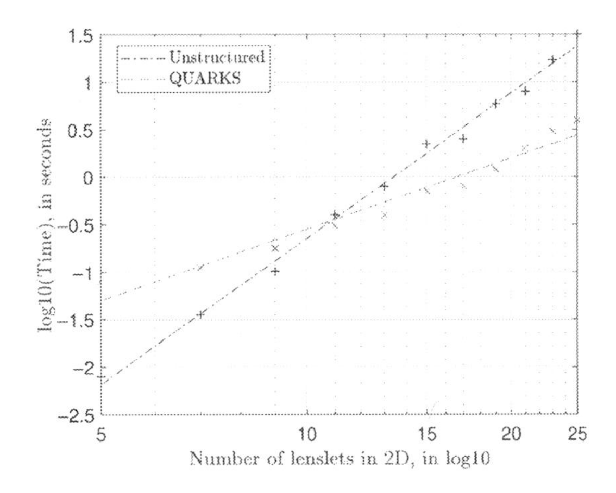

Figure 6.2 Evolution of the computation time for estimating Quarks models or FIR models with unstructured coefficient matrices versus the size of the 2D array. ©2019 IEEE Reprinted, with permission, from Sinquin and Verhaegen (2019a)

6.3 Tensor VARX and FIR Models

There are not many differences in the strategy when dealing with tensors. Optimization problems are multilinear, in general, instead of only bilinear. ALS remains the cornerstone of the algorithms for estimating tensor VARX models.

6.3.1 Problem Formulation and Cost Function

The generalization of the Quarks model was discussed in Section 3.6 and led to the definition of the tensor VAR model in Equation (3.25). This representation is recalled in tensor format. For this purpose, we define a dummy data vector $z(k)$ that is stored in the tensor $\mathcal{Z}(k)$. Similarly, we store the input and output data, respectively, in the tensors $\mathcal{U}(k) \in \mathbb{R}^{I_1 \times I_2 \times \cdots \times I_d}$ and $\mathcal{Y}(k) \in \mathbb{R}^{J_1 \times J_2 \times \cdots \times J_d}$ with I_ℓ and J_ℓ integers. Then, the following tensor relationship is defined:

$$\mathcal{Y}(k) = \sum_{i=1}^{n} \sum_{j=1}^{r_i} \left(\mathcal{Z}(k-i) \times_1 M_{i,j,1} \times_2 \cdots \times_d M_{i,j,d} \right) + \mathcal{E}(k), \tag{6.18}$$

where $\mathcal{E}(k)$ defines an additive perturbation term that is statistically uncorrelated with $\mathcal{Z}(k-i)$ for $i = 1:n$.

REMARK 6.3 (Tensor VAR and tensor FIR) Equation (6.18) can be interpreted as either

1. a tensor VAR model if $\mathcal{Z}(k)$ is replaced by $\mathcal{Y}(k)$,
2. or a tensor VFIR model if $\mathcal{Z}(k)$ is replaced by $\mathcal{U}(k)$.

This interchangeability between VAR and FIR models allows us to design a single algorithm that can deal with both situations. In the same way, the algorithm could (with slight modifications) also deal with VARX models. For that reason in the sequel of this section, the VAR case is treated only.

The estimation of the factor matrices of tensor VAR models follows the same lines as presented in Section 6.2.2 in spite of the tensor-specific notation. Without loss of generality and taking Remark 6.3 into consideration, we develop an algorithmic solution for estimating the coefficients of a tensor VAR model that have Kronecker rank one with the least-squares cost function,

$$\min_{\{M_{i,j}\}_{i=1:n}^{j=1:d}} \sum_{k=n+1}^{N_t} \left\| \mathcal{Y}(k) - \sum_{i=1}^{n} \Big(\mathcal{Y}(k-i) \times_1 M_{i,1} \times_2 \cdots \times_d M_{i,d} \Big) \right\|_2, \quad M_{i,j} \in \mathbb{R}^{J_j \times J_j}. \tag{6.19}$$

The norm used is the sum of the squared elements of the tensor that extends from the Frobenius norm for matrices. Using Proposition B.1 and writing $\operatorname{vec}\Big([\mathcal{Y}(k)]_{(1)}\Big)$ with some abuse of notation of $y(k)$, we can re-write this equation as:

$$\min_{\{M_{i,j}\}_{i=1:n}^{j=1:d}} \sum_{k=n+1}^{N_t} \left\| y(k) - \sum_{i=1}^{n} \Big(M_{i,d} \otimes \cdots \otimes M_{i,1} \Big) y(k-i) \right\|_2^2. \tag{6.20}$$

This problem is multi-linear in the unknowns $M_{i,j}$. The alternating least squares are standardly used in applications with tensor calculus (Kolda and Bader, 2009) and despite the popularity, more sophisticated alternatives are recommended in case of ill-conditioning. Such alternatives include quasi-Newton and non-linear least-squares algorithms. The latter may be combined with block-Jacobi preconditioners for reducing the number of conjugate gradient iterations and to improve the overall convergence of the iterative calculations (Boussé, 2019). In the context of the identification of Quarks models, the ill-conditioning has not appeared to be an issue when the data batch is large enough and persistently exciting. The Quarks identification is outlined in Algorithm 6.2 when the input and output signals are recast into tensors. We use the jth mode reshuffling of the tensors $\mathcal{Y}(k)$, see Appendix B, and denote these as $Y_{(j)}(k)$. Further, if we assume that the factor matrices $M_{\ell,\mu}$ are known for all $(\ell, \mu) \in \{1, \ldots, n\} \times \{1, \ldots, j-1, j+1, \ldots, d\}$, then such a linear least-squares problem in terms of $M_{i,j}$ only reads, for fixed j and $i = 1:n$,

$$\min_{\{M_{i,j}\}_{i=1:n}} \sum_{k=n+1}^{N_t} \left\| Y_{(j)}(k) - \sum_{i=1}^{n} M_{i,j} F_{i,j}(k-i) \right\|_F^2, \tag{6.21}$$

with,

$$F_{i,j}(k-i) = Y_{(j)}(k-i) \Big(M_{i,d} \otimes \cdots \otimes M_{i,j+1} \otimes M_{i,j-1} \otimes \cdots \otimes M_{i,1} \Big)^T. \tag{6.22}$$

Algorithm 6.2: Estimation of tensorized QUARKS model

Input : $\{y(k)\}_{k=1:N_t}, n$, and the sizes $\{J_j\}_{j=1...d}$
Output: $\{\widehat{M}_{i,j}\}_{(i,j)\in\{1,\dots,n\}\times\{1,\dots,d\}}$

```
/* Prepare the given data */
```
1 **for** $j = 1 \dots d$ **do**
2 **for** $k = n+1 \dots N_t$ **do**
3 Form the unfoldings $Y_{(j)}(k)$
4 **end**
5 Form $\bar{Y}_j := \begin{bmatrix} Y_{(j)}(n+1) & \dots & Y_{(j)}(N_t) \end{bmatrix}$
```
/* Initial guesses */
```
6 **for** $i = 1 \dots n$ **do**
7 $M_{i,j}^{(0)} = \text{rand}(J_j)$
8 **end**
9 **end**
```
/* Use the following default values */
```
10 $\ell = 1, \ell_{max} = 30, \epsilon = \infty, \epsilon_{min} = 10^{-3}, c^{(0)} = 0$
```
/* Start ALS */
```
11 **while** $\ell < \ell_{max}$ *and* $\epsilon > \epsilon_{min}$ **do**
12 **for** $j = 1 \dots d$ **do**
13 Form $\bar{F}_j^{(\ell)}$ with Equation (6.24), and solve the least squares problem (6.25); denote its solution as
14 $M_j^{(\ell)} = \bar{Y}_j (\bar{F}_j^{(\ell)})^T \left(\bar{F}_j^{(\ell)} (\bar{F}_j^{(\ell)})^T\right)^{-1}$
15 **end**
16 $c^{(\ell)} = ||\bar{Y}_d - M_d^{(\ell)} \bar{F}_d||_F^2$
17 $\epsilon = |c^{(\ell)} - c^{(\ell-1)}|$
18 $\ell = \ell + 1$
19 **end**
20 Set $\widehat{M}_{i,j}$ to the optimal values

To denote this as a more recognizable least-squares problem, we introduce the following notation:

$$M_j = \begin{bmatrix} M_{1,j} & \dots & M_{n,j} \end{bmatrix}, \tag{6.23}$$

$$\overline{F}_j = \begin{bmatrix} F_{1,j}(n) & \dots & F_{1,j}(N_t - 1) \\ \vdots & \ddots & \vdots \\ F_{n,j}(1) & \dots & F_{n,j}(N_t - n) \end{bmatrix} \in \mathbb{R}^{nJ_j \times (N_t - n) \prod_{\ell=1,\ell\neq j}^{d} J_\ell}. \tag{6.24}$$

Then, by storing the matrices $Y_{(j)}(k)$ next to each other for $k = n+1 : N_t$ to define the matrix $\bar{Y}_j$, the least-squares Equation (6.21) can be rewritten as:

$$\min_{M_j} \quad \|\bar{Y}_j - M_j \bar{F}_j\|_F^2. \tag{6.25}$$

Table 6.2 Computational complexity for the most expensive operations in the Quarks model.

Operation	Flops	With $J_i = J_j$
Form $\bar{F}_j$	$(N_t - n)nJ\sum_{j=1}^{d} J_j$	$\mathcal{O}(N_t J^{(d+1)/d})$
Compute $\bar{F}_j\bar{F}_j^T$	$(J_j n)^2 (N_t - n)\prod_{i=1,i\neq j}^{d} J_i$	$\mathcal{O}(N_t J^{(d+1)/d})$
Invert $\bar{F}_j\bar{F}_j^T$	$(J_j n)^3$	$\mathcal{O}(J^{3/d})$

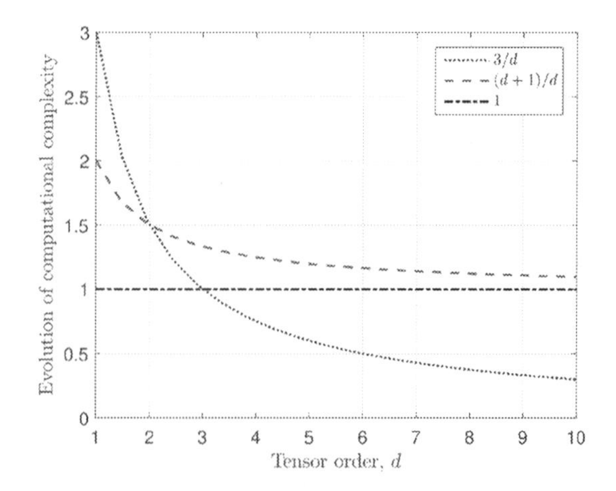

Figure 6.3 Exponent of the computation complexity as indicated by the last column in Table 6.2 as a function of the tensor order. Here, the total number of nodes $J = \prod_{j=1}^{d} J_j$. The horizontal line corresponds to a linear complexity with respect to the number of nodes in the array for $d = 1$.

We assume the matrix $\bar{F}_j$ to be of full row rank.The latter condition is satisfied by making the matrix $\bar{F}_j$ flat and by selecting an initial guess randomly. When d gets larger, comparably few temporal samples N_t are needed as the smallest dimension of M_j decreases.

6.3.2 Computational Complexity

The computational complexity for the most expensive parts is summarized in Table 6.2. The most expensive part is the formation of the matrix $\bar{F}_j$. With respect to the size of the network (indicated by the number of sensor nodes $J = \prod_{j=1}^{d} J_j$) the computational complexity is asymptotically linear in d. Figure 6.3 displays the exponent of the number of sensors J in Table 6.2 for $J_i = J_j = J$ as a function of the order of the tensor d.

When $d = 1$, the cost for the identification is cubic in J. This corresponds to the unstructured identification case. When $d > 2$, the most expensive algorithmic component is the formation of the matrix F_i and its multiplication with its transpose. When dealing with two-dimensional sensor arrays, it is expected that d cannot be increased to very large values without losing too much accuracy in the identification (unless increasing the Kronecker rank). However, most of the improvements are obtained for small tensor orders. For example, for $d = 4$ the complexity of Quarks case is $\mathcal{O}\left(J^{1.25}\right)$ compared to $\mathcal{O}\left(J^{3}\right)$ for the full unstructured case.

REMARK 6.4 The complexity is cubic with n. When considering the Quarks model coefficient matrices with Kronecker rank strictly larger than one, the complexity is cubic with r. This corresponds to the cubic complexity with the local system order in underlying SSS matrix operations (Rice and Verhaegen, 2009).

6.4 Dynamic Network/Structure Function Models

DNF and DSF models were introduced in Section 2.2.2. Here, the identification of the transfer functions in these models is restricted to the identification of DSF models for the sake of brevity as similar methods can be used for DNF models with minor modification. In this section, we focus on the following three identification problems, also considered in Weerts (2018):

1. The MISO direct method to identify individual (block)-rows of a DSF model within the prediction error framework.
2. The joint-direct method to identify the full network within the prediction error framework.
3. The identification of a single module or single transfer function in the transfer matrix $Q(q)$ of the DSF model (3.10).

A key restrictive requirement in these identification problems and their analysis is that the network topology is known and/or that parametrization of the transfer function is given. In this section, we start with considering the DSF model (2.13), briefly recalled here:

$$y(k) = Q(q)y(k) + P(q)u(k) + R(q)e(k), \tag{6.26}$$

where $y(k) \in \mathbb{R}^p$ is the collection (i.e. storing in a particular order[1]) of all recorded node signals in the network, and the same for the external inputs $u(k)$ and white-noise perturbation $e(k)$. Further, it is assumed that the DSF is well posed such that the inverse of $(I - Q(q))$ exists.

6.4.1 The MISO Direct Method

This method is described under the following restrictions:

1. The external input transfer function $P(q)$ in Equation (6.26) is equal to the identity matrix.

[1] It should be remarked that the order in which this storage is done can have an influence on the sparsity of the transfer function matrices.

2. The noise term $H(q)e(k)$ is assumed to be of the following diagonal form:

$$H(q)e(k) = \text{diag}(H_{i,i}(q)e_i(k)),$$

   and $e_i(k)$ is uncorrelated with $e_j(k)$ for $i \neq j$.
3. The entries of the transfer function matrix $Q(q)$ may be proper, but algebraic loops are excluded.

For this special case the DSF can be denoted explicitly as:

$$\begin{bmatrix} y_1(k) \\ y_2(k) \\ \vdots \\ y_p(k) \end{bmatrix} = \begin{bmatrix} 0 & G_{1,2}(q) & \cdots & G_{1,p}(q) \\ G_{2,1}(q) & 0 & \cdots & G_{2,p}(q) \\ \vdots & & \ddots & \vdots \\ G_{p,1}(q) & G_{p,2}(q) & \cdots & 0 \end{bmatrix} \begin{bmatrix} y_1(k) \\ y_2(k) \\ \vdots \\ y_p(k) \end{bmatrix} + \begin{bmatrix} u_1(k) \\ u_2(k) \\ \vdots \\ u_p(k) \end{bmatrix}$$
$$+ \begin{bmatrix} H_{1,1}(q) & 0 & \cdots & 0 \\ 0 & H_{2,2}(q) & \cdots & 0 \\ \vdots & & \ddots & \vdots \\ 0 & 0 & \cdots & H_{p,p}(q) \end{bmatrix} \begin{bmatrix} e_1(k) \\ e_2(k) \\ \vdots \\ e_p(k) \end{bmatrix}. \tag{6.27}$$

REMARK 6.5 The specific structure of the transfer function matrices $Q(q), P(q), R(q)$ of the general DSF form in Equation (6.27) makes the latter structure identifiable (see Lemma 3.1).

When the entries of the transfer functions $G_{i,j}(q)$ and $H_{i,i}(q)$ in Equation (6.27) have been parametrized by a parameter vector θ taking into consideration that $H_{i,i}(q)$ is invertible, then the one-step ahead predictor of this local output is given by Ljung (1999):

$$\hat{y}_i(k|k-1;\theta) = y_i(k) - H_{i,i}^{-1}(q,\theta)\Big(y_i(k) - u_i(k) - \sum_{t \in \mathcal{N}_i} G_{i,t}(q,\theta) y_t(k) \Big), \tag{6.28}$$

where $\mathcal{N}_i$ denotes the set of integers t for which $G_{i,t}(q)$ is non-zero. This model has one output and many inputs and is therefore called a multi-input, single-output (MISO) model. The related one-step ahead prediction error then follows from:

$$\epsilon_i(k|k-1;\theta) = y_i(k) - \hat{y}_i(k|k-1;\theta). \tag{6.29}$$

Identification is then generally framed by solving the following optimization problem (Ljung, 1999):

$$\hat{\theta}_{N_t} = \text{argmin}_\theta \frac{1}{N} \sum_{k=1}^{N_t} \epsilon_i(k|k-1;\theta)^2. \tag{6.30}$$

An analysis of such standard identification problems is made in Ljung (1999), where conditions for consistency of the estimates and their uncertainty are analysed.

6.4.2 The Joint-Direct Method

The one-step ahead predictor for the DSF in Equation (6.26) is summarized in the following lemma.

LEMMA 6.1 (One-step ahead predictor for output of Equation (6.26)) *For the DSF in Equation* (6.26) *with $R(q)$ square and invertible and the entries of the tranfer function matrices $Q(q)$ and $P(q)$ strictly proper, then the one-step ahead predictor is given by*

$$\hat{y}(k|k-1) = y(k) - R(q)^{-1}(I - G(q))y(k) + R(q)^{-1}P(q)u(k). \tag{6.31}$$

Proof The proof is similar to the derivation of the one-step ahead predictor discussed for the MISO case in Section 6.4.1. It can also be found in the proof of proposition 4.4 in Weerts (2018). □

When parametrizing the transfer functions $Q(q)$, $P(q)$ and $R(q)$ in Equation (6.26), the one-step prediction error is given as:

$$\epsilon(k|k-1;\theta) = R(q,\theta)^{-1}(I - G(q,\theta))y(k) + R(q,\theta)^{-1}P(q,\theta)u(k). \tag{6.32}$$

Estimating the parameter vector can typically be done via the solution of the following weighted least squares (WLS) problem Ljung (1999):

$$\hat{\theta}_{N_t}^{\mathrm{WLS}} = \mathrm{argmin}_{\theta\in\Theta}\frac{1}{N}\sum_{k=1}^{N_t}\epsilon(k|k-1;\theta)^T\Lambda\epsilon(k|k-1;\theta), \tag{6.33}$$

with $\Lambda > 0$ and Θ a set that constrains the parameter vector θ based on prior information, such as the rank deficiency of the power spectrum of the noise $R(q)e(k)$ (Weerts, 2018). Further, and similar to the MISO case, the prediction error framework developed in Ljung (1999) can be used to analyse the consistency and covariance properties of the estimate of this WLS minimization problem. See Chapter 4 of Weerts (2018) for more details.

REMARK 6.6 (Sparsity in DNF/DSF models) The method presented in this section and the previous one are global methods that aim to identify full network structures. When consistent estimates can be obtained, such an approach may highlight afterwards the sparsity that is present within the network. However, because of the facts that it is necessary to work with an identifiable network, the computational complexity related to full network identification can make this approach very restricted in terms of the size of the networks.

6.4.3 Identifying a Single Transfer Function in a DSF Model

The question of identifying a local system or a single transfer function in an DNF or DSF model is motivated in two ways:

(a) The fact that in networks often not all relevant signals in the network can be measured. This is partly due to the lack of sensors in particular spots of the network or because relevant signals that are communicated within a network are unmeasured (state) quantities. Many examples of this are given in Section 4.2, such as the α-decomposable systems of Section 4.2.2 or the block-tridiagonal systems of Section 4.2.3. Also, the systems with SSS system matrices demonstrate that many crucial quantities are usually not measured within a network.

(b) The computational complexity with identifying full networks when the latter have a large number of (signal) nodes.

These two restrictions have stimulated research in the areas of DNF and DSF to attempt to deal with these problems. A key factor is to deal with the unmeasurable signals in dynamic networks. This topic is discussed next.

To deal with unmeasurable signals in dynamic networks, so-called principles of *abstraction* and *immersion* have been developed (Weerts, 2018). These principles are summarized by a general transformation considered in Lemma 6.2. This Lemma introduces the following partitioning of the special DSF model (Weerts, 2018):

$$\begin{bmatrix} y_{\tilde{\mathcal{S}}}(k) \\ y_{\mathcal{L}}(k) \\ y_{\mathcal{V}}(k) \\ y_{\tilde{\mathcal{Z}}}(k) \end{bmatrix} = \begin{bmatrix} Q_{\tilde{\mathcal{S}}\tilde{\mathcal{S}}}(q) & Q_{\tilde{\mathcal{S}}\mathcal{L}}(q) & Q_{\tilde{\mathcal{S}}\mathcal{V}}(q) & Q_{\tilde{\mathcal{S}}\tilde{\mathcal{Z}}}(q) \\ Q_{\mathcal{L}\tilde{\mathcal{S}}}(q) & Q_{\mathcal{L}\mathcal{L}}(q) & Q_{\mathcal{L}\mathcal{V}}(q) & Q_{\mathcal{L}\tilde{\mathcal{Z}}}(q) \\ Q_{\mathcal{V}\tilde{\mathcal{S}}}(q) & Q_{\mathcal{V}\mathcal{L}}(q) & Q_{\mathcal{V}\mathcal{V}}(q) & Q_{\mathcal{V}\tilde{\mathcal{Z}}}(q) \\ Q_{\tilde{\mathcal{Z}}\tilde{\mathcal{S}}}(q) & Q_{\tilde{\mathcal{Z}}\mathcal{L}}(q) & Q_{\tilde{\mathcal{Z}}\mathcal{V}}(q) & Q_{\tilde{\mathcal{Z}}\tilde{\mathcal{Z}}}(q) \end{bmatrix} \begin{bmatrix} y_{\tilde{\mathcal{S}}}(k) \\ y_{\mathcal{L}}(k) \\ y_{\mathcal{V}}(k) \\ y_{\tilde{\mathcal{Z}}}(k) \end{bmatrix} + \begin{bmatrix} u_{\tilde{\mathcal{S}}}(k) \\ u_{\mathcal{L}}(k) \\ u_{\mathcal{V}}(k) \\ u_{\tilde{\mathcal{Z}}}(k) \end{bmatrix} + \begin{bmatrix} v_{\tilde{\mathcal{S}}}(k) \\ v_{\mathcal{L}}(k) \\ v_{\mathcal{V}}(k) \\ v_{\tilde{\mathcal{Z}}}(k) \end{bmatrix}, \tag{6.34}$$

with the following conventions:

- $y_{\mathcal{V}}(k)$ is the vector of *unmeasurable nodes* that are not directly measured but that are directly connected to other nodes that are measured.
- $y_{\mathcal{L}}(k)$ are the *measured* nodes that are directly connected to one element of $y_{\mathcal{V}}(k)$.
- $y_{\tilde{\mathcal{S}}}(k)$ are the remaining measured nodes, i.e. all the measured nodes minus $y_{\mathcal{L}}(k)$.
- $y_{\tilde{\mathcal{Z}}}(k)$ are the remaining unmeasured nodes, i.e. all the unmeasured nodes minus $y_{\mathcal{V}}(k)$.

The goal of abstraction and immersion is to derive a sub-network that only consists of measurable quantities. For that purpose, one seeks an invertible transformation $T(q)$ that is applied to the left of the DSF (6.26) such that the unmeasurable nodes are 'removed' from the sub-network of measurable nodes.

REMARK 6.7 (Equivalent DSF) An invertible transformation applied to the left of the DSF (6.26) results in an equivalent DSF in the sense that it leaves the transfer function from the external input $u(k)$ to the output $y(k)$ and that from the noise $v(k)$ to the output $y(k)$ invariant. To see this, consider the DSF given as:

$$y(k) = Q(q)y(k) + P(q)u(k) + R(q)e(k), \tag{6.35}$$

and the transformed variant:

$$T(q)y(k) = T(q)Q(q)y(k) + T(q)P(q)u(k) + T(q)R(q)e(k). \tag{6.36}$$

The transfer function from the second DNF from $u(k)$ to $y(k)$ reads:

$$T_{uy}(q) = (I - Q(q))^{-1}T(q)^{-1}T(q)P(q) = (I - Q(q))^{-1}P(q). \tag{6.37}$$

The same holds for the transfer function from $e(k)$ to $y(k)$:

$$T_{ey}(q) = (I - Q(q))^{-1}T(q)^{-1}T(q)R(q)e(k) = (I - Q(q))^{-1}R(q)e(k). \tag{6.38}$$

LEMMA 6.2 (Removing unmeasurable nodes from a DSF) *Given the partitioned DSF in Equation* (6.34) *and assuming this DSF is well posed, then there exists a transformation $T(q)$ applied to the left of this DSF such that a sub-network is extracted that only contains measurable nodes and that is governed by a DSF.*

Proof Let the output vector of the partitioned DSF in Equation (6.34) be explicitly split up in measured and unmeasured nodes as follows:

$$\left[\begin{array}{c} y_{\tilde{\mathcal{S}}}(k) \\ y_{\mathcal{L}}(k) \\ \hline y_{\mathcal{V}}(k) \\ y_{\tilde{\mathcal{Z}}}(k) \end{array}\right] = \left[\begin{array}{c} y_{\text{meas}}(k) \\ \hline y_{\text{umeas}}(k) \end{array}\right]. \tag{6.39}$$

Accordingly, we can write Equation (6.34) compactly as,

$$\left[\begin{array}{c} y_{\text{meas}}(k) \\ \hline y_{\text{umeas}}(k) \end{array}\right] = \left[\begin{array}{c|c} Q_{11}(q) & Q_{12}(q) \\ \hline Q_{21}(q) & Q_{22}(q) \end{array}\right] \left[\begin{array}{c} y_{\text{meas}}(k) \\ \hline y_{\text{umeas}}(k) \end{array}\right] + \begin{bmatrix} u_1(k) \\ u_2(k) \end{bmatrix} + \begin{bmatrix} v_1(k) \\ v_2(k) \end{bmatrix}. \tag{6.40}$$

This can be denoted as,

$$\begin{bmatrix} I - Q_{11}(q) & -Q_{12}(q) \\ -Q_{21}(q) & I - Q_{22}(q) \end{bmatrix} \left[\begin{array}{c} y_{\text{meas}}(k) \\ \hline y_{\text{umeas}}(k) \end{array}\right] = \begin{bmatrix} u_1(k) \\ u_2(k) \end{bmatrix} + \begin{bmatrix} v_1(k) \\ v_2(k) \end{bmatrix}. \tag{6.41}$$

Now apply the following transformation $T(q)$ from the left:

$$T(q) = \begin{bmatrix} I & Q_{12}(q)(I - Q_{22}(q))^{-1} \\ 0 & I \end{bmatrix}. \tag{6.42}$$

This results in,

$$\begin{aligned} &\begin{bmatrix} (I - (Q_{11}(q) + Q_{12}(q)(I - Q_{22}(q))^{-1}Q_{21}(q)) & 0 \\ -Q_{21}(q) & (I - Q_{22}(q)) \end{bmatrix} \left[\begin{array}{c} y_{\text{meas}}(k) \\ \hline y_{\text{umeas}}(k) \end{array}\right] \\ &= \begin{bmatrix} u_1(k) + Q_{12}(q)(I - Q_{22}(q))^{-1}u_2(k) \\ u_2(k) \end{bmatrix} + \begin{bmatrix} v_1(k) + Q_{12}(q)(I - Q_{22}(q))^{-1}v_2(k) \\ v_2(k) \end{bmatrix}. \end{aligned} \tag{6.43}$$

The DNF only containing $y_{\text{meas}}(k)$ reads as,

$$y_{\text{meas}}(k) = (Q_{11}(q) + Q_{12}(q)(I - Q_{22}(q))^{-1}Q_{21}(q))y_{\text{meas}}(k) + u_1(k) \\ + Q_{12}(q)(I - Q_{22}(q))^{-1}u_2(k) + v_1(k) + Q_{12}(q)(I - Q_{22}(q))^{-1}v_2(k). \quad (6.44)$$

To transform this to a DSF, we perform an hollow abstraction as outlined in the derivation of Equation (2.13). □

Lemma 6.2 shows how to remove the unmeasurable nodes using an equivalence transformation on the original DSF. However, the resulting DSF derived in the proof may no longer be identifiable as the weighting matrix of the external inputs is no longer the identity matrix. Further, the noise properties may be altered such that the power spectrum is no longer full rank. As such, the removal of nodes may require special attention in the identification of such reduced-size networks. See Weerts (2018) and Woodbury (2019) for more details.

REMARK 6.8 (Invariance of a single transfer function in an equivalent DSF) A number of conditions have been derived in Weerts (2018), see proposition 6.15, such that the procedure to remove unmeasurable outputs signals in a DSF in order to keep the transfer functions that one is interested in invariant.

6.5 Conclusions

This chapter reviewed algorithmic strategies for identifying signal flow dynamic networks. Sparsity in the coefficient matrices of VAR(X) models is sought by formulating a regularized least-squares cost function, which, however, favourably scales to large network dimensions only if the position of the (non-)zero entries is known. Further assumptions such as spatial invariance may then be useful.

Quarks models allow us to represent the spatial-temporal dynamics for systems with a sensor array. The workhorse for identifying the Quarks and its extension to higher tensor order is the alternating least-squares approach, which does not provide theoretical guarantees of global convergence although the latter is observed in practice.

The identification of DSF models was proposed following three main approaches: either a MISO direct approach, for which the transfer function of each module is identified solving a linear least-squares, or a joint-direct method or an approach that removes the unobservable nodes to focus on the estimation of a reduced-size network.

7 Subspace-Like Identification of Multi-Dimensional Systems

7.1 Introduction

Two-dimensional (2D) state-space models were introduced in Section 4.3.2 as a special case of multi-dimensional state-space models. In this chapter, we show that the identification of this special case is far from trivial. Therefore, one might expect that the identification of more general classes of models will also be challenging.

A subspace-like solution to the identification of 2D state-space models is presented. The notion 'subspace-like' is used to reflect that apart from a (limited) number of non-convex optimization problems in a small number of parameters, the major parts of the algorithmic solution are convex.

By this subspace-like nature, the identification solution presented in this chapter differs from those presented in Chapter 8 as these methods are restricted to convex or rank constrained optimization problems only. A further difference is that the identification methods in this chapter consider the global identification, while in Chapter 8 special emphasis is put on the identification of subsystems in the network using local measurements only.

The current chapter is based on Sinquin and Verhaegen (2019a).

7.2 Kronecker-Structured State-Space Models

7.2.1 Problem Formulation

The simplest tensor state-space model (4.31) has the system matrices (A, B, C) given as $(A_2 \otimes A_1, B_2 \otimes B_1, C_2 \otimes C_1)$. The special Kronecker structure of the system matrices allows us to write such state-space models in terms of signal matrices instead of signal vectors. To see this, let the 2D state matrix $X(k) \in \mathbb{R}^{nN \times N}$ be defined as:

$$X(k) = \begin{bmatrix} x_{1,1}(k) & x_{1,2}(k) & \cdots & x_{1,N}(k) \\ x_{2,1}(k) & x_{2,2}(k) & & x_{2,N}(k) \\ \vdots & & \ddots & \vdots \\ x_{N,1}(k) & & \cdots & x_{N,N}(k) \end{bmatrix}. \tag{7.1}$$

In the same way, define the 2D input matrix $U(k)$ from the inputs $u_{i,j}(k) \in \mathbb{R}^m$ recorded on a 2D grid with coordinates (i, j) and the 2D output matrix $Y(k)$ from

$y_{i,j}(k) \in \mathbb{R}^p$. Then, the simplest tensor state-space model referred to as the Kronecker matrix state-space model (MSSM) has the following form:

$$\begin{aligned} X(k+1) &= A_1 X(k) A_2^T + B_1 U(k) B_2^T, \\ Y(k) &= C_1 X(k) C_2^T + E(k), \end{aligned} \tag{7.2}$$

with,

$$\begin{aligned} &A_1 \in \mathbb{R}^{n_1 \times n_1}, \quad B_1 \in \mathbb{R}^{n_1 \times mN}, \quad C_1 \in \mathbb{R}^{pN \times n_1}, \\ &A_2 \in \mathbb{R}^{n_2 \times n_2}, \quad B_2 \in \mathbb{R}^{n_2 \times N}, \quad C_2 \in \mathbb{R}^{N \times n_2}. \end{aligned} \tag{7.3}$$

DEFINITION 7.1 (Generators of 2D state-space model) The set of generators $\mathcal{S}$ for the Kronecker MSSM (7.2) is defined from the factors of the state-space matrices by the following sextuple:

$$\mathcal{S} = \{A_1, A_2, B_1, B_2, C_1, C_2\},$$

where the dimensions of the corresponding matrices are given in (7.3).

For a discussion on the relationship between the stability, observability and controllability of the local system matrices in the generator $\mathcal{S}$ and the global system matrices $(A_2 \otimes A_1, B_2 \otimes B_1, C_2 \otimes C_1)$, we refer to Lemmas 4.1, 4.2 and 4.3. Here, we refine the analysis of equivalence between two different sets of generators as outlined for the general tensor state-space model in Lemma 4.4 to the 2D case. This is summarized in Lemma 7.1.

DEFINITION 7.2 (Persistency of excitation for Kronecker MSSM models) The input matrix $U(k)$ to the Kronecker MSSM (7.2) is persistently exciting of order s if and only if there exists an integer N_t such that the (block-)Hankel matrix:

$$\bar{U}_s = \begin{bmatrix} \text{vec}(U(1)) & \text{vec}(U(2)) & \dots & \text{vec}(U(N_t - s + 1)) \\ \text{vec}(U(2)) & & & \vdots \\ \vdots & & \ddots & \vdots \\ \text{vec}(U(s)) & & \dots & \text{vec}(U(N_t)) \end{bmatrix}, \tag{7.4}$$

has full rank smN^2.

DEFINITION 7.3 (Equivalence of Kronecker MSSM models) Let N_t be the number of temporal samples. Then, two Kronecker MSSM models with generators $\mathcal{S}_1$ and $\mathcal{S}_2$, respectively, are said to be equivalent if the input-output behaviour of the corresponding state-space model (7.2), for $E(k) \equiv 0$, is identical for all $k = 1, \dots, N_t$.

LEMMA 7.1 (Equivalence of Kronecker MSSM models) *The sets of generators $\mathcal{S}_1$ and $\mathcal{S}_2$ for the Kronecker MSSM equivalently model (7.2), for $E(k) \equiv 0$, if there exist non-singular transformation matrices $T_1 \in \mathbb{R}^{n_1 \times n_1}$ and $T_2 \in \mathbb{R}^{n_2 \times n_2}$ as well as non-zero scalars α, β, γ that satisfy:*

$$\begin{cases} A_2^{(1)} = \alpha T_2^{-1} A_2^{(2)} T_2, \\ B_2^{(1)} = \beta T_2^{-1} B_2^{(2)}, \\ C_2^{(1)} = \gamma C_2^{(2)} T_2, \end{cases} \tag{7.5}$$

$$\begin{cases} A_1^{(1)} = \frac{1}{\alpha} T_1^{-1} A_1^{(2)} T_1, \\ B_1^{(1)} = \frac{1}{\beta} T_1^{-1} B_1^{(2)}, \\ C_1^{(1)} = \frac{1}{\gamma} C_1^{(2)} T_1. \end{cases} \tag{7.6}$$

Proof The Lemma readily follows from Lemma 4.4, the definition of the Kronecker product and the input-output equation as governed by (5.19) (for $E(k) \equiv 0$). □

Before we can state the identification problem for the class of Kronecker MSSM models, we need the definition of minimality.

DEFINITION 7.4 (Minimality Kronecker MSSM model) A minimal realization of the state-space model (7.2) defined by a set $\mathcal{S}$ given in Definition 7.1 such that the extended observability matrix $\mathcal{O}_s$ and controllability matrix $\mathcal{C}_s$ are denoted as:

$$\mathcal{O}_s = \begin{bmatrix} C_2 \otimes C_1 \\ C_2 A_2 \otimes C_1 A_1 \\ \vdots \\ C_2 A_2^{s-1} \otimes C_1 A_1^{s-1} \end{bmatrix},$$

$$\mathcal{C}_s = \begin{bmatrix} B_2 \otimes B_1 & B_2 A_2 \otimes B_1 A_1 & \dots & B_2 A_2^{s-1} \otimes B_1 A_1^{s-1} \end{bmatrix}, \tag{7.7}$$

both have minimal rank $n_2 n_1$.

COROLLARY 7.1 *If the set of generators* $\mathcal{S} = \{A_1, A_2, B_1, B_2, C_1, C_2\}$ *corresponds to a minimal realization of the MSSM* (7.2)*, then both triplets (of local state-space models)* (A_1, B_1, C_1) *and* (A_2, B_2, C_2) *represent a minimal realization.*
The reverse is not true in general.

Proof The proof readily follows from Lemmas 4.2 and 4.3. □

REMARK 7.1 When an unstructured similarity transform is applied to the triplet $(A_2 \otimes A_1, B_2 \otimes B_1, C_2 \otimes C_1)$ of the global system, the Kronecker rank one structure of this global model is lost.

As stated in Remark 7.1, and from a computational perspective, it is not wise to first identify the global system matrices up to a similarity transformation, e.g. making use of standard subspace identification methods (Verhaegen and Verdult, 2007) and then try to impose the Kronecker structure on the system matrices by finding an appropriate similarity transformation. The latter problem fits in the framework of structured model identification (Yu et al., 2018a).

In order to derive a scalable identification algorithm for identifying the system matrices of the state-space model (4.31) with system matrices belonging to the class of Kronecker MSSM models, the following assumptions on the data and system matrices in (7.2) are made:

- **A1:** System (7.2) is minimal.
- **A2:** The eigenvalues of both A_1 and A_2 all lie strictly within the unit circle.
- **A3:** The input is persistently exciting of any order.
- **A4:** The measurement noise $e(k) = \text{vec}(E(k))$ is zero-mean white gaussian.
- **A5:** The measurement noise is uncorrelated from all past inputs, that is:

$$\text{for all } k \leq j, \mathbb{E}\left[u(k)e(j)^T\right] = 0$$

Assumptions **A1–A3** and **A5** are linked to the global system properties and are commonly used in subspace identification (Verhaegen and Verdult, 2007). Assumption **A4** ensures our attention is focused on the essential steps of deriving subspace-like identification methods and not the complicated noise scenarios. The latter should be considered in future research activities.

PROBLEM 7.1 Assuming **A1–A5**, and given the input-output data matrices $U(k), Y(k)$ of the state-space model *(7.2)* for $k = 1, \ldots, N_t$, estimate, up to the similarity transformations T_1, T_2 and ambiguity scaling factors α, β, γ defined in Lemma 7.1, a minimal set of generators $A_1, A_2, B_1, B_2, C_1, C_2$ of an MSSM model that correspond to a minimal realization. The main difficulty is in deriving an algorithm with $\mathcal{O}(N^3 N_t)$ computational complexity.

The constraint on the computational cost, as stated in Problem 7.1, prevents us from first using standard subspace methods like MOESP or SSARX (Verhaegen and Verdult, 2007) and then imposing the MSSM structure on the (large-scale) identified system matrices. These methods fail for three main reasons. First, these standard subspace identification methods perform a QR decomposition of concatenated block-Hankel matrices that are constructed from the input-output data. The size of this compound matrix is $(p+m)sN^2 \times (N_t - s + 1)$, for some user-selected dimension parameter $s \geq M$. The triangular matrix of such a QR decompostion is only square if $N_t \geq psN^2$ and this constraint implies the need to collect a large number of data samples. Second, for a global system order of $n_2 n_1$, the calculation of this QR decomposition and the subsequent SVD performed on $N^2 \times N^2$ matrices scales with $\mathcal{O}(N^6)$. Even in the best case that the system order is known a priori, a rank-$n_2 n_1$ SVD can be computed at a cost of $\mathcal{O}(n_2 n_1 N^4)$. More efficient methods for computing SVD, as presented in the paper of Halko et al. (2011), do not break down the curse of dimensionality that appears with multi-dimensional systems and still require $\mathcal{O}(log(n_2 n_1) N^4)$ flops. Final drawbacks are the storage (and formation) of the global matrices as well as doing operations on these system matrices such as using them to design a controller. For example, doing a matrix-vector multiplication with the dense unstructured matrix requires $\mathcal{O}(N^4)$ instead of $\mathcal{O}(N^3)$ in the MSSM format. In order to avoid this computational bottleneck, one could opt to impose this MSSM on the first identfied large-scale system matrices with standard subspace methods, approximating these matrices with matrices of lower Kronecker rank. This only adds to the problems already illustrated.

The algorithm PBSID provides an alternative route that first estimates a high-order VARX and then computes the SVD of a large-size data matrix (Chiuso, 2007). The computational cost associated with the latter operation, along with the estimation of unstructured and dense estimates of the state-space matrices, reaches $\mathcal{O}(N^6)$ and is reduced in this chapter by working rather with the factor Markov parameters $C_1 A_1^i B_1$ and $C_2 A_2^i B_2$.

7.2.1.1 Generic Algorithmic Strategy

The main idea of the algorithm is to work only with factor parameters and never with the full Kronecker products. Although convexity is lost when formulating optimization problems, the number of unknowns is far less. This allows us to reach a solution that is only an approximation of the one that would have been obtained by solving without such a bilinear parametrization of the state-space matrices. The subspace-like method is composed of three major steps, two of them closely following PBSID: once an estimate of the state sequence is available, it is used for retrieving the state-space matrices solving a least-squares problem. The strategy here differs only because of the bilinear nature of the problem, which implies an additional step to estimate an ambiguity parameter.

More precisely, in the following three subsections, we start with the identification of the factor Markov parameters. For that purpose, we will make use of a globally convergent algorithm. These factor Markov parameters are, however, estimated up to an unknown scaling factor that differs for each estimated Markov parameter. Therefore, in a second step, a low-rank optimization problem with bilinear constraints is formulated to remove the unequal scaling of these factor Markov parameter estimates, that is to equalize these estimates. The third and final step is the identification of the factor state-space model matrices that are estimated from the estimated state sequence. This is done in Subsection 7.2.4.

7.2.2 Estimating the Factor Markov Parameters

Based on Assumption **A2**, the vectorization of the output $Y(k)$ of the MSSM (7.2) can be approximated with the tensor FIR model (6.20) for $d = 2$. If we denote $M_{i,1} := C_1 A_1^{i-1} B_1 \in \mathbb{R}^{N \times N}$ and $M_{i,2} := C_2 A_2^{i-1} B_2 \in \mathbb{R}^{pN \times mN}$, then this approximation can be written for all $k > s$ as,

$$y(k) \approx \sum_{i=1}^{s} C A^{i-1} B u(k-i) + e(k)$$
$$\approx \sum_{i=1}^{s} \left(C_2 A_2^{i-1} B_2 \otimes C_1 A_1^{i-1} B_1\right) u(k-i) + e(k). \tag{7.8}$$

We further denote $\begin{bmatrix} M_{1,1} & \dots & M_{s,1} \end{bmatrix}$ by the matrix M_1 as the matrix containing the left-factor impulse responses $C_1A_1^{i-1}B_1$ for $i = 1 : s$. Similarly, the matrix M_2 is built from the factor matrices $M_{i,2}$ and this matrix is constructed from the right-factor impulse responses $C_2A_2^{i-1}B_2$ for $i = 1 : s$.

Choosing the parameter s in this model to be big enough, as in the standard subspace identification literature (Verhaegen and Verdult, 2007), we can make the approximation error with model (7.8) arbitrarily small (Knudsen, 2001).

A computationally efficient and globally convergent alternating least-squares algorithm was presented in Algorithm 6.2. This algorithm optimizes with respect to the factor matrices $M_{i,1}, M_{i,2}$. Additional constraints, such as stability of the impulse responses via kernel methods, can be imposed on these factor matrices directly, as shown in Chapter 6. To demonstrate this, let the kernel matrix $P_t \in \mathbb{R}^{s\times s}$ be introduced and let its square root be denoted by Q_t, then stability is induced by expanding the cost function in Equation (6.20) with regularization terms as follows,

$$\min_{M_{i,1},M_{i,2}} \sum_{k=s+1}^{N_t} \left\| Y(k) - \sum_{i=1}^{s} M_{i,1}U(k-i)M_{i,2}^T \right\|_F^2 + \lambda_{ALS} \left\| Q_t \begin{bmatrix} \text{vec}(M_{1,2})\text{vec}(M_{1,1})^T \\ \vdots \\ \text{vec}(M_{s,2})\text{vec}(M_{s,1})^T \end{bmatrix} \right\|_F^2, \tag{7.9}$$

where λ_{ALS} is a regularization parameter.

Though the alternating least-squares is globally convergent, the FIR approximation and the presence of the noise (that in the limit $N_t \to \infty$ can be correlated out), approximate estimates of the impulse response factor matrices will be obtained. Further, due to the Kronecker product bilinear nature, these impulse response factor matrices are related to the original as follows:

$$\widehat{M}_{i,1} \approx v_i M_{i,1}, \quad \widehat{M}_{i,2} \approx t_i M_{i,2}, \tag{7.10}$$

where $t_i v_i \approx 1$. The non-zero ambiguity constants t_i that differ for each factor Markov parameter are unknown. This difference prevents us from realizing the individual local system matrices (A_j, B_j, C_j) for $j = 1, 2$ (up to the ambiguity outlined in Lemma 7.1) from the corresponding estimates $\widehat{M}_{i,j}$.

This problem is tackled in Subsection 7.2.3. We now propose to study how we could reduce the different scalings with respect to the index i in Equation (7.10) to the scaling of the factor Markov parameters given as $wC_2(\eta A_2)^{i-1}B_2$ for w, η non-zero scalars.

7.2.3 Rescaling the Estimated Factor Markov Parameters

The analysis is performed for the left factor matrices $\widehat{M}_{i,2}$ and can easily be extended to the right factor matrices $\widehat{M}_{i,1}$.

Let the local observability matrix $\mathcal{O}_{2,s_1}$ and controllability matrix $\mathcal{C}_{2,s_1}$ be defined as:

$$\mathcal{O}_{2,s_1} = \begin{bmatrix} C_2 \\ C_2 A_2 \\ \vdots \\ C_2 A_2^{s_1-1} \end{bmatrix}, \quad \mathcal{C}_{2,s_1} = \begin{bmatrix} B_2 & A_2 B_2 & \dots & A_2^{s_1-1} B_2 \end{bmatrix}. \tag{7.11}$$

The goal is to construct a block-Hankel matrix from the left factor Markov parameters $\widehat{M}_{i,2}$ such that it equals $\mathcal{O}_{2,s_1}\mathcal{C}_{2,s_1}$ for some integer $s_1 \geq n_2$. An SVD could then follow according to the classical Ho–Kalman realization theory (Ho and Kalman, 1966), for example. The i-dependent scaling of the left factor Markov parameter estimates will, in general, cause the block-Hankel matrix constructed from them to be full rank. A way to compensate for this unwanted scaling is to try to rescale the estimated left factor Markov parameter such that the rank of the block-Hankel matrix constructed from these rescaled variants (again) equals n_2. This is analysed in Theorem 7.1, which requires Lemma 7.2 and for us to make use of the following notation:

$$\mathcal{H}(x_i) = \begin{bmatrix} x_1 & x_2 & \dots & x_{s_1} \\ \vdots & & \ddots & \vdots \\ x_{s_1} & x_{s_1+1} & \dots & x_s \end{bmatrix}. \tag{7.12}$$

LEMMA 7.2 (Partial realization problem) (Gragg and Lindquist, 1983) *Let $\delta \in \mathbb{N}$. Let $s = 2s_1 - 1$ with s_1 an integer strictly larger than δ. For $i = 1 \dots s$, let $x_i \in \mathbb{R}$ such that rank $\left(\mathcal{H}(x_i)\right) = \delta$. Then there exists a realization (a, b, c) of minimal degree δ with $a \in \mathbb{R}^{\delta\times\delta}, b \in \mathbb{R}^{\delta\times 1}, c \in \mathbb{R}^{1\times\delta}$ such that for $i = 1 \dots s, x_i = ca^{i-1}b$. This decomposition is unique up to a similarity transformation.*

The triplet (a, b, c) defines a partial realization of the sequence $\{x_i\}_{i=1\dots s}$.

THEOREM 7.1 (Rescaling the estimated factor Markov parameters $\widehat{M}_{i,2}$) (Sinquin and Verhaegen, 2019a) *Let $s = 2s_1 - 1$ with s_1 an integer such that $s_1 N \min(p,m) \geq \max(n_2, n_1)$. For $i \in \{1, \dots, s\}$, let (α_i, t_i) be non-zero scalars and let the matrices $\widehat{M}_{i,2}$ satisfy:*

$$\widehat{M}_{i,2} = t_i M_{i,2}, \tag{7.13}$$

with rank$(\mathcal{H}(M_{i,2})) = n_2$.

- *If rank $\left(\mathcal{H}(\alpha_i \widehat{M}_{i,2})\right) = n_2$, then rank$\left(\mathcal{H}(\alpha_i t_i)\right) = 1$.*
- *If $\alpha_i t_i = \eta^{i-1}$ for a non-zero scalar η, then:*

$$\text{rank}\left(\mathcal{H}(\alpha_i \widehat{M}_{i,2})\right) = n_2. \tag{7.14}$$

Proof The proof is based on Sinquin and Verhaegen (2019a). We derive the proof using contraposition. In the following, we denote $x_i = \alpha_i t_i$ and $X_i = M_{i,2}$ and drop the index of the system matrices (A_2, B_2, C_2). Let $\delta \in \mathbb{N}$ such that $1 < \delta \leq s_1$ and suppose that:

$$\text{rank}\left(\mathcal{H}(x_i)\right) = \delta. \tag{7.15}$$

From Lemma 7.2, there exists a realization (a, b, c) of minimal degree δ with $a \in \mathbb{R}^{\delta\times\delta}, b \in \mathbb{R}^{\delta\times 1}, c \in \mathbb{R}^{1\times\delta}$ such that for $i = 1 \dots s, x_i = ca^{i-1}b$. If every eigenvalue of a is 0, then a is nilpotent (via Cayley–Hamilton). This would violate the assumption $x_i \neq 0$ for all i. Therefore, the matrix a has to have at least one non-zero eigenvalue.

We continue the proof by considering two cases. First, a is considered diagonalizable, such that the following holds $a = P\Lambda P^{-1}$ for P invertible and Λ a diagonal matrix containing the eigenvalues. All eigenvalues λ_i are distinct and, letting $k \in \mathbb{N}$, then:

$$\begin{aligned} ca^k bCA^k B &= C\left(ca^k b\right) A^k B \\ &= CcP\Lambda^k P^{-1} bA^k B. \end{aligned} \tag{7.16}$$

Denote $\tilde{c} = cP$, $\tilde{b} = P^{-1}b$ and $r_i = \tilde{c}_i\tilde{b}_i \neq 0$. Then we write:

$$\begin{aligned} ca^k bCA^k B &= C\tilde{c}\Lambda^k\tilde{b}A^k B \\ &= C\sum_{i=1}^{\delta} \tilde{c}_i\lambda_i^k\tilde{b}_i A^k B \\ &= \sum_{i=1}^{\delta} r_i C(\lambda_i A)^k B. \end{aligned} \tag{7.17}$$

Without loss of generality, we restrict ourselves to the case $\delta = 2$. Then the Markov parameters are associated with the following state-space matrices:

$$\tilde{A} = \begin{bmatrix} \lambda_1 A & 0 \\ 0 & \lambda_2 A \end{bmatrix}, \quad \tilde{B} = \begin{bmatrix} B \\ B \end{bmatrix}, \quad \tilde{C} = \begin{bmatrix} r_1 C & r_2 C \end{bmatrix}. \tag{7.18}$$

For this case, if W_i is an eigenvector of A, then both $\left[\begin{smallmatrix} W_i \\ 0 \end{smallmatrix}\right]$ and $\left[\begin{smallmatrix} 0 \\ W_i \end{smallmatrix}\right]$ are eigenvectors of $\tilde{A}$. The condition $\tilde{C}\left[\begin{smallmatrix} W_i \\ 0 \end{smallmatrix}\right] = 0$ or $\tilde{C}\left[\begin{smallmatrix} 0 \\ W_i \end{smallmatrix}\right] = 0$ can then be written as:

$$r_i C W_i = 0, \tag{7.19}$$

for $i \in \{1, 2\}$. The Popov–Belevitch–Hautus (PBH) observability test and the assumption that the pair (A, C) is observable, implies that $W_i = 0$. Therefore, the pair $(\tilde{A}, \tilde{C})$ is observable. It follows that the rank of $\mathcal{H}(x_i X_i)$ is strictly larger than n_2 and we have a contradiction.

For the second case, we assume that the matrix a is not diagonalizable. In that case there is an invertible matrix $P^{-1}aP = J$ with J a Jordan matrix (Golub and Van Loan, 1996). This Jordan matrix has square blocks on its diagonal of size equal to the multiplicity of the corresponding eigenvalue.

Without loss of generality, we restrict λ_i to multiplicity 2. This corresponds to a Jordan block J_i of the following form:

$$J_i = \begin{bmatrix} \lambda_i & 1 \\ 0 & \lambda_i \end{bmatrix}. \tag{7.20}$$

By induction it can be shown that J_i^k equals:

$$J_i^k = \begin{bmatrix} \lambda_i^k & k\lambda_i^{k-1} \\ 0 & \lambda_i^k \end{bmatrix}. \tag{7.21}$$

The expression in Equation (7.17) reads:

$$\begin{aligned} ca^k bCA^k B &= C \sum_{i=1}^{q} \tilde{c}_i \begin{bmatrix} \lambda_i^k & k\lambda_i^{k-1} \\ 0 & \lambda_i^k \end{bmatrix} \tilde{b}_i A^k B \\ &= \sum_{i=1}^{q} \tilde{C}_i \tilde{A}_i^k \tilde{B}_i, \end{aligned} \tag{7.22}$$

where,

$$\tilde{A}_i = \begin{bmatrix} \lambda_i A & A \\ 0 & \lambda_i A \end{bmatrix}, \quad \tilde{B}_i = \begin{bmatrix} \tilde{b}_{i,1} B \\ \tilde{b}_{i,2} B \end{bmatrix}, \quad \tilde{C}_i = \begin{bmatrix} \tilde{c}_{i,1} C & \tilde{c}_{i,2} C \end{bmatrix}. \tag{7.23}$$

From the assumption that all $x_i \neq 0$, it follows that the matrix a has at least one non-zero eigenvalue. Hence we may assume that $\lambda_i \neq 0$.

The observability matrix associated to the triplet $(\tilde{A}_i, \tilde{B}_i, \tilde{C}_i)$ is written as:

$$\mathcal{V}_{n,i} = \begin{bmatrix} \tilde{c}_{i,1} C & \tilde{c}_{i,2} C \\ \tilde{c}_{i,1} C(\lambda_i A) & \tilde{c}_{i,1} CA + \tilde{c}_{i,2} C(\lambda_i A) \\ \vdots & \vdots \end{bmatrix}. \tag{7.24}$$

The rank of this matrix is equal to the rank of $\frac{1}{\tilde{c}_{i,1}} \mathcal{V}_n$ for $\lambda_i \neq 0, \tilde{c}_{i,1} \neq 0$:

$$\frac{1}{\tilde{c}_{i,1}} \mathcal{V}_{n,i} = \begin{bmatrix} C & 0 \\ C(\lambda_i A) & C(\lambda_i A) \\ C(\lambda_i A)^2 & 2C(\lambda_i A)^2 \\ \vdots & \vdots \end{bmatrix} \begin{bmatrix} I & \frac{\tilde{c}_{i,2}}{\tilde{c}_{i,1}} I \\ 0 & \frac{1}{\lambda_i} I \end{bmatrix}. \tag{7.25}$$

Sylvester's inequality (Golub and Van Loan, 1996) shows that the rank of $\frac{1}{\tilde{c}_{i,1}} \mathcal{V}_{n,i}$ is equal to the rank of the following matrix:

$$\begin{bmatrix} C & 0 \\ C(\lambda_i A) & C(\lambda_i A) \\ C(\lambda_i A)^2 & 2C(\lambda_i A)^2 \\ \vdots & \vdots \end{bmatrix}. \tag{7.26}$$

When the pair $(C, \lambda_i A)$ is observable, both matrices:

$$\begin{bmatrix} C \\ C(\lambda_i A) \\ C(\lambda_i A)^2 \\ \vdots \end{bmatrix}, \quad \begin{bmatrix} I & & & \\ & 2I & & \\ & & \ddots & \\ & & & (s-1)I \end{bmatrix} \begin{bmatrix} C \\ C(\lambda_i A) \\ C(\lambda_i A)^2 \\ \vdots \end{bmatrix}. \tag{7.27}$$

have full column rank. The zero-block in the upper right part of matrix (7.26) shows that the rank of $\mathcal{V}_{n,i}$ is strictly larger than n_2 and this is again a contradiction.

The proof for the second bullet of the theorem goes as follows. When $\alpha_i t_i = \eta^{i-1}$ for all i, then:

$$\begin{aligned}\mathcal{H}(\alpha_i \widehat{M}_{i,2}) &= \mathcal{H}(C_2(\eta A_2)^{i-1} B_2) \\ &= \mathcal{O}_{2,s_1} \mathcal{C}_{2,s_1},\end{aligned} \tag{7.28}$$

with $\mathcal{O}_{2,s_1}$ and $\mathcal{C}_{2,s_1}$ being the observability and controllability matrices of the pairs $(C_2, \eta A_2)$ and $(\eta A_2, B_2)$, respectively. Using Sylvester's inequality, the rank of $\mathcal{O}_{2,s_1}\mathcal{C}_{2,s_1}$ equals n_2. This proves that the rank of $\mathcal{H}(\alpha_i \widehat{M}_{i,2})$ is n_2 under the conditions specified in the theorem. □

COROLLARY 7.2 (Sinquin and Verhaegen (2019a)) *Using the notation in Theorem 7.1, when the* $\text{rank}\big(\mathcal{H}(\alpha_i \widehat{M}_{i,2})\big) = n_2$*, there exist non-zero scalars* $(a_\alpha, b_\alpha, c_\alpha)$ *such that* $\alpha_i = \frac{c_\alpha a_\alpha^{i-1} b_\alpha}{t_i}$.

Proof It follows from Theorem 7.1 by using $x_i = \alpha_i t_i$. □

COROLLARY 7.3 (Sinquin and Verhaegen (2019a)) *Using the notation introduced in Theorem 7.1, when the rank* $\big(\mathcal{H}(\beta_i \widehat{M}_{i,1})\big) = n_1$*, there exist* $(a_\beta, b_\beta, c_\beta)$ *non-zero scalars such that* $\beta_i = c_\beta a_\beta^{i-1} b_\beta t_i$.

Proof It follows from Theorem 7.1 adapted to $\mathcal{H}\big(\beta_i \widehat{M}_{i,r}\big)$ and by using $x_i = \frac{\beta_i}{t_i}$. □

In summary, Theorem 7.1 and Corollaries 7.2 and 7.3 show that the matrix $\mathcal{H}\big(\widehat{M}_{i,2}\big)$ does not, in general, have low rank (equal to the order of the local triplet (A_2, B_2, C_2)). This is due to the scaling factors t_i that are different for each factor Markov parameter. From the properties of the block-Hankel $\mathcal{H}\big(\alpha_i \widehat{M}_{i,2}\big)$ that were studied in this theorem, we conclude that if the rank of the latter matrix is minimal, the following rank condition has to hold: $\text{rank}\big(\mathcal{H}(\alpha_i t_i)\big) = 1$. The number of α_i's for which this rank condition holds is, however, infinite. In the following we consider the low-rank property of the block-Hankel matrices $\mathcal{H}\big(\alpha_i \widehat{M}_{i,2}\big)$ and the block-Hankel matrix $\mathcal{H}\big(\beta_i \widehat{M}_{i,1}\big)$. Theorem 7.2 tries to obtain insight into the uniqueness of the scalings α_i, β_i when they are related by a bilinear constraint. The theorem shows that they are not unique.

THEOREM 7.2 (Non-uniqueness of the rank constraints in Theorem 7.1) (Sinquin and Verhaegen, 2019a) *The multi-criteria feasibility problem:*

$$\begin{aligned} \textit{find} \quad & (\alpha_i, \beta_i) \quad \forall\, i \in \{1, \ldots, s\} \\ \textit{such that} \quad & \left\{\textit{rank}\left(\mathcal{H}(\alpha_i \widehat{M}_{i,2})\right) = n_2, \textit{rank}\left(\mathcal{H}(\beta_i \widehat{M}_{i,1})\right) = n_1\right\}, \\ & \alpha_i \beta_i = 1, \end{aligned} \tag{7.29}$$

has a non-unique set of feasible solutions for all α_i, β_i*, as described in Corollaries 7.2 and 7.3, (even) when adding the additional constraints* $a_\alpha a_\beta = 1$ *and* $c_\alpha c_\beta b_\alpha b_\beta = 1$.

Proof From Corollaries 7.2 and 7.3, we observe that the rank conditions in (7.29) are satisfied for all $i \in \{1, \ldots, s\}$ when

$$\alpha_i = \frac{c_\alpha a_\alpha^{i-1} b_\alpha}{t_i}, \qquad \beta_i = c_\beta a_\beta^{i-1} b_\beta t_i. \tag{7.30}$$

Replacing these expressions in the bilinear constraint (7.29) yields:

$$\alpha_i \beta_i = c_\alpha c_\beta (a_\alpha a_\beta)^{i-1} b_\alpha b_\beta = 1, \tag{7.31}$$

which implies $a_\alpha a_\beta = 1$ and $c_\alpha c_\beta b_\alpha b_\beta = 1$. □

REMARK 7.2 With the bilinear constraint $\alpha_i \beta_i = 1$, both α_i and β_i cannot be zero. Moreover, the scalars t_i, v_i are related with $t_i v_i = 1$. Therefore, both sequences $\alpha_i t_i$ and $\frac{\beta_i}{t_i}$ differ from zero and this fulfills the condition imposed on the scaling coefficients in Theorem 7.2.

The results from Lemma 7.2 to the above last remark can be summarized in the definition of a bilinear constrained low-rank optimization problem to rescale the factor Markov parameters.

From problem (7.29), when not knowing the system orders (n_2, n_1) a priori, we formulate the following bilinear rank optimization problem:

$$\begin{aligned} \min_{\alpha, \beta} \quad & \operatorname{rank}\left(\mathcal{H}(\alpha_i \widehat{M}_{i,2})\right) + \operatorname{rank}\left(\mathcal{H}(\beta_i \widehat{M}_{i,1})\right) \\ \text{such that} \quad & \forall\, i \in \{1, \ldots, s\}, \quad \alpha_i \beta_i = 1. \end{aligned} \tag{7.32}$$

The minimization problem (7.32) is bilinear in the scaling parameters α_i, β_i. Further, the presence of the rank operator in this problem makes it non-convex. Relaxing this rank operator to the nuclear norm turns the problem in to the class of multi-convex optimization problems discussed in Nocedal and Wright (2006). A possible solution to such problems, as worked out in Xu and Yin (2013) and Doelman and Verhaegen (2016), is via the solution of alternating convex problems. The work of Xu and Yin (2013) uses a block-coordinate update (BCU) algorithm with slack variables. The slack variables are used to relax the bilinear constraints that are gradually tightened by increasing a regularization parameter. The latter reformulates each bilinear term to create a 2×2 matrix whose rank is one when the variables are set to the correct values, leading to a sequential optimization of a cost function featuring nuclear norms.

Let $\|.\|_\star$ denote the nuclear norm, then the optimization (7.32) is relaxed into:

$$\begin{aligned} \min_{\alpha_i, \beta_i, q_i} \quad & \|\mathcal{H}(\alpha_i \widehat{M}_{i,2})\|_\star + \|\mathcal{H}(\beta_i \widehat{M}_{i,1})\|_\star + \mu \sum_{i=1}^{s} q_i^2, \\ \text{such that} \quad & \forall\, i \in \{1, \ldots, s\}, \quad \alpha_i \beta_i - 1 = q_i, \end{aligned} \tag{7.33}$$

where μ is a regularization parameter to be selected by the user. For that, use can be made of the following intuition. The larger the parameter μ is, the more emphasis is put on setting q to 0. The optimization problem is solved by alternating with respect to (α_i, q_i) and (β_i, q_i), respectively, in iterative manner. It starts with a non-zero initial guess and with a low value of μ. In this sense, the first iteration(s) will generally violate the bilinear equality. The number of variables is, moreover, only $2s$. Algorithm 7.1 details all the steps. Here, we use the notation $\mathcal{B}_{\widehat{M}_2,\alpha_i,\mu}(\beta_i,q_i)$ for $i = 1 \dots s$ to indicate the cost function (7.33) when the decision variables are the pair (β_i,q_i) while keeping the parameters $\{\alpha_i\}_{i=1\dots s}$ fixed, and with the regularization parameter μ. This regularization parameter is to be increased gradually during the iterations in order to increasingly emphasize the bilinear constraint in (7.33) (Xu and Yin, 2013). The prominent role of the initial value $\mu^{(0)}$ for μ is highlighted next. If it is set too large when optimizing $\mathcal{B}_{\widehat{M}_1,\beta_i,\mu}(\alpha_i,q_i)$ with respect to the decision pair (α_i,q_i) (respectively $\mathcal{B}_{\widehat{M}_2,\alpha_i,\mu}(\beta_i,q_i)$ with respect to (β_i,q_i)), the variable α_i is fixed to $1/\beta_i$ by the constraint. If $\mu^{(0)}$ is set too low, α_i goes to 0 and q_i to -1. In that respect, $\mu^{(0)}$ plays the role of a regularization parameter whose optimal value is determined by a grid search.

Standard experience with ADMM optimization techniques are helpful in such grid search and more details can be found for example in Verhaegen and Hansson (2016) using the linear operator framework to enforce the block-Hankel structure. The nuclear

Algorithm 7.1: Summary of the algorithm to solve (7.32).

Input : $\widehat{M}_{2,i}, \widehat{M}_{1,i}, \mu^{(0)}$
Output: $\widehat{\alpha}_i, \widehat{\beta}_i$ for $i = 1 \dots s$

```
/* Default values                                                  */
```
1 $\ell = 5, \tau = 5, \kappa_{max} = 40, \epsilon_{min} = 10^{-3}$
2 $\kappa \leftarrow 0$
3 $\beta^{(\kappa)} \leftarrow \begin{bmatrix} \beta_1^{(\kappa)} \\ \vdots \\ \beta_s^{(\kappa)} \end{bmatrix}$ and $\alpha^{(\kappa)} \leftarrow \begin{bmatrix} \alpha_1^{(\kappa)} \\ \vdots \\ \alpha_s^{(\kappa)} \end{bmatrix}$
4 **for each** $i \leq s$ **do**
5 | $\alpha_i^{(\kappa)} \leftarrow 1$
6 **end**
7 **while** $\kappa \leq \kappa_{max}$ and $\epsilon > \epsilon_{min}$ **do**
8 | $\beta^{(\kappa+1)} \leftarrow \text{argmin } \mathcal{B}_{\widehat{M}_2,\alpha^{(\kappa)},\mu^{(\kappa)}}(\beta,q)$.
9 | $\alpha^{(\kappa+1)} \leftarrow \text{argmin } \mathcal{B}_{\widehat{M}_1,\beta^{(\kappa+1)},\mu^{(\kappa)}}(\alpha,q)$.
10 | **if** $mod(\kappa,\ell) = 0$ **then**
11 | | $\mu^{(\kappa+1)} \leftarrow \tau\mu^{(\kappa)}$
12 | **end**
13 | $\epsilon \leftarrow \sum_{i=1}^{s}(\alpha_i\beta_i - 1)^2$
14 | $\kappa \leftarrow \kappa + 1$
15 **end**
16 Set $\widehat{\alpha}$ and $\widehat{\beta}$ to the optimal values.

norm minimization can simply be done by soft-thresholding the singular values in the SVD step. The details are not reproduced here in order to focus on the subspace algorithm.

REMARK 7.3 (Computational complexity) The computational complexity of Algorithm 7.1 is determined by the solution of each ADMM problem. When the number of iterations is independent of N, the computation of a Gramian matrix is computed based on the sequences $\{\widehat{M}_{i,1}\}, \{\widehat{M}_{i,2}\}$ prior to performing the ADMM update costs $\mathcal{O}(N^2)$ flops. Two operations that appear in each of the above ADMM steps are detailed below. As the number of unknowns is only $2s$, the cost of the primal variable update in each of the ADMM steps is not dominated by the matrix inversion, but rather by forming the matrices prior to solving the least-squares, which scales with $\mathcal{O}(N^2)$ only. However, in each iteration, an SVD of the block-Hankel matrix $\mathcal{H}(\alpha_i^{(\kappa)}\widehat{M}_{i,2})$ (respectively, $\mathcal{H}(\beta_i^{(\kappa)}\widehat{M}_{i,1})$) is to be calculated by soft-thresholding its a singular-values. This is the bottleneck in Algorithm 7.1 because this operation scales with $\mathcal{O}(N^3)$.

7.2.4 Estimating the Factor Matrices of the State-Space Matrices

7.2.4.1 A Block-Hankel Low-Rank Matrix

The theoretical insights presented in the previous subsections show that a solution can be obtained, as in Lemma 7.1, which can be given as:

$$\alpha_i \widehat{M}_{i,2} = \alpha_i t_i M_{i,2} = c_\alpha a_\alpha^{i-1} b_\alpha M_{i,2}. \tag{7.34}$$

The block-Hankel matrix $\mathcal{H}\left(\alpha_i \widehat{M}_{i,2}\right)$ is then:

$$\mathcal{H}\left(\alpha_i \widehat{M}_{i,2}\right) = \begin{bmatrix} c_\alpha b_\alpha M_{1,2} & c_\alpha a_\alpha b_\alpha M_{2,2} & \cdots & c_\alpha a_\alpha^{s_1-1} b_\alpha M_{s_1,2} \\ c_\alpha a_\alpha b_\alpha M_{2,2} & c_\alpha a_\alpha^2 b_\alpha M_{3,2} & \cdots & c_\alpha a_\alpha^{s_1} b_\alpha M_{s_1+1,2} \\ \vdots & & & \vdots \\ c_\alpha a_\alpha^{s_1-1} b_\alpha M_{s_1,2} & \cdots & \cdots & c_\alpha a_\alpha^{s-1} b_\alpha M_{s,2} \end{bmatrix}. \tag{7.35}$$

An SVD of this matrix delivers its rank $\hat{n}_2$,

$$\mathcal{H}(\alpha_i \widehat{M}_{i,2}) = \begin{bmatrix} c_\alpha C_2 \\ c_\alpha C_2(a_\alpha A_2) \\ \vdots \\ c_\alpha C_2(a_\alpha A_2)^{s_1-1} \end{bmatrix} \begin{bmatrix} b_\alpha B_2 & (a_\alpha A_2) b_\alpha B_2 & \cdots & (a_\alpha A_2)^{s_1-1} b_\alpha B_2 \end{bmatrix}. \tag{7.36}$$

After selecting the system order $\widehat{n}_2$, the matrices $\widehat{A_2}, \widehat{B_2}, \widehat{C_2}$ are estimated via the standard realization algorithm as summarized in Algorithm 7.2.

Algorithm 7.2: Realization steps

Input : $\widehat{\alpha}, \{\widehat{M}_{i,2}\}_{i=1\ldots s}$
Output: $\widehat{A}_2, \widehat{B}_2, \widehat{C}_2$

1 Compute a SVD, $\mathcal{H}_2(\alpha_i \widehat{M}_{i,2}) = U_2\Sigma_2 V_2^T$
2 Select the system order, $\widehat{n}_2$
3 Denote: $U_{2,\widehat{n}_2} = U_2(:,1:\widehat{n}_2)$ and $V_{2,\widehat{n}_2} = \Sigma_2(1:\widehat{n}_2, 1:\widehat{n}_2)V_2(:,1:\widehat{n}_2)^T$
/* Estimate B_2 and C_2 */
4 Extract $\widehat{B_2} = V_{2,\widehat{n}_2}(:,1:N)$ and $\widehat{C_2} = U_{2,\widehat{n}_2}(1:N,:)$
/* Estimate A_2 */
5 $\mu = 10^{-6}$, $\widehat{A}_j = I_{\widehat{n}_2}$
6 **while** $\widehat{A}_2$ *is not strictly stable* **do**
7 Solve the following least-squares problem denote its solution as $\widehat{A}_2$,

$$\min_{A_2} \|U_{2,\widehat{n}_2}(N+1:s_1N,:) - U_{2,\widehat{n}_2}(1:(s_1-1)N,:)A_2\|_F^2 + \mu\|A_2\|_F^2$$

$\mu = 10 \cdot \mu$
8 **end**

LEMMA 7.3 (Sinquin and Verhaegen (2019a)) *Let the regularization parameter μ be zero in Algorithm 7.2, then the factor matrices estimated by Algorithm 7.2 are* not *an equivalent realization of the MSSM.*

Proof Let $T_2 \in \mathbb{R}^{n_2 \times n_2}$ denote a similarity transformation such that:

$$\widehat{A_2} = T_2 a_\alpha A_2 T_2^{-1}, \quad \widehat{B_2} = T_2 b_\alpha B_2, \quad \widehat{C_2} = c_\alpha C_j T_2^{-1}. \tag{7.37}$$

Using the fact that $a_\alpha a_\beta = 1$, we write:

$$\widehat{A_2} \otimes \widehat{A_1} = a_\alpha a_\beta T A T^{-1} = T A T^{-1}. \tag{7.38}$$

However, it is not true that $b_\alpha b_\beta = 1$ or that $c_\alpha c_\beta = 1$, and therefore,

$$\widehat{C_2} \otimes \widehat{C_1} \neq C T^{-1}, \quad \widehat{B_2} \otimes \widehat{B_1} \neq T B. \tag{7.39}$$

Nonetheless, $c_\alpha b_\alpha c_\beta b_\beta = 1$ implying:

$$\widehat{C_2}\widehat{B_2} \otimes \widehat{C_1}\widehat{B_1} = CB. \tag{7.40}$$

There are infinite possibilities for choosing c_α, c_β such that the Hankel matrix built from $\{c_\alpha a_\alpha^{i-1} b_\alpha\}_{i \in \{1,\ldots,s\}}$ is rank one. □

What remains now is to comment on whether the non-uniqueness highlighted in Lemma 7.3 can be dealt with. In general, it does, and it might also depend on the purpose of the identified model. The input-output relationship can be given by the convolution of an infinite impulse response with the input sequence because the parameters of this impulse response do not change by the scaling of the factor matrices of the state-space model. However, for realizing a state-space model, none of the terms, like B or C, appear separately. As such, they cannot be determined individually.

7.2.4.2 A Data-Equation in Matrix Form

In order to estimate a similarly equivalent state-space model, we now propose to estimate a similarly equivalent state-sequence and use that sequence to estimate the system factor matrices of the system matrices (7.2) via a bilinear least-squares optimization. For that purpose, we start with the following data equation, first written in matrix form:

$$\mathcal{Y} = \mathcal{V} + \mathcal{T}_u + \mathcal{E}, \tag{7.41}$$

where the block-Hankel matrix $\mathcal{Y} \in \mathbb{R}^{pNs \times NM}$ is as follows:

$$\mathcal{Y} = \begin{bmatrix} Y(1) & Y(2) & \dots & Y(M) \\ Y(2) & \ddots & & \vdots \\ \vdots & & \ddots & \vdots \\ Y(s) & Y(s+1) & \dots & Y(N_t) \end{bmatrix}, \tag{7.42}$$

with $M = N_t - s + 1$. The block-Hankel matrix denoted by $\mathcal{E}$ is built in a similar manner to that of $\mathcal{Y}$ from the noise matrices $E(k)$. The matrix $\mathcal{V}$ is defined with:

$$\mathcal{V} = \begin{bmatrix} C_1 X(1) C_2^T & \dots & C_1 X(M) C_2^T \\ C_1 A_1 X(1)(C_2 A_2)^T & \dots & \vdots \\ \vdots & \dots & \vdots \\ C_1 A_1^{s-1} X(1)(C_2 A_2^{s-1})^T & \dots & C_1 A_1^{s-1} X(M)(C_2 A_2^{s-1})^T \end{bmatrix}, \tag{7.43}$$

and $\mathcal{T}_u$ with:

$$\begin{bmatrix} 0 & \dots & 0 \\ M_{1,1} U(1) M_{1,2}^T & \dots & M_{1,1} U(M) M_{1,2}^T \\ \sum_{i=0}^{1} M_{i+1,1} U(2-i) M_{i+1,2}^T & & \vdots \\ \vdots & \ddots & \vdots \\ \sum_{i=0}^{s-2} M_{i+1,1} U(s-1-i) M_{i+1,2}^T & \dots & \sum_{i=0}^{s-2} M_{i+1,1} U(N_t - i) M_{i+1,2}^T \end{bmatrix}. \tag{7.44}$$

Data equation (7.41) gives the relationship between matrices of sizes of order N rather than N^2. However, the composing matrices do not have key structural properties like low-rankness that have been exploited in standard subspace identification (Verhaegen and Verdult, 2007). For example, the matrix $\mathcal{V}$ is, in general, not of low rank and hence estimating the state sequence is not possible.

In order to proceed, we construct a new matrix having low rank property form entries of the matrix $\mathcal{V}$, but also containing unknowns to be constructed from the derived data. For that purpose, we embed the entries of the matrix $\mathcal{V}$ into a structured third-order tensor denoted by $\mathcal{A}$. Let $\varphi \in \mathbb{N}$ such that $\varphi p N > n_1$ and $\varphi N > n_2$. For all $k = 1 \dots M$, a slice $\mathcal{A}(:,:,k) \in \mathbb{R}^{p\varphi N \times \varphi N}$ of the tensor $\mathcal{A}$ is given as:

$$\begin{bmatrix} C_1 X(k) C_2^T & \dots & C_1 X(k) \left(C_2 A_2^{\varphi-1}\right)^T \\ \vdots & & \vdots \\ C_1 A_1^{\varphi-1} X(k) C_2^T & \dots & C_1 A_1^{\varphi-1} X(k) \left(C_2 A_2^{\varphi-1}\right)^T \end{bmatrix}. \tag{7.45}$$

Before estimating the entries of the tensor $\mathcal{A}$, we first focus on the rank properties of matricizations of this tensor, pretending we have this tensor available.

DEFINITION 7.5 Let $\mathcal{A} \in \mathbb{R}^{p\varphi N \times \varphi N \times M}$ be a third-order tensor. The unfolding $\mathcal{A}_{(1)}$ is defined as:

$$\mathcal{A}_{(1)} = \begin{bmatrix} \mathcal{A}(:,:,1) & \dots & \mathcal{A}(:,:,M) \end{bmatrix} \in \mathbb{R}^{p\varphi N \times \varphi N M}. \tag{7.46}$$

Then using the definition in (7.45) provides the following matrix factorization of $\mathcal{A}_{(1)}$:

$$\mathcal{A}_{(1)} = \begin{bmatrix} C_1 \\ C_1 A_1 \\ \vdots \\ C_1 A_1^{\varphi-1} \end{bmatrix} \begin{bmatrix} X(1) C_2^T & \dots & X(M) \left(C_2 A_2^{\varphi-1}\right)^T \end{bmatrix}. \tag{7.47}$$

The rank of $\mathcal{A}_{(1)}$ is equal to n_1:

$$\operatorname{rank}\left(\mathcal{A}_{(1)}\right) = n_1 < p\varphi N. \tag{7.48}$$

Hence, it has an SVD as:

$$\begin{aligned} \mathcal{A}_{(1)} &= U_1 V_1, \\ U_1 &= \mathcal{O}_{\varphi,1} T_1, \\ V_1 &= T_1^{-1} \begin{bmatrix} X(1) C_2^T & \dots X(M) \left(C_2 A_2^{\varphi-1}\right)^T \end{bmatrix}, \end{aligned} \tag{7.49}$$

with the matrix T_1 non-singular. From this, and more precisely from U_1, the matrices A_1 and C_1 can be estimated. With the entries of the matrix V_1, we can define the following low-rank matrix H:

$$\begin{aligned} H &= \begin{bmatrix} T_1^{-1} X(1) C_2^T & \dots & T_1^{-1} X(1) \left(C_2 A_2^{\varphi-1}\right)^T \\ \vdots & & \vdots \\ T_1^{-1} X(M) C_2^T & \dots & T_1^{-1} X(M) \left(C_2 A_2^{\varphi-1}\right)^T \end{bmatrix} \\ &= \begin{bmatrix} T_1^{-1} X(1) \\ \vdots \\ T_1^{-1} X(M) \end{bmatrix} \begin{bmatrix} C_2^T & \dots & \left(C_2 A_2^{\varphi-1}\right)^T \end{bmatrix}, \end{aligned} \tag{7.50}$$

with rank given as:

$$\operatorname{rank}(H) = n_2 < \varphi N. \tag{7.51}$$

An SVD of this matrix H is denoted by:

$$H = U_2 V_2, \tag{7.52}$$
$$U_2 = \begin{bmatrix} T_1^{-1} X(1) T_2 \\ \vdots \\ T_1^{-1} X(M) T_2 \end{bmatrix},$$
$$V_2 = T_2^{-1} \begin{bmatrix} C_2^T & \dots & (C_2 A_2^{\varphi-1})^T \end{bmatrix}.$$

Consequently, the matrix U_2 provides an estimate for the state-sequence up to the similarity transformations T_1 and T_2 as presented in Lemma 7.1. From the matrix V_2, equal to the extended observability matrix $\boldsymbol{O}_{\varphi,2}$ up to a similarity transformation T_2, the matrices A_2 and C_2 can be estimated.

REMARK 7.4 The SVD (7.49) is performed on the tensor unfolding $\mathcal{A}_{(1)}$ while the second SVD (7.52) deals with the matrix H of reduced size and obtained from the right singular vectors in (7.49). A canonical polyadic decomposition (CPD) of the tensor $\mathcal{A}$ does provide an estimate for the state-sequence as such CPD would provide sets of matrices of size $p\varphi N \times r$, $\varphi N \times r$, and one of size $M \times r$ (instead of $Mr \times r$).

REMARK 7.5 The factor matrices derived from the feasibility problem (7.29), as well as the rank minimization (7.32), may lead to unstable factored models. Either the matrix A_1 or the matrix A_2 may have eigenvalues outside the unit circle, while the other can compensate for that such that the Kronecker product A is still stable. For α_i, as described in Corollary 7.3, the scaled factor matrices $\alpha_i \widehat{M}_{i,2}$ of the Markov parameters become:

$$\alpha_i \widehat{M}_{i,2} = c_\alpha \eta^{i-1} b_\alpha C_2 A_2^{i-1} B_2$$
$$= c_\alpha C_2 (\eta A_2)^{i-1} b_\alpha B_2.$$

The stability of ηA_1 is not guaranteed due to the presence of η. Although this does not affect the estimation of α, β (defined in Algorithm 7.1), when we know that the underlying system is stable, it is recommendable to derive two stable factor matrices. This can be realized by computing an eigenvalue of the unstable matrix, say $\widehat{A}_1$, and dividing its entries by the number $(1+\epsilon) \cdot \max_{i=1,2}(\lambda_{max}(A_i))$ (for some small ϵ) while counter-scaling the other matrix.

7.2.4.3 Completing the Tensor $\mathcal{A}$

We consider the noise-free case $\mathcal{E} = 0$. The submatrices $C_1 A_1^i X(k) (C_2 A_2^j)^T$ for $i = j$ are located on the main block-diagonal of $\mathcal{A}(:,:,k)$. These entries can be read from:

$$\mathcal{V} \approx \mathcal{Y} - \widehat{\mathcal{T}}_{\boldsymbol{u}}, \tag{7.53}$$

where $\widehat{\mathcal{T}}_{\boldsymbol{u}}$ is obtained from $\mathcal{T}_{\boldsymbol{u}}$ by replacing all $M_{i,2}, M_{i,1}$ with $\widehat{M}_{i,2}, \widehat{M}_{i,1}$.

Next, we consider the block entries away from the main block-diagonal of $\mathcal{A}(:, :, k)$. These contain cross-terms such as $C_1 A_1^i X(k)\big(C_2 A_2^j\big)^T$ for $i \neq j$. These cross-terms can be estimated by simulating the following MSSM:

$$\begin{cases} X(k+1) = A_1 X(k) A_2^T + B_1 U(k) B_2^T, \\ Y_{g,h}^{\sharp}(k) = C_{1,g}^{\sharp} X(k) \big(C_{2,h}^{\sharp}\big)^T \end{cases} \tag{7.54}$$

This MSSM has virtual output $Y_{g,h}^{\sharp}(k)$, for $g, h \in \mathbb{N}$. The output matrices in this MSSM are $C_{1,g}^{\sharp} = C_1 A_1^g$ and $C_{2,h}^{\sharp} = C_2 A_2^h$. This means that the MSSM has the following generator:

$$\mathcal{S}_{g,h} = \big\{A_1, A_2, B_1, B_2, C_1 A_1^g, C_2 A_2^h\big\}. \tag{7.55}$$

The virtual outputs $\big\{Y_{g,h}^{\sharp}(k)\big\}_{k=1\ldots N_t}$ are unknown when both g, h are non-zero. For such cases they are approximated from the following high-order FIR models.

$$Y_{g,h}^{\sharp}(k) \approx \sum_{i=0}^{z-1} C_1 A_1^{i+g} B_1 U(k-i-1) \left(C_2 A_2^{i+h} B_2\right)^T, \tag{7.56}$$

$$Y_{g,h}^{\sharp}(k) \approx \sum_{i=0}^{z-1} M_{i+g+1,1} U(k-i-1) M_{i+h+1,2}^T, \tag{7.57}$$

for $z \in \mathbb{N}$ and for all $k \geq z$.

Using the estimates $\widehat{M}_{i,1}, \widehat{M}_{i,2}$ along with the estimates $\widehat{\alpha}, \widehat{\beta}$, Equation (7.56) reads:

$$Y_{g,h}^{\sharp}(k) \approx \sum_{i=0}^{z-1} \widehat{\beta}_{i+g+1} \widehat{M}_{i+g+1,1} U(k-i-1) \widehat{\alpha}_{i+h+1} \widehat{M}_{i+h+1,2}^T. \tag{7.58}$$

The retrieval of $Y_{0,0}^{\sharp}(k)$ does not require knowledge of the coefficients $\widehat{\alpha}, \widehat{\beta}$.

Let us now turn to some rules of thumb to specify the indices g, h to fill the tensor $\mathcal{A}$, as in (7.45). Having estimated the scaling coefficients $\widehat{\alpha}_i$ and $\widehat{\beta}_i$ for $i = 1 : s$, Equation (7.58) implies the following ranges for choosing the triplet (z, g, h):

$$z + g \leq s, \quad z + h \leq s. \tag{7.59}$$

In order to satisfy the rank inequalities (7.48) and (7.51), the index φ needs to be strictly smaller than s and larger than $\frac{n}{p}$. Hence the allowable combinations of the pair (g, h) that determine the filling of the tensor $\mathcal{A}$ are $g = 0, h \in \{1, \ldots, \varphi - 1\}$, $g = 0$, $h = 0$, $g \in \{1, \ldots, \varphi - 1\}, h = 0$. The maximum value of g is $g = \varphi - 1$, which implies $z + \varphi - 1 = s$. A total number of $2\varphi - 1$ virtual outputs are available within the temporal range $\{z, \ldots, N_t\}$. For each of the associated virtual MSSM (7.54), we can write a data equation in matrix form similar to (7.53) as:

$$\mathcal{Y}_{g,h}^{\sharp} = \mathcal{V}_{g,h} + \mathcal{T}_{g,h}, \tag{7.60}$$

where, for $M_z = N_t - \varphi + 1$:

$$\mathcal{Y}^{\sharp}_{g,h} = \begin{bmatrix} Y^{\sharp}_{g,h}(z) & Y^{\sharp}_{g,h}(z+1) & \dots & Y^{\sharp}_{g,h}(M_z) \\ Y^{\sharp}_{g,h}(z+1) & \ddots & & \vdots \\ \vdots & & \ddots & \vdots \\ Y^{\sharp}_{g,h}(z+\varphi-1) & \dots & \dots & Y^{\sharp}_{g,h}(N_t) \end{bmatrix},$$

$$\mathcal{V}_{g,h} = \begin{bmatrix} C^{\sharp}_{1,g} X(z) C^{\sharp\,T}_{2,h} & \dots & C^{\sharp}_{1,g} X(M_z) C^{\sharp\,T}_{2,h} \\ \vdots & \ddots & \vdots \\ C^{\sharp}_{1,g} A_1^{\varphi-1} X(z) (C^{\sharp}_{2,h} A_2^{\varphi-1})^T & \dots & C^{\sharp}_{1,g} A_1^{\varphi-1} X(M_z) (C^{\sharp}_{2,h} A_2^{\varphi-1})^T \end{bmatrix}.$$

$\mathcal{T}_{g,h}$ is as follows:

$$\begin{bmatrix} 0 & \dots & 0 \\ C^{\sharp}_{1,g} B_1 U(z) (C^{\sharp}_{2,h} B_2)^T & \dots & C^{\sharp}_{1,g} B_1 U(M_z) (C^{\sharp}_{2,h} B_2)^T \\ \sum_{i=0}^{1} C^{\sharp}_{1,g} A_1^i B_1 U(z+1-i) (C^{\sharp}_{2,h} A_2^i B_2)^T & \dots & \vdots \\ \vdots & & \vdots \\ \sum_{i=0}^{\varphi-2} C^{\sharp}_{1,g} A_1^i B_1 U(z+\varphi-2-i) (C^{\sharp}_{2,h} A_2^i B_2)^T & \dots & \sum_{i=0}^{\varphi-2} C^{\sharp}_{1,g} A_1^i B_1 U(N_t-i) (C^{\sharp}_{2,h} A_2^i B_2)^T \end{bmatrix}. \tag{7.61}$$

The matrices $\mathcal{V}_{g,h}$ are estimated from:

$$\mathcal{V}_{g,h} \approx \mathcal{Y}^{\sharp}_{g,h} - \widehat{\mathcal{T}}_{g,h}. \tag{7.62}$$

These matrices are contained as the diagonal slices of the tensor $\mathcal{A}$ given as follows:

$$\mathcal{A}((i-1)pN + 1 : ipN, (i-1)N + 1 : iN, :). \tag{7.63}$$

This containment is illustrated in Figures 7.1 and 7.2.

For example, in Figure 7.1 the block-diagonal terms of each $\mathcal{A}(:\,,\,:\,,k)$ are contained in $\mathcal{V}_{0,0}$, while the block-subdiagonal terms are contained in $\mathcal{V}_{1,0}$ and finally the block-superdiagonal terms are contained in $\mathcal{V}_{0,1}$. One diagonal slice is provided by one data equation (for $\mathcal{E} = 0$) that is derived from one MSSM (7.54). The main block-diagonal can be computed without knowledge of the scalings $(\widehat{\alpha}, \widehat{\beta})$. This is not the case for all the other slices.

From the estimates of the state sequence $\widehat{X}_T(k) = T_1^{-1} X(k) T_2$, as outlined above, the following bilinear least-squares can be formulated to recover the matrices B_2 and B_1:

$$\min_{B_2, B_1} \sum_{k=z+1}^{M_z-1} \left\| \left(\widehat{X}_T(k+1) - A_1 \widehat{X}_T(k) A_2^T \right) - B_1 U(k) B_2^T \right\|_F^2. \tag{7.64}$$

The minimization problem (7.64) can again be solved via alternating least squares that is started with a random initial value.

Algorithm 7.3 summarizes the steps to estimate the state sequence.

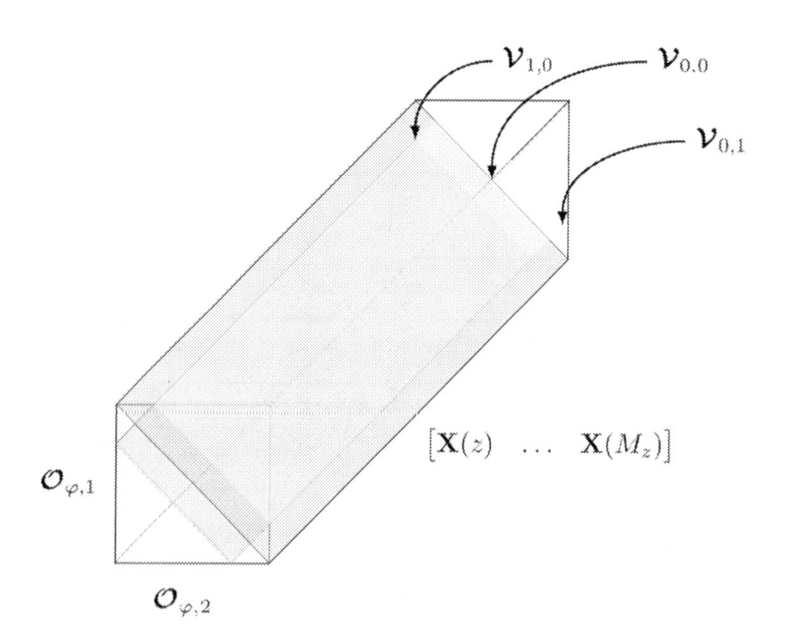

Figure 7.1 Schematic representation of the tensor $\mathcal{A}$ in terms of the diagonal slices $\mathcal{V}_{g,h}$ for the index pair (g, h) equal to $(0, 1), (0, 0), (1, 0)$. The figure also illustrates the position of the observability matrices $\mathcal{O}_{\varphi,1}$ and $\mathcal{O}_{\varphi,2}$. ©2019 IEEE Reprinted, with permission, from Sinquin and Verhaegen (2019a)

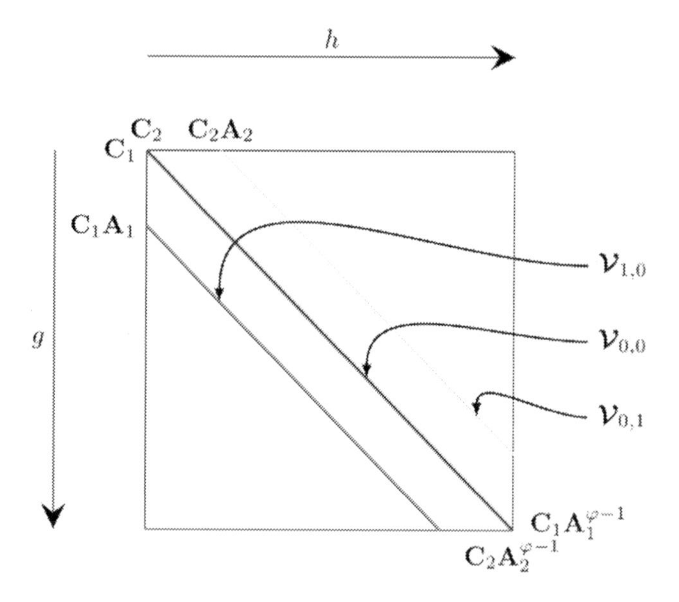

Figure 7.2 Schematic of a slice $\mathcal{A}(:, :, k)$. ©2019 IEEE Reprinted, with permission, from Sinquin and Verhaegen (2019a)

Algorithm 7.3: Estimation of the state-sequence $X(k)$ up to similarity transformation. ©2019 IEEE Reprinted, with permission, from Sinquin and Verhaegen (2019a).

Input : $\{u(k)\}_{1:N_t}, \{y(k)\}_{1:N_t}, \widehat{\alpha}, \widehat{\beta}, \widehat{M}_{i,2}, \widehat{M}_{i,1}, z, \varphi$
Output: $\widehat{A}_1, \widehat{A}_2, \widehat{B}_1, \widehat{B}_2, \widehat{C}_1, \widehat{C}_2$

```
1  for each η = 1 : 2φ − 1 do
2      if η < φ then
3          g = 0, h = η
4      else if η = φ then
5          g = 0, h = 0
6      else if η > φ then
7          g = η − φ, h = 0.
8      foreach k = z + 1 : N_t do
9          Compute the virtual outputs with (7.58)
10     end
11     Compute V_{g,h} with (7.62)
       /* Fill the tensor A                                  */
12     foreach k = z : N_t do
13         Fill each block (sub- or super) diagonal A(: , : ,k), η − φ) of the tensor
            A with the kth block column of the matrix V_{g,h} as indicated in (7.45).
14     end
15 end
   /* Compute the state-sequence                              */
16 Form the unfolded matrix A_(1) and compute the SVD (7.49)
17 Estimate Â1, Ĉ1
18 Form the matrix H and compute its SVD as in (7.52)
19 Estimate Â2, Ĉ2
   /* Compute the input state-space matrices                  */
20 Solve (7.64) iteratively with alternative least squares to estimate B̂2, B̂1
```

REMARK 7.6 (Computational complexity) Computing the virtual outputs requires $(2\varphi - 1)N^3 N_t z$ flops and, therefore, scales with $\mathcal{O}(N^3 N_t)$. The first and second SVD cost respectively $\mathcal{O}(N^3 M)$ and $\mathcal{O}(\widehat{n_1} N^2 M)$. Last, solving the bilinear least-squares in (7.64) requires $N^3 M_z$ flops. The overall computational complexity for Algorithm 7.3 is $\mathcal{O}(N^3 N_t)$.

7.2.5 Numerical Example

For now, we apply the algorithm on deterministic systems with the Kronecker rank-one structure for all state-space matrices and vary the size of the sensor array from 5×5 to 30×30. The algorithm is applied on control for adaptive optics using datasets collected on-sky at the William Herschel Telescope, as discussed in Chapter 10.

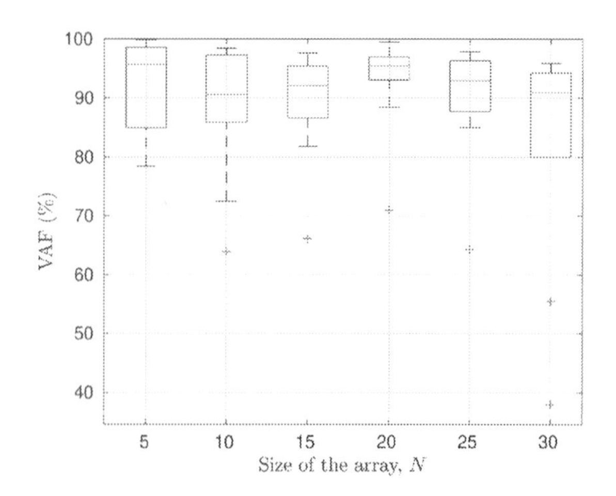

Figure 7.3 Variance Accounted For (%) on a validation set as a function of the size of the array, N. The algorithm was applied on 20 different realizations. The linear model for the Quarks has a regression coefficient of 1.15, whereas it is 1.59 for K4SID.

For each size of the array, 20 systems are randomly generated. The spectral radius of the matrix $A_2 \otimes A_1$ is set between 0.95 and 0.97. The state-space matrices are estimated using Algorithm 7.3. The model orders n_1 and n_2 are set equal to the first index of the singular values vector from (7.49) and (7.52), respectively, whose corresponding value is larger than 99% of the accumulated sum of the singular values. The parameter s was set to 11.

Figure 7.3 shows the VAF as a function of the size of the array. An identification with PI-MOESP or N4SID reaches values with a much smaller variance thanks to the convex optimization scheme they rely on. The low-rank optimization with bilinear constraints does not, in general, converge to the global minimum and, as a consequence, the spread in the VAF results is not negligible. However, the mean remains above 90%. It may happen that the algorithm gives poor identification results: one case out of 20 gave unsatisfactory VAF.

Figure 7.4 shows the scalability of the algorithms Quarks and K4SID. The computational cost scales much more favourably using Quarks and K4SID than standard methods. The regression coefficients for both fitting lines differ: K4SID suffers from the increased number of iterations in the BCU algorithm to estimate the sequence of ambiguity parameters.

7.2.6 Conclusions

In this chapter, we presented a new subspace-like identification methodology to identify MSSM for the dynamics between large-scale multi-dimensional sensor/actuator arrays with $\mathcal{O}(N^3 N_t)$ complexity. This complexity is achieved for the case the global state-space system matrices are of Kronecker rank. Though a number of key steps reveal information about the local orders of the state-space model via SVDs, such as in

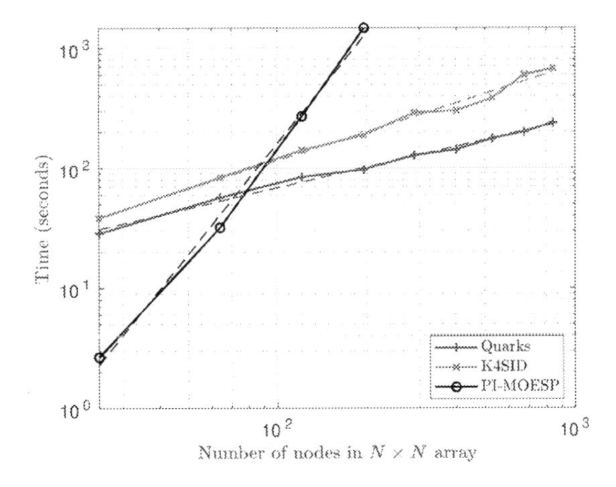

Figure 7.4 Time (seconds) as a function of the number of nodes in the array, N^2, for computing the QUARKS and K4SID. The values were averaged for the 20 times the algorithm was applied on each size of the array.

classical subspace identification methods, the solution to identify MSSMs is far more complex and consists of a number of bilinear optimization problems. These problems are, however, of much smaller scale as they are defined in terms of the factor matrices or scaling factors. This is why the presented solution is, in essence, not a subspace identification method, but referred to as subspace-like identification as it inherits a number of key steps.

The benefits of large-scale modelling with the Kronecker structure are mainly twofold. First, the Kronecker-based subspace algorithm handles larger systems than centralized algorithms in general, such as the N4SID or PI-MOESP subspace algorithms. Second, although the method we propose leads to less accurate models, the number of time samples required no longer increases with N^2, hence there is a significant reduction in the required memory storage and a decreasing burden laid on the application for collecting data is realized.

The estimates may be further refined with a non-linear output-error algorithm provided the computational cost does not overcome that of standard subspace identification methods, the latter converging globally. A preliminary step in the derivation of such algorithms for the present class of state-space models is made in Sinquin and Verhaegen (2017).

In real-life applications, the Kronecker rank of the global state-space matrices may be larger than one. Such problems need further consideration. One such extension may focus on the fact that only the matrices B (and/or C) have such higher Kronecker rank.

Networks often replicate a similar interconnection pattern at different scales. Although the ordering of the nodes to be able to express the matrices with a Kronecker product is straightforward when dealing with a regular sensor grid, this is no longer the case for random networks. The (weighted) adjacency matrix of such graphs may be modelled with Kronecker products (see e.g. figures 1 and 2 in Leskovec et al. (2010)

that are enlightening in this respect), which paved the way for analysing the network properties despite the large dimensions. From an identification perspective, the first approach to deal with random graphs having clusters that share interconnection links is to recast these for applying the methods discussed in Chapter 8. That is, identify the subsystems in each cluster using only local input-output data. If there exists a re-ordering of the nodes such that the state matrix of dynamical models is a Kronecker product, this then fits into the class of multi-dimensional state-space models. The disadvantage is that data from the whole network is required, hence the call for a centralised computing unit. The main asset is that it is able to capture the interconnection links between clusters immediately provided the factors of the Kronecker matrices are also parametrized according to the *known* topology. In an attempt to formulate an identification problem for self-replicating networks, it would be of interest to re-cast the identification of Kronecker-structured models within the framework of structured identification that is described in Chapter 9.

8 Subspace Identification of Local Systems in an NDS

8.1 Introduction

This chapter focuses on the identification of a local dynamic system operating in a network when the direct interaction with the neighbouring systems cannot be measured.

In accordance with the classification of NDS in Figure 5.1, we restrict ourselves *mainly* to a system defined on a regular lattice. The architypal scenario that is considered in this chapter is a finite 1D network of locally connected LTI systems, as depicted in Figure 8.1. The LTI systems Σ_i in this network communicate information with their neighbours that is assumed not to be measured directly. For systems operating in a regular lattice network configuration, we will consider:

1. *Homogeneous systems*: All systems Σ_i in the network are assumed to be identical with the restriction that the systems at the beginning and end do not have a missing input signal from either left or right.
2. *Heterogeneous systems*: In principle, similar to the homogeneous case but now the systems Σ_i can change with index i.

The chapter will consists of two main blocks. First we will consider the case that the communication between local systems in the network is *restricted* to the exchange of state information between two adjacent local systems *only*. This gives rise to the class of *block-tridiagonal systems*, defined in Section 4.2.3. This will be analysed in Section 8.3. Second, we will consider the case that the inter-system communication is generic in a spatial sense, allowing a transfer of information from one system in the network far down the line beyond the immediately adjacent system. This will be analysed in Section 8.4 for sequentially semi-separable (SSS) systems.

In this chapter we will present *subspace identification* methods. This means that only information about the *partial topology NDS for state-space models*, as defined in Definition 4.2, will be required. Apart from this partial structural network information, the methods presented allow the retrieval of crucial information about the size of the submatrices of the considered system matrices (or dimension of related state vectors) via *convex* optimization. The latter will also, like existing subspace identification methods (Verhaegen and Verdult, 2007), enable the retrieval of estimates of the system matrices of a cluster of local systems. Such estimates often form an intermediate step in retrieving the system matrices related to one particular local system.

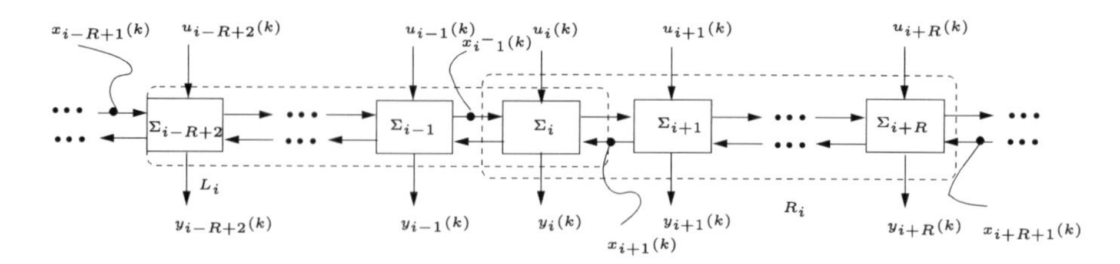

Figure 8.1 Local system Σ_i in a connected network with neighbourhoods L_i and R_i, each containing R systems. The index i indicates the spatial index, while k indicates the dependence on time.

The organization of this chapter is as follows. In Section 8.2 we consider the identification of a local LTI system in a NDS within the framework of *blind system identification* (Yu and Verhaegen, 2016). Here, the (essential) missing information in a state-space context is that external non-white inputs cannot be measured. Such external inputs could represent the missing communication signals in between the local systems considered in this chapter. This first section analyses that for (local) state-space models that have a generic unknown input. Full knowledge of the (local) system state is not sufficient to find the system matrices uniquely (up to a similarity transformation), so there is an inherent ambiguity for such 'blind' identification problems. This highlights two things. First the difficulty, or challenge, of these type, of blind identification problems, and second, that additional information is necessary, such as the input-output measurements of the local system to be identified.

An example of such additional information is the *block-tridiagonal* structure. Therefore, in Section 8.3 this class of system is considered first. Section 8.3.1 considers homogeneous systems in the network. The method presented uses the idea of rank-constrained subspace identification as presented for lumped parameter systems in Verhaegen and Hansson (2016). For this method, both the block-tridiagonal structure as well as the system homogeneity are essential. This approach can be generalized to the class of *decomposable* systems, defined in Section 4.2.2. Section 8.3.2 presents a second subspace identification based on estimating the state sequence by subspace intersection, as was considered for lumped parameter systems in Moonen et al. (1989). For this method only the block-tridiagonal system structure of the system matrices is essential. Therefore, this method is presented for a network of heterogeneous systems.

A second type of additional information is considered in Section 8.4, namely that a cluster of local systems are to be *strongly observable* (Sundaram, 2012). Though, in this case, more general inter-network communication can be considered, we start this section again with block-tridiagonal systems. This clearly highlights the connection and difference between the methods presented in the previous section. The generalization of the method of Section 8.4 is highlighted by presenting its essential applicability to the identification of the class of SSS state-space models. This is done in Section 8.4.2.

Some sections presenting new subspace identification methods contain an illustration of the derived algorithm(s) to an illustrative case study.

8.2 Ambiguity in Blind Subspace Identification with Full State Information

The identification of a local system Σ_i in a network such as that shown in Figure 8.1 is related to blind system identification (Yu and Verhaegen, 2016). This paper addressed the goal of how to identify LTI systems while only having access to the output of the system while the input, that in general is non-white, cannot be measured. In this section we demonstrate that there is an unresolvable ambiguity that shows that additional information is necessary to address this identification problem. Examples of such additional information in the scope of identifying local systems operating in a network are addressed in the next three sections.

The following LTI state-space model is considered:

$$\begin{aligned} x(k+1) &= Ax(k) + Bu(k) + Fv(k), \\ y(k) &= Cx(k), \end{aligned} \tag{8.1}$$

with $x(k) \in \mathbb{R}^n, u(k) \in \mathbb{R}^m, v(k) \in \mathbb{R}^q$ and $y(k) \in \mathbb{R}^p$. A blind identification problem would consider only the pair $\{u(k), y(k)\}$ to be measurable, and the input $v(k)$ to be unmeasurable (and non-white). We analyse the potential difficulties of identifying the system matrices of this model when the state sequence $\{x(k)\}$ (up to similarity transformation) is available.

For this we have the following two Lemmas.[1]

LEMMA 8.1 *The state equation of the system* (8.1) *is equivalent to the following state equation:*

$$x(k+1) = (A + \delta_A)x(k) + (B + \delta_B)u(k) + Fw(k), \tag{8.2}$$

where $w(k) \in \mathbb{R}^q$*, and the matrices* δ_A*,* δ_B *satisfy*

$$Range\begin{pmatrix} \delta_A & \delta_B \end{pmatrix} \subset Range(F). \tag{8.3}$$

Proof State equation (8.1) is a special case of Equation (8.2) obtained by setting $\delta_A = \delta_B = 0$ and $w(k) = v(k)$ and noting that Condition (8.3) holds trivially.

Conversely, let us prove that the state sequence (8.2) under Condition (8.3) can also be covered by state equation (8.1). Indeed, Equation (8.2) can be rewritten as follows:

$$x(k+1) = Ax(k) + Bu(k) + \delta_A x(k) + \delta_B u(k) + Fw(k). \tag{8.4}$$

In view of Condition (8.3), it holds that

$$\delta_A x(k) + \delta_B u(k) \in \text{Range}\begin{pmatrix} \delta_A & \delta_B \end{pmatrix} \subset \text{Range}(F).$$

[1] These Lemmas are the result of a discussion with Dr. Thao Nguyen from the Delft Center for Systems and Control.

Then there exists an $z(k) \in \mathbb{R}^\ell$ such that

$$\delta_A x(k) + \delta_B u(k) = F z(k).$$

Plugging this into Equation (8.4) and setting $v(k) := z(k) + w(k)$, we obtain

$$\begin{aligned} x(k+1) &= Ax(k) + Bu(k) + Fz(k) + Fw(k) \\ &= Ax(k) + Bu(k) + Fv(k). \end{aligned}$$

This is exactly the state equation (8.1) and hence the proof is complete. □

When the matrix F is full column rank, Lemma 8.1 can be read as below.

LEMMA 8.2 *Suppose that F is full column rank. Then the state equation of system* (8.1) *is equivalent to the state equation* (8.2) *with $w(k) \in \mathbb{R}^\ell$ and*

$$\begin{bmatrix} \delta_A & \delta_B \end{bmatrix} = F \begin{bmatrix} L_a & L_b \end{bmatrix}, \tag{8.5}$$

for some $L_a \in \mathbb{R}^{\ell \times n}$ and $L_b \in \mathbb{R}^{\ell \times m}$.

Proof In view of Lemma 8.1, all we need is to show that Conditions (8.3) and (8.5) are interchangeable in order to guarantee the equivalence between Equations (8.2) and (8.1). It is clear that Equation (8.5) implies Condition (8.3). Conversely, Condition (8.3) also implies the following:

$$\text{Range}(\delta_A), \text{Range}(\delta_B) \subset \text{Range}(F). \tag{8.6}$$

This in turn implies Condition (8.5), where the matrices L_a and L_b are explicitly given by

$$L_a = F^\dagger \delta_A, \; L_b = F^\dagger \delta_B.$$

Here, $F^\dagger$ is the Moore–Penrose pseudo-inverse of F, which exists by the full column rank of F. □

One can formally include an ambiguity of F into the state equation (8.2) as follows:

$$x(k+1) = (A + \delta_A)x(k) + (B + \delta_B)u(k) + (F + \delta_F)w(k), \tag{8.7}$$

where $w(k) \in \mathbb{R}^q$. Then the condition under which Equation (8.7) is equivalent to the state equation (8.1) is:

$$\text{Range}\begin{pmatrix} \delta_A & \delta_B & \delta_F \end{pmatrix} \subset \text{Range}(F).$$

This condition implies that $\text{Range}(\delta_F) \subset \text{Range}(F)$. In particular, there exists $z(k) \in \mathbb{R}^q$ such that $\delta_F w(k) = Fz(k)$. Then

$$(F + \delta_F)w(k) = Fw(k) + Fz(k) = Fw'(k),$$

where $w'(k) := w(k) + z(k)$. From the above, Equation (8.7) reduces to Equation (8.2) with $w'(k)$ in place of $w(k)$. As a result, the state equation (8.7) is also equivalent to Equation (8.2).

REMARK 8.1 It is seen from the proof of Lemma 8.1 that the role of Condition (8.3) is to guarantee

$$\delta_A x(k) + \delta_B u(k) \in \text{Range}(F) \quad \forall\, k \in \mathbb{N}.$$

Lemmas 8.1 and 8.2, and Remark 8.1 indicate that even knowledge of the state sequence $x(k)$ does not enable the unique identification of the system matrices. Therfore, in the following three sections we seek additional information to enable us to identify the local system within a network.

8.3 Tri-diagonal Systems

In this section, we will present two subspace identification methods for which the tri-diagonal structure of the system matrices of the global network, as defined in Section 4.2.3, is *essential*. First, a method is presented that is based on the rank constrained subspace identification for lumped parameter systems in Verhaegen and Hansson (2016). For this method, both the block-tridiagonal structure as well as the homogeneity of the systems in the network are essential. A second method is presented in Section 8.3.2 and is based on estimating the state sequence by subspace intersection, as was considered for lumped parameter systems in Moonen et al. (1989), which only requires the block-tridiagonal matrix structure.

8.3.1 Rank Constrained Subspace Identification

This section is based on Yu et al. (2018b). The method outlined can be applied to more general classes of large-scale identification problems beyond the archetypal 1D chain of identical LTI systems. For example, Yu and Verhaegen (2015, 2017) considered using the outlined methodology for the identification of a cluster of decomposable systems that communicate with states of neighbouring systems in the following structured way:

$$\begin{aligned} x(k+1) &= (I \otimes A_a + P \otimes A_b)x(k) + (R \otimes A_b)v(k) + (I \otimes B)u(k), \\ y(k) &= (I \otimes C_a + P \otimes C_b)x(k) + (R \otimes C_b)w(k) + e(k), \end{aligned} \tag{8.8}$$

where $v(k)$ and $w(k)$ are the collection of state vectors of external states of systems of other clusters that interact with the cluster of decomposable systems. That external interaction pattern is described by the pattern matrix R. The internal pattern matrix is denoted by P and corresponds to the pattern matrix introduced for decomposable systems in Section 4.2.2. Such a system could be part of the α-decomposable NDS mentioned in Section 4.2.2.

8.3.1.1 Problem Formulation and Notation

We consider the case of homogeneous block-tridiagonal systems. Then Equation (4.18) simplifies to,

$$\begin{aligned}
\Sigma_1 \;:\; x_1(k+1) &= Ax_1(k) + A_r x_2(k) + Bu_1(k),\\
y_1(k) &= Cx_1(k) + e_1(k),\\
\Sigma_i \;:\; x_i(k+1) &= Ax_i(k) + A_\ell x_{i-1}(k) + A_r x_{i+1}(k) + Bu_i(k),\\
y_i(k) &= Cx_i(k) + e_i(k),\\
\Sigma_N \;:\; x_N(k+1) &= Ax_1(k) + A_\ell x_{N-1}(k) + Bu_N(k),\\
y_N(k) &= Cx_N(k) + e_N(k),
\end{aligned} \tag{8.9}$$

where $x_i(k) \in \mathbb{R}^n, u_i(k) \in \mathbb{R}^m, y_i(k) \in \mathbb{R}^p$, $e_i(k)$ is a zero-mean white-noise sequence with unknown covariance matrix and $e_i(k), e_j(k)$ are mutually independent for all $i \neq j$. The spatial index i satisfies $1 \leq i \leq N$ with $N \gg n$ for $n > max(p,m)$.

Lifting these local models by stacking all the local state, input and output vectors in one state, input and output vector, e.g. indicated for the state vector as $x(k) = \begin{bmatrix} x_1^T(k) & \cdots & x_N^T(k) \end{bmatrix}^T$, the standard (A, B, C) model is obtained, and is denoted as:

$$\begin{aligned}
x(k+1) &= \mathcal{A}x(k) + \mathcal{B}u(k),\\
y(k) &= \mathcal{C}x(k) + e(k),
\end{aligned} \tag{8.10}$$

with the system matrices having the following form:

$$\mathcal{A} = \begin{bmatrix} A & A_r & & \\ A_\ell & A & \ddots & \\ & \ddots & \ddots & A_r \\ & & A_\ell & A \end{bmatrix}, \quad \begin{aligned} &\mathcal{B} = \text{diag}(B|i = 1:N) \text{ and}\\ &\mathcal{C} = \text{diag}(C|i = 1:N). \end{aligned} \tag{8.11}$$

The state-space model in Equation (8.10) can, in principle, be identified (up to a similarity transformation) with standard subspace identification methods (Verhaegen and Verdult, 2007). However, this has two major disadavantages:

1. The similarity transformation destroys the block-tridiagonal or block-diagonal structure of the system matrices. Finding a similarity transformation that restores the block-(tri)diagonal structure in the system matrices is a non-trivial problem. See e.g. Yu et al. (2018a). This constraint excludes the use of first finding a global state-space model even for small scale networks, i.e. for small N.
2. For large scale networks, such that $N \gg n$, first estimating a global state-space model via subspace identification is to be avoided as such methods scale cubically with N.

To better deal with both disadvantages of classical subspace identification methods, in this section we focus on the identification of a local system Σ_i. The key problem in identifying a local system is addressed via the identification of a cluster of local systems $\{\Sigma_j\}_{j=i-R}^{i+R}$ in the neighbourhood of Σ_i. The radius R satisfies $R < i < N-R$ and $N \gg R$. The lifted state-space model of this cluster is represented as,

$$\begin{aligned}\mathbf{x}_i(k+1) &= \mathbf{A}_i\mathbf{x}_i(k) + \mathbf{B}_i\mathbf{u}_i(k) + \mathbf{E}_i\mathbf{v}_i(k),\\ \mathbf{y}_i(k) &= \mathbf{C}_i\mathbf{x}_i(k) + \mathbf{e}_i(k),\end{aligned} \tag{8.12}$$

where $\mathbf{A}_i \in \mathbb{R}^{(2R+1)n\times(2R+1)n}$ is the $2R+1$ *leading principal block submatrix* of $\mathcal{A}$ in Equation (8.10) (from block row 1 to $2R+1$ and similarly for the columns), and $\mathbf{B}_i$ and $\mathbf{C}_i$ are the $2R+1$ leading principal block submatrices of $\mathcal{B}$ respectively $\mathcal{C}$ in Equation (8.10). The unknown inputs to this cluster of local models are given by:

$$\mathbf{E}_i\mathbf{v}_i(k) = \begin{bmatrix} A_\ell & 0 \\ 0 & 0 \\ \vdots & \vdots \\ 0 & 0 \\ 0 & A_r \end{bmatrix} \begin{bmatrix} x_{i-R-1}(k) \\ x_{i+R+1}(k) \end{bmatrix}. \tag{8.13}$$

The identification of the local system matrices is split into the solution of the following two problems:

PROBLEM 8.1 The Markov-parameter estimation for the cluster of LTI systems in Σ_i in Equation (8.12) using only its local inputs $\mathbf{u}_i(k)$ and outputs $\mathbf{y}_i(k)$ for $k = 1 : N_t$.

PROBLEM 8.2 Using the results from the solution to Problem 8.1, the estimation of the order n of the local system Σ_i and the system matrices up to a similarity transformation. That is, in this case:

$$A^Q = Q^{-1}AQ, \;\; A_\ell^Q = Q^{-1}A_\ell Q, \;\; A_r^Q = Q^{-1}A_r Q, \;\; B^Q = Q^{-1}B, \;\; C^Q = CQ,$$

where Q is an invertible $n \times n$ matrix.

These two problems will be addressed in this section where the following assumption is made.

ASSUMPTION 8.1 (minimality) *The global triplet of system matrices* $(\mathcal{A},\mathcal{B},\mathcal{C})$ *of the model in Equation* (8.10) *as well as the triplet* $(\mathbf{A}_i,\mathbf{B}_i,\mathbf{C}_i)$ *of the local cluster* (8.12) *are minimal.*

The notion of persistency of excitation was introduced in Definition 7.2 for MSSM models. It could, however, also be used in a slightly modified form for the NDS considered in Equation (8.10), as summarized in the following definition.

DEFINITION 8.1 (Persistency of excitation) A time sequence $u(k) \in \mathbb{R}^{Nm}$ is persistently exciting of order s if there exists an integer N_t such that the following block-Hankel matrix is of full row rank:

$$\begin{bmatrix} u(k) & u(k+1) & \cdots & u(k+N_\mathrm{t}-1) \\ u(k+1) & u(k+2) & & u(k+N_\mathrm{t}) \\ \vdots & & \ddots & \vdots \\ u(k+s-1) & u(k+s) & \cdots & u(k+N_\mathrm{t}+s-2) \end{bmatrix}.$$

For this chapter, we introduce further common notation. This is done for now given the relationship of a simple LTI system of the form:

$$\begin{aligned} x(k+1) &= Ax(k) + Bu(k), \\ y(k) &= Cx(k) + Du(k). \end{aligned}$$

Then the input signal $u(k)$ (but basically any other signal in this model) can be stored in the block-Hankel matrix (Definition 8.1) re-introduced here with the following notation:

$$\mathcal{H}_{s,N_t}[u(k)] = \begin{bmatrix} u(k) & u(k+1) & \cdots & u(k+N_\mathrm{t}-1) \\ u(k+1) & u(k+2) & & u(k+N_\mathrm{t}) \\ \vdots & & \ddots & \vdots \\ u(k+s-1) & u(k+s) & \cdots & u(k+N_\mathrm{t}+s-2) \end{bmatrix}. \tag{8.14}$$

Given the system matrices (A, B, C, D) of the above LTI system, the Toeplitz matrix $\mathcal{T}_s(A, B, C, D)$ is defined as:

$$\mathcal{T}_s(A,B,C,D) = \begin{bmatrix} D & 0 & \cdots & 0 & 0 \\ CB & 0 & & & \\ \vdots & \ddots & \ddots & & 0 \\ CA^{s-2}B & & \cdots & CB & D \end{bmatrix}. \tag{8.15}$$

Finally, for the above state-space model the extended observability matrix $\mathcal{O}_s(A, C)$ is defined as:

$$\mathcal{O}_s(A,C) = \begin{bmatrix} C \\ CA \\ \vdots \\ CA^{s-1} \end{bmatrix}. \tag{8.16}$$

Sometimes this matrix, for the sake of brevity, is simply denoted as $\mathcal{O}_s$.

8.3.1.2 Failure of Existing Identification Methods

Consider the lifted state-space model in Equation (8.12) for generic system matrices and generic input signals $u_i(k)$ and $v_i(k)$. For such systems it will be shown that the unknown input term $\mathbf{E}_i\mathbf{v}_i(k)$ causes existing subspace methods (Verhaegen, 2015) and the low-rank optimization approach (Liu et al., 2013) to fail, to yield accurate identification results.

The data equation for the local system model in Equation (8.12) can be written as

$$\mathbf{Y}^i_{s,N_\mathrm{t}} = \mathcal{O}_s\mathbf{x}^i_{N_\mathrm{t}} + \mathcal{T}_U\mathbf{U}^i_{s,N_\mathrm{t}} + \mathcal{T}_V\mathbf{V}^i_{s,N_\mathrm{t}} + \mathbf{E}^i_{s,N_\mathrm{t}}, \tag{8.17}$$

where the (block-)Hankel matrices $\mathbf{Y}^i_{s,N_t} = \mathcal{H}_{s,N_t}[\mathbf{y}_i(k)]$ and $\mathbf{U}^i_{s,N_t}$, $\mathbf{V}^i_{s,N_t}$, $\mathbf{E}^i_{s,N_t}$ are defined similarly from the corresponding lower case signals; the block Toeplitz matrix $\mathcal{T}_U = \mathcal{T}_s(\mathbf{A}_i, \mathbf{B}_i, \mathbf{C}_i, \mathbf{0})$, while $\mathcal{T}_V = \mathcal{T}_s(\mathbf{A}_i, \mathbf{E}_i, \mathbf{C}_i, \mathbf{0})$; the extended observability matrix $\mathcal{O}_s = \mathcal{O}_s(\mathbf{A}_i, \mathbf{C}_i)$ and the initial state related sequence $\mathbf{x}^i_{N_t}$ is defined as

$$\mathbf{x}^i_{N_t} = [\mathbf{x}_i(k) \cdots \mathbf{x}_i(k + N_t - 1)].$$

First, it will be shown that the existing subspace identification methods, see e.g. Verhaegen (2015), fail to retrieve a matrix with a column subspace like $\mathcal{O}_s$ or a row subspace like $\mathbf{x}^i_{N_t}$ from the data matrices $\mathbf{Y}^i_{s,N_t}$ and $\mathbf{U}^i_{s,N_t}$ in Equation (8.17).

Let

$$\Pi_U^{\perp} = I - \mathbf{U}^{i,T}_{s,N_t} \left(\mathbf{U}^i_{s,N_t} U^{i,T}_{s,N_t} \right)^{-1} \mathbf{U}^i_{s,N_t}.$$

In the absence of measurement noise, i.e. $\mathbf{e}_i(k) = 0$ (or $\mathbf{E}^i_{s,N_t} = 0$), the subspace revealing matrix (Verhaegen, 2015) is $\mathbf{Y}^i_{s,N_t}\Pi_U^{\perp}$, or $\left[\mathbf{Y}^{i,T}_{s,N_t} \;\; \mathbf{U}^{i,T}_{s,N_t}\right]^T$ if we consider the approach in Verhaegen and Hansson (2016).

When we assume the unknown input to be zero, that is $\mathbf{v}_i(k) = 0$ (or $\mathbf{V}^i_{s,N_t} = 0$), the column space of $\mathcal{O}_s$ can be calculated from that of $\mathbf{Y}^i_{s,N_t}\Pi_U^{\perp}$ by the matrix SVD operation, while the row space of $\mathbf{x}^i_{N_t}$ can be computed from that of the matrix $\left[\mathbf{Y}^{i,T}_{s,N_t} \;\; \mathbf{U}^{i,T}_{s,N_t}\right]^T$. However, the unknown bilinear term $\mathcal{T}_V \mathbf{V}^i_{s,N_t}$ prevents the retrieval of the column space of $\mathcal{O}_s$ or the row space of $\mathbf{x}^i_{N_t}$.

Next, it will be shown that the identification method in Liu et al. (2013) is incapable of dealing with the concerned identification problem. By treating the unknown input $\mathbf{v}_i(k)$ in Equation (8.12) as missing variables, and by denoting $\mathbb{H}_v$ as the set of block-Hankel matrices of the same structure as the matrix $\mathbf{V}^i_{s,N_t}$, the associated low rank optimization problem of Liu et al. (2013) can be written as

$$\min_{V \in \mathbb{H}_v} \quad \text{rank} \begin{bmatrix} \mathbf{Y}^i_{s,N_t} \\ \mathbf{U}^i_{s,N_t} \\ V \end{bmatrix}. \tag{8.18}$$

By the following inequality:

$$\text{rank} \begin{bmatrix} \mathbf{Y}^i_{s,N_t} \\ \mathbf{U}^i_{s,N_t} \\ V \end{bmatrix} \geq \text{rank} \begin{bmatrix} \mathbf{Y}^i_{s,N_t} \\ \mathbf{U}^i_{s,N_t} \end{bmatrix},$$

it can easily be seen that $V = 0$ is an optimal solution to Equation (8.18), indicating that the missing input sequence cannot be wholly restored by the rank minimization method presented in Liu et al. (2013). From the above analysis, we can conjecture that a few missing entries might be recovered using the rank minimization method. However, it is impossible to recover the entire missing sequence using the rank minimization method.

From the above analysis, it can be summarized that the failure of the existing subspace methods is caused by the unknown and non-white input signal $\mathbf{v}_i(k)$. In the presence of unknown inputs, it is difficult to obtain an accurate (or consistent) estimate of the column space of $\mathcal{O}_s$ or the row space of $\mathbf{x}^i_{N_t}$ using the traditional subspace identification methods, nor an accurate estimate of structured system matrices $\{\mathbf{A}_i, \mathbf{B}_i, \mathbf{C}_i\}$ in Equation (8.12).

8.3.1.3 General Algorithmic Strategy

To deal with the challenge caused by the unknown input sequence, a new subspace identification approach will be developed by using the low-rank property of the term $\mathcal{O}_s\mathbf{x}^i_{N_t}+\mathcal{T}_V\mathbf{V}^i_{s,N_t}$ in Equation (8.17) and the block Toeplitz structure of $\mathcal{T}_U$ in Equation (8.17).

The *generic strategy* of the solution to Problems 8.1 and 8.2 is as follows. First, to tackle Problem 8.1 a combination of the low rank property of the term $\mathcal{O}_s\mathbf{x}^i_{N_t} + \mathcal{T}_V\mathbf{V}^i_{s,N_t}$ is used, under mild conditions, to determine (part of) the true Markov parameters of the lifted state-space model in Equation (8.12) (in the noise free setting) via a rank constrained subspace identification step. The subsequent challenge is then to derive the local system matrices of the system Σ_i from these (partial) Markov parameters. This subsequent challenge tackles Problem 8.2. These two steps will be described in the following subsections.

The Algorithmic outline is based on (Yu et al., 2018b).

8.3.1.4 Estimating the Markov Parameters $\mathbf{C}_i\mathbf{A}_i^j\mathbf{B}_i$ of the Cluster (8.12)

Estimation of the Markov parameters $\mathbf{C}_i\mathbf{A}_i^j\mathbf{B}_i$ of the cluster in Equation (8.12) starts with the exploration of the required dimension selection parameters s and R in the determination of the size of the matrices in the data equation (8.17) such that the term $\mathcal{O}_s\mathbf{x}^i_{N_t} + \mathcal{T}_V\mathbf{V}^i_{s,N_t}$ is of low rank. This is done in Lemma 8.3 and Remark 8.2.

LEMMA 8.3 (Rank property of the term $\mathcal{O}_s\mathbf{x}^i_{N_t}+\mathcal{T}_V\mathbf{V}^i_{s,N_t}$) *Consider the data equation (8.17) when $N_t > ps$ or $\mathbf{Y}^i_{s,N_t}$ is a fat matrix, the rank of $\mathcal{O}_s\mathbf{x}^i_{N_t} + \mathcal{T}_V\mathbf{V}^i_{s,N_t}$ satisfies*

$$rank\left(\mathcal{O}_s\mathbf{x}^i_{N_t} + \mathcal{T}_V\mathbf{V}^i_{s,N_t}\right) \leq (2R+1)n + \min\{(s-1)sp, 2(s-1)n\}, \tag{8.19}$$

where $(s-1)sp$ and $2(s-1)n$ represent, respectively, the numbers of non-zero rows and columns of $\mathcal{T}_U$.

Proof According to the shifting property of $\mathcal{O}_s$ and the block Toeplitz structure of $\mathcal{T}_U$, it can be obtained that $\text{rank}(\mathcal{O}_s\mathbf{x}^i_{N_t}) \leq \text{rank}(\mathcal{O}_s) \leq (2R+1)n$ and $\text{rank}(\mathcal{T}_U\mathbf{V}^i_{s,N_t}) \leq \text{rank}(\mathcal{T}_U) \leq \min\{(s-1)sp, 2(s-1)n\}$. This completes the proof. □

REMARK 8.2 Lemma 8.3 provides a criterion to select the dimension parameters s and R such that $\mathcal{O}_s\mathbf{x}^i_{N_t} + \mathcal{T}_V\mathbf{V}^i_{s,N_t}$ has a low rank. This criterion is explicitly given as

$$(2R+1)sp > (2R+1)n + \min\{(s-1)sp, 2(s-1)n\}. \tag{8.20}$$

This inequality implies that the number of the rows of the matrix $\mathcal{O}_s\mathbf{x}^i_{N_t} + \mathcal{T}_V\mathbf{V}^i_{s,N_t}$ is larger than an upper bound of its rank. In practice, by fixing a value of s satisfying that $s > \frac{n}{p}$, we can always find a value of R such that this inequality holds. Therefore, we will assume that the matrix sum $\mathcal{O}_s \mathbf{x}^i_{N_t} + \mathcal{T}_V \mathbf{V}^i_{s,N_t}$ has a low rank.

It would be interesting to show that the block Toeplitz matrix $\mathcal{T}_U$ has a two-layer block Toeplitz structure, where the upper layer is the block Toeplitz structure of $\mathcal{T}_U$ with respect to the block entries $\mathbf{C}_i \mathbf{A}_i^j \mathbf{B}_i$ and the lower layer is the partial block Toeplitz structure inside the block entries $\mathbf{C}_i \mathbf{A}_i^j \mathbf{B}_i$.

Example 8.1 (Partial Toeplitz structures embedded in Markov parameters) Let $R = 3$ and suppose that each block entry in $\mathbf{A}_i, \mathbf{B}_i, \mathbf{C}_i$ has size 2×2. Then, the grey image in Figure 8.2 illustrates the structures embedded in the matrices $\{\mathbf{M}_j = \mathbf{C}_i \mathbf{A}_i^j \mathbf{B}_i\}_{j=1}^3$.

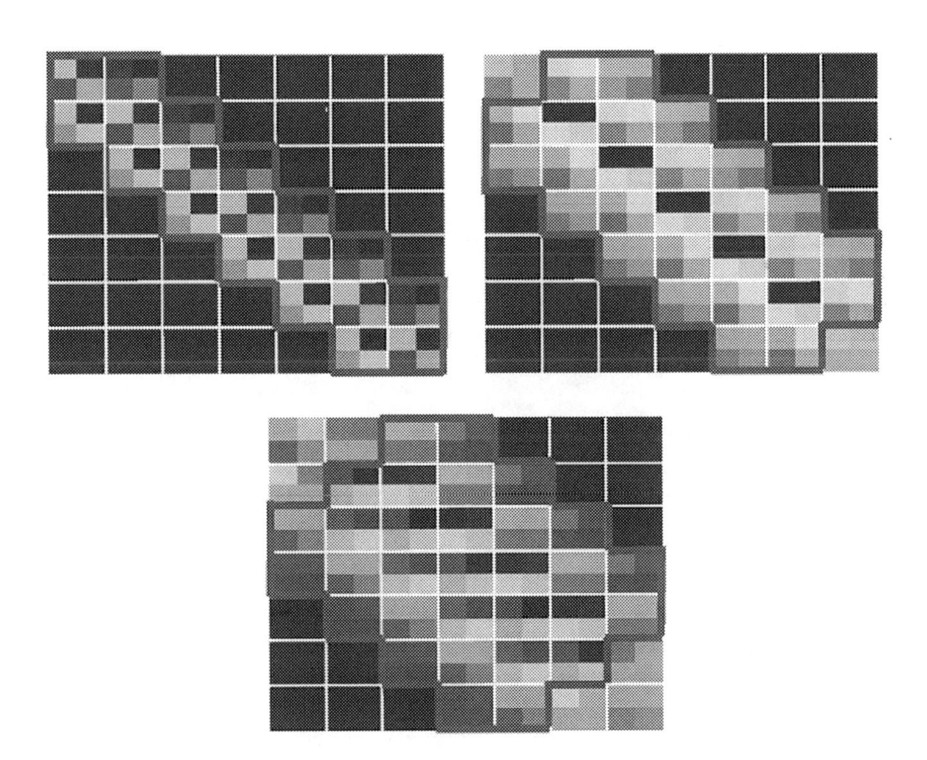

Figure 8.2 Grey images to show the partial Toeplitz structures of the matrices $\{\mathbf{M}_j = \mathbf{C}_i \mathbf{A}_i^j \mathbf{B}_i\}_{j=1}^3$ for: (Top left) $\mathbf{C}_i \mathbf{A}_i \mathbf{B}_i$ with block-bandwidth 1; (Top right) $\mathbf{C}_i \mathbf{A}_i^2 \mathbf{B}_i$ with block-bandwidth 2; (Bottom) $\mathbf{C}_i \mathbf{A}_i^3 \mathbf{B}_i$ with block-bandwidth 3. The outer zero entries are dark grey and the inner non-zero entries exhibit partial block Toeplitz structures. ©2018 IEEE Reprinted, with permission, from Yu et al. (2018b)

LEMMA 8.4 (Partial Toeplitz structure embedded in $\mathbf{M}_j = \mathbf{C}_i \mathbf{A}_i^j \mathbf{B}_i$) *According to the definitions of the matrices* $\mathbf{A}_i, \mathbf{B}_i, \mathbf{C}_i$ *in Equation* (8.12), *the matrix product* $\mathbf{M}_j = \mathbf{C}_i \mathbf{A}_i^j \mathbf{B}_i$ *satisfies that:*

1. $\mathbf{M}_j$ *is a banded block matrix, with block-bandwidth* j.
2. *The submatrices of* $\mathbf{M}_j$ *consisting of the block entries with row index* l *and column index* q *satisfying* $j+1 \leq l+q \leq 4R+3-i$ *has a partial block Toeplitz structure.*

Proof Since the block diagonal matrices $\mathbf{B}_i$ and $\mathbf{C}_i$ have constant diagonal blocks, the matrix $\mathbf{M}_j$ has the same block structure as $\mathbf{A}_i^j$. The matrix $\mathbf{A}_i$ can be represented by

$$\mathbf{A}_i = I \otimes A + J_- \otimes A_l + J_+ \otimes A_r,$$

with J_- and J_+ block-columns being, respectively, the left and right of the shifting matrices.

The proof of this lemma is carried out by induction. For $j = 1$ and $j = 2$, the results of the lemma are obvious. Suppose that $\mathbf{A}_i^j$ has a block Toeplitz structure consisting of block entries with the row and column index pair (l, q) satisfying $j+1 \leq l+q \leq 4R+3-j$. Using the expression for $\mathbf{A}_i$ above, the matrix $\mathbf{A}_i^{j+1}$ can be written as

$$\mathbf{A}_i^{j+1} = \mathbf{A}_i^j \mathbf{A}_i = \mathbf{A}_i^j (I \otimes A + J_- \otimes A_l + J_+ \otimes A_r). \tag{8.21}$$

Then, it can be verified that $\mathbf{A}_i^{j+1}$ has a partial Toeplitz structure consisting of block entries with the index pair (l, q) satisfying $(j+1)+1 \leq l+q \leq 4R+3-(j+1)$. The lemma is then proven. □

After the above two preparatory steps, we are ready to tackle Problem 8.1. In outlining this solution, we will use the notation $\mathbb{T}_U$ and $\mathbb{H}_Y$ to denote the sets of matrices having the same structures as the matrices $\mathcal{T}_U$ and $\mathbf{Y}^i_{s,N_t}$, respectively. By exploiting the low-rank property of the term $\mathcal{O}_s \mathbf{x}^i_{N_t} + \mathcal{T}_V \mathbf{V}^i_{s,N_t}$ and the two-layer block Toeplitz structure of $\mathcal{T}_U$, a low-rank regularized optimization problem will be given.

Denote by $\hat{\mathbf{y}}_i(k) = \mathbf{y}_i(k) - \mathbf{e}_i(k)$ the noise-free output and its related block Hankel matrix $\hat{\mathbf{Y}}^i_{s,N_t}$. Then, a low-rank regularized optimization problem is given as

$$\min_{\Theta_U \in \mathbb{T}_U, \hat{\mathbf{Y}}^i_{s,N_t} \in \mathbb{H}_Y} \sum_{t=1}^{N_t+s-1} \|\hat{\mathbf{y}}_i(t) - \mathbf{y}_i(t)\|^2 + \lambda \cdot \operatorname{rank}\left[\hat{\mathbf{Y}}^i_{s,N_t} - \Theta_U \mathbf{U}^i_{s,N_t}\right], \tag{8.22}$$

where the regularization parameter λ can be adjusted to yield a reasonable optimization result.

Next, the globally optimal solution to Equation (8.22) without noise perturbation will be analyzed, and it will be shown that the partial Toeplitz regions embedded in the Markov parameters $\mathbf{M}_j = \mathbf{C}_i \mathbf{A}_i^j \mathbf{B}_i$ can be correctly recovered in the noise-free case. For that purpose, the following time-varying observability matrices will be defined.

DEFINITION 8.2 (Time-varying observability matrix) Define the block Toeplitz matrix G_j of the size $j \times (j+2)$ as

$$G_j = \begin{bmatrix} A_l & A & A_r & & & \\ & A_l & A & A_r & & \\ & & \ddots & \ddots & \ddots & \\ & & & A_l & A & A_r \end{bmatrix}. \tag{8.23}$$

Denote $\Xi_j = I_j \otimes C$ and define a time-varying observability matrix $\mathbf{O}_{j,k}$ in terms of the matrix pair (Ξ_j, G_j) as Dewilde and van der Veen (1998, chapter 3):

$$\mathbf{O}_{j,k} = \begin{bmatrix} \Xi_j \\ \Xi_{j-2} G_{j-2} \\ \Xi_{j-4} G_{j-4} G_{j-2} \\ \vdots \\ \Xi_{j-2(k-1)} G_{j-2(k-1)} \cdots G_{j-2} \end{bmatrix},$$

where $j > 2(k-1)$.

THEOREM 8.1 (Exact recovery of partial Markov parameters) *Suppose that the following assumptions are satisfied:*

1. *Assumption 8.1 holds, and the augmented matrices* $[A_l \quad B]$ *and* $[A_r \quad B]$ *are of full row rank;*
2. *The input signal* $u(k)$ *in Equation* (8.10) *is persistently exciting of order* $Nn + s$ *with the dimension parameter* s *defined as in Equation* (8.17)*;*
3. *The dimension parameters* R *in Equation* (8.12) *and* s *in Equation* (8.17) *satisfy*

$$s > \nu_o, \quad R \geq s - 1,$$

where ν_o *is defined as the observability index of the matrix pair* $(\mathbf{A}_i, \mathbf{C}_i)$*;*
4. *The observability matrix* $\mathbf{O}_{2R+1,s-1}$ *in Definition 8.2 is of full column rank;*
5. *There is no measurement noise, i.e.* $\hat{\mathbf{y}}_i(k) \equiv \mathbf{y}_i(k)$.

Then the second layer block Toeplitz structured submatrices of the Markov parameters $\mathbf{M}_j = \mathbf{C}_i \mathbf{A}_i^j \mathbf{B}_i$, *for* $j = 0, 1, \ldots, s-2$, *can be computed uniquely by solving the following low-rank optimization problem:*

$$\min_{\Theta_U \in \mathbb{T}_U} \operatorname{rank} \left[\mathbf{Y}^i_{s,N_t} - \Theta_U \mathbf{U}^i_{s,N_t} \right] \quad \text{for } R + s < i < N - R - s. \tag{8.24}$$

The proof of the above theorem has been provided in Yu et al. (2018b). It can be seen that the global optimal solution can yield unique recovery of the partial Markov parameters $\mathbf{M}_j = \mathbf{C}_i \mathbf{A}_i^j \mathbf{B}_i$, for $j = 0, 1, \ldots, s-2$. Additionally, Theorem 8.1 implies that the second-layer Toeplitz structure is crucial in yielding a (unique) solution to Problem 8.1.

Since the low rank minimization problem in Equation (8.22) is non-convex in nature, it is impossible to obtain an optimal solution using traditional optimization methods. To deal with this problem, the re-weighted nuclear norm optimization method (Mohan and Fazel, 2010) is used, which is inherently an iterative heuristic for the exact rank optimization problem (8.22).

8.3.1.5 Identifying the Systems Matrices of a Single Subsystem

Algorithmic Strategy

Based on the estimate of Markov parameters $\mathbf{M}_j = \mathbf{C}_i\mathbf{A}_i^j\mathbf{B}_i$ for $j = 0, 1, \ldots, s-2$, the system matrices $\{C, A, A_l, A_r, B\}$ are to be identified up to a similarity transformation. Special care is necessary in retrieving these local system matrices because a similarity transformation on the lifted matrices $\mathbf{A}_i, \mathbf{B}_i, \mathbf{C}_i$ preserves the Markov parameters but destroys the (spatial) structure in these matrices. Furthermore, as indicated in Theorem 8.1, only partial Toeplitz regions of $\mathbf{M}_j = \mathbf{C}_i\mathbf{A}_i^j\mathbf{B}_i$ can be reliably estimated by solving the optimization problem in Equation (8.22).

The *rational* of the algorithmic solution to estimate the local system matrices $\{C, A, A_l, A_r, B\}$ can be done by first constructing a Hankel matrix from the estimated part of the Markov parameters $\mathbf{M}_j$ and then factorizing this Hankel matrix into the product of a time-varying observability matrix and a controllability matrix. The inner shifting structure of the observability matrix is then used to estimate the local system matrices by the least-squares method.

Therefore, to work out the above algorithmic strategy we start relating the block entries inside the partial Toeplitz regions to the basic system matrices $\{C, A, A_l, A_r, B\}$. This is done in Lemma 8.5. This is followed by the presentation of a structured matrix factorization of a Hankel-type of matrix. Its definition is given in Equation (8.27) and the structured factorization is presented in Equation (8.28).

LEMMA 8.5 *According to the definitions of the matrices* $\mathbf{A}_i, \mathbf{B}_i$ *and* $\mathbf{C}_i$ *in Equation* (8.12), *the non-zero block entries of the* $(j+1)$*th block row of* $\mathbf{C}_i\mathbf{A}_i^j\mathbf{B}_i$ *are denoted by* $\{F_{j,-j}, F_{j,1-j}, \ldots, F_{j,j-1}, F_{j,j}\}$, *which satisfy that*

$$\sum_{k=-j}^{j} F_{j,k} z^{-k} = C(A_l z^{-1} + A + A_r z)^j B, \tag{8.25}$$

where $z \in \mathbb{C}$.

This lemma can be proved straightforwardly using the filter bank theory in Strang (2009).

The block entries inside the partial Toeplitz regions of $\mathbf{M}_j = \mathbf{C}_i\mathbf{A}_i^j\mathbf{B}_i$ are denoted by $\{F_{j,k}\}$, as shown in Lemma 8.5, and are assumed to be known in the following derivation of an estimation algorithm for the system matrices $\{C, A, A_l, A_r, B\}$.

Using an analogous approach to that for the time-varying observability matrix in Definition 8.2, the corresponding time-varying controllability matrix is defined below.

DEFINITION 8.3 (Time-varying controllability matrix) Define the block Toeplitz matrix Γ_j of size $(j+2) \times j$ as

$$\Gamma_j = \begin{bmatrix} A_r & & & \\ A & A_r & & \\ A_l & A & \ddots & \\ & A_l & \ddots & A_r \\ & & \ddots & A \\ & & & A_l \end{bmatrix}. \quad (8.26)$$

Denote $\Pi_j = I_j \otimes B$ and define the time-varying controllability matrix $\mathbf{C}_{j,k}$ in terms of the matrix pair (Γ_j, Π_j) as

$$\mathbf{C}_{j,k} = \left[\Pi_j \mid \Gamma_{j-2}\Pi_{j-2} \mid \cdots \mid \Gamma_{j-2}\cdots\Gamma_{j-2(k-1)}\Pi_{j-2(k-1)}\right],$$

where $j > 2(k-1)$.

Estimation of the system matrices $\{C, A_l, A, A_r, B\}$ from the Markov parameters will be done in two phases. First, the time-varying observability and controllability matrices $\{\mathbf{O}_{2R+1,s/2}, \mathbf{C}_{2R+1,s/2}\}$ are estimated from the available matrix values $\{F_{i,k}\}$. Second, the system matrices $\{C, A, A_l, A_r, B\}$ are estimated from the estimated matrices of $\{\mathbf{O}_{2R+1,s/2}, \mathbf{C}_{2R+1,s/2}\}$. For notational simplicity, the subscript $_{s/2}$ is assumed to be an integer.

As illustrated in Example 8.2, the product of $\mathbf{O}_{2R+1,s/2}$ and $\mathbf{C}_{2R+1,s/2}$ can be represented by the available matrix values $\{F_{j,k}\}_{k=-j}^{j}$ for $j = 0, 1, \ldots, s-2$.

Example 8.2 (Product of a time-varying observability matrix and a time-varying controllability matrix) When $R = 1$ and $s = 4$, the product of $\mathbf{O}_{3,2}$ and $\mathbf{C}_{3,2}$ can be expressed as

$$\mathbf{O}_{3,2}\mathbf{C}_{3,2} = \left[\begin{array}{ccc} C & 0 & 0 \\ 0 & C & 0 \\ 0 & 0 & C \\ \hline CA_l & CA & CA_r \end{array}\right] \left[\begin{array}{ccc|c} B & 0 & 0 & A_rB \\ 0 & B & 0 & AB \\ 0 & 0 & B & A_lB \end{array}\right]$$

$$= \left[\begin{array}{ccc|c} F_{0,0} & 0 & 0 & F_{1,1} \\ 0 & F_{0,0} & 0 & F_{1,0} \\ 0 & 0 & F_{0,0} & F_{1,-1} \\ \hline F_{1,-1} & F_{1,0} & F_{1,1} & F_{2,0} \end{array}\right].$$

Suppose that the product of $\mathbf{O}_{2R+1,s/2}$ and $\mathbf{C}_{2R+1,s/2}$ is expressed as

$$\mathbf{O}_{2R+1,s/2}\mathbf{C}_{2R+1,s/2} = \mathbf{H}_{2R+1,2R+1}, \quad (8.27)$$

where the $(2R+1) \times (2R+1)$ block matrix $\mathbf{H}_{2R+1,2R+1}$ is assumed to be known. It will be interesting to estimate $\mathbf{O}_{2R+1,s/2}$ and $\mathbf{C}_{2R+1,s/2}$ from this equation. Denote by $\mathbb{O}_{2R+1,s/2}$ and $\mathbb{C}_{2R+1,s/2}$ the sets of matrices having the same structures of $\mathbf{O}_{2R+1,s/2}$

and $\mathbf{C}_{2R+1,s/2}$, respectively. Then, the structured low-rank matrix factorization problem is formulated as

$$\begin{aligned} \min_{\mathbf{O},\mathbf{C}} \quad & \|\mathbf{H}_{2R+1,2R+1} - \mathbf{O}\mathbf{C}\|_F^2, \\ \text{such that} \quad & \mathbf{O} \in \mathbb{O}_{2R+1,s/2}, \mathbf{C} \in \mathbb{C}_{2R+1,s/2}. \end{aligned} \tag{8.28}$$

In Theorem 8.2 it is shown that the matrices $\{\mathbf{O},\mathbf{C}\}$ can be estimated up to a block-diagonal ambiguity matrix with identical block diagonal entries.

THEOREM 8.2 *Consider the optimization problem in Equation* (8.28). *Suppose that the following conditions are satisfied:*

1. *The dimension parameters R and s satisfy $R \geq s - 2$;*
2. *The system order n is known;*
3. *The matrix $\mathbf{O}_{j,s/2}$ has full column rank and the matrix $\mathbf{C}_{j,s/2}$ has full row rank;*
4. *The matrix $\mathbf{H}_{2R+1,2R+1}$ in Equation* (8.27) *is available.*

Then, the optimal solution $\{\hat{\mathbf{O}},\hat{\mathbf{C}}\}$ to Equation (8.28) *satisfies that*

$$\begin{aligned} \hat{\mathbf{O}} &= \mathbf{O}_{2R+1,s/2}\mathbf{Q}, \\ \hat{\mathbf{C}} &= \mathbf{Q}^{-1}\mathbf{C}_{2R+1,s/2}, \end{aligned} \tag{8.29}$$

where the ambiguity matrix $\mathbf{Q}$ has the form $\mathbf{Q} = I_{2R+1} \otimes Q$ and $Q \in \mathbb{R}^{n\times n}$ is a constant matrix.

This theorem was proven in Yu et al. (2018b).

Since the optimization problem (8.28) is bilinear and non-convex, it is usually difficult to handle. Here, this bilinear optimization problem will be reformulated as a rank-constrained optimization problem. First, the parametrization of the time-varying observability and controllability matrices will be provided. For $j = 0, 1, \ldots, s/2$ and $z \in \mathbb{C}$, the matrices $\{W_{j,l}\}_{l=-j}^{j}$ and $\{E_{j,l}\}_{l=-j}^{j}$ are defined as

$$\begin{aligned} \sum_{l=-j}^{j} W_{j,l} z^{-l} &= C(A_l z^{-1} + A + A_r z)^j, \\ \sum_{l=-j}^{j} E_{j,l} z^{-l} &= (A_l z^{-1} + A + A_r z)^j B. \end{aligned} \tag{8.30}$$

According to the quantities $W_{j,l}$ and $E_{j,l}$ given in Equation (8.30), the matrices $\mathbf{W}_j$ and $\mathbf{E}_j$ are defined as

$$\mathbf{W}_j = \begin{bmatrix} W_{0,0} \\ W_{1,-1} \\ W_{1,0} \\ W_{1,1} \\ W_{2,-2} \\ \vdots \\ W_{j-1,j-1} \end{bmatrix}, \quad \mathbf{E}_j = \begin{bmatrix} E_{0,0}^T \\ E_{1,-1}^T \\ E_{1,0}^T \\ E_{1,1}^T \\ E_{2,-2}^T \\ \vdots \\ E_{j-1,j-1}^T \end{bmatrix}^T. \tag{8.31}$$

Then the matrix $\mathbf{O}_{j,k}$ (or $\mathbf{C}_{j,k}$) can be linearly represented in terms of $\mathbf{W}_k$ (or $\mathbf{E}_k$), and the product $\mathbf{O}_{2R+1,s/2}\mathbf{C}_{2R+1,s/2}$ can be linearly represented in terms of block entries of $\mathbf{W}_{s/2}\mathbf{E}_{s/2}$ (Choudhary and Mitra, 2014). This linear representation operator $\mathcal{H}(\cdot)$ is denoted by

$$\mathbf{O}_{2R+1,s/2}\mathbf{C}_{2R+1,s/2} = \mathcal{H}(\mathbf{W}_{s/2}\mathbf{E}_{s/2}). \tag{8.32}$$

The above linear operator $\mathcal{H}(\cdot)$ will be assumed to be known.

Instead of treating $\mathbf{O}_{2R+1,s/2}$ and $\mathbf{C}_{2R+1,s/2}$ as two separate variables, the whole matrix product $\mathbf{X} = \mathbf{W}_{s/2}\mathbf{E}_{s/2}$ is regarded as a variable. Then, the linear operator $\mathcal{H}(\mathbf{X})$ relies on the block entries of $\mathbf{X}$. Since the matrix product $\mathbf{W}_{s/2}\mathbf{E}_{s/2}$ has a low-rank property, a rank-constrained optimization problem is proposed:

$$\begin{aligned} \min_{\mathbf{X}} \quad & \|\mathbf{H}_{2R+1,2R+1} - \mathcal{H}(\mathbf{X})\|_F^2, \\ \text{such that} \quad & \text{rank}(\mathbf{X}) = n, \end{aligned} \tag{8.33}$$

where n is the system order and $\mathbf{X}$ is a matrix variable having the same size as $\mathbf{W}_{s/2}\mathbf{E}_{s/2}$. It is remarked that when the order n of a single system is unavailable, it can be estimated using the N2SID idea (Verhaegen and Hansson, 2016) by transforming the above rank-constrained optimization problem as a nuclear-norm regularized optimization problem.

The rank-constrained optimization problem in Equation (8.33) is non-convex and NP-hard, which can be addressed using the re-weighted nuclear-norm method (Mohan and Fazel, 2010) or the difference-of-convex programming method (Yu et al., 2018a).

Denote by $\hat{\mathbf{X}}$ an optimal solution to Equation (8.33). The SVD decomposition of $\hat{\mathbf{X}}$ is given by

$$\hat{\mathbf{X}} = \begin{bmatrix} U_1 & U_2 \end{bmatrix} \begin{bmatrix} \Sigma_1 & \\ & \Sigma_2 \end{bmatrix} \begin{bmatrix} V_1^T \\ V_2^T \end{bmatrix}, \tag{8.34}$$

where $\Sigma_1 \in \mathbb{R}^{n \times n}$ is a diagonal matrix with its diagonal entries larger than Σ_2. The estimates of $\mathbf{W}_{s/2}$ and $\mathbf{E}_{s/2}$ are given below:

$$\hat{\mathbf{W}}_{s/2} = U_1, \qquad \hat{\mathbf{E}}_{s/2} = \Sigma_1 V_1^T.$$

Since $\mathbf{O}_{2R+1,s/2}$ and $\mathbf{C}_{2R+1,s/2}$ are linearly parametrized in terms of $\mathbf{W}_{s/2}$ and $\mathbf{E}_{s/2}$, their estimates can be obtained accordingly. For the sake of brevity, the estimates $\hat{\mathbf{O}}_{2R+1,s/2}$ and $\hat{\mathbf{C}}_{2R+1,s/2}$ are assumed to satisfy

$$\begin{aligned} \hat{\mathbf{O}}_{2R+1,s/2} &= \mathbf{O}_{2R+1,s/2}\mathbf{Q}, \\ \hat{\mathbf{C}}_{2R+1,s/2} &= \mathbf{Q}^{-1}\mathbf{C}_{2R+1,s/2}, \end{aligned} \tag{8.35}$$

where $\mathbf{Q} = I_{2R+1} \otimes Q$ and $Q \in \mathbb{R}^{n \times n}$ is a constant matrix. According to the above equalities, the system matrices $\{A, A_l, A_r, B, C\}$ are to be determined up to a similarity transformation.

First, in order to exploit the shifting structure of the matrix $\mathbf{O}_{2R+1,s/2}$, we denote

$$\mathbf{O}_{j,k_1:k_2} = \begin{bmatrix} \Xi_{j-2k_1} G_{j-2k_1} \cdots G_{j-2} \\ \Xi_{j-2(k_1+1)} G_{j-2(k_1+1)} \cdots G_{j-2} \\ \vdots \\ \Xi_{j-2k_2} G_{j-2k_2} \cdots G_{j-2} \end{bmatrix},$$

where $0 \leq k_1 < k_2 \leq s/2 - 2$ and $2k_2 \leq j \leq 2R+1$. The matrix $\mathbf{O}_{j,k_1:k_2}$ above is generated by stacking the block rows of $\mathbf{O}_{j,k}$ with block-row indices from k_1 to k_2. Then, the *structure-shifting property* of $\mathbf{O}_{2R+1,s/2}$ can be mathematically represented as

$$\mathbf{O}_{2R-1,0:s/2-2} G_{2R-1} = \mathbf{O}_{2R+1,1:s/2-1}, \tag{8.36}$$

where $\mathbf{O}_{2R-1,0:s/2-2}$ and $\mathbf{O}_{2R+1,1:s/2-1}$ are sub-matrices of $\mathbf{O}_{2R+1,s/2}$, and G_{2R-1} is defined as in Definition 8.2. The structure-shifting property in Equation (8.36) will be illustrated via a simple example.

Example 8.3 (Structure-shifting property) Given the dimension parameters $R = 2$ and $s = 6$, the structure-shifting property in Equation (8.36) can be explicitly shown as

$$\left[\begin{array}{ccc} C & & \\ & C & \\ & & C \\ \hline CA_l & CA & CA_r \end{array}\right] \begin{bmatrix} A_l & A & A_r & & \\ & A_l & A & A_r & \\ & & A_l & A & A_r \end{bmatrix}$$

$$= \left[\begin{array}{ccccc} CA_l & CA & CA_r & & \\ & CA_l & CA & CA_r & \\ & & CA_l & CA & CA_r \\ \hline CA_l^2 & C(AA_l + A_lA) & C(A_lA_r + A^2 + A_rA_l) & C(AA_r + A_rA) & CA_r^2 \end{array}\right].$$

In view of Equation (8.36), the least-squares optimization method is adopted to identify the matrices A_l, A, A_r from the estimate $\hat{\mathbf{O}}_{2R+1,s/2}$:

$$\begin{aligned} \min_{G} \quad & \|\hat{\mathbf{O}}_{2R-1,0:s/2-2} G - \hat{\mathbf{O}}_{2R+1,1:s/2-1}\|_F^2, \\ \text{such that} \quad & G \in \mathbb{G}_{2R-1}, \end{aligned} \tag{8.37}$$

where $\mathbb{G}_{2R-1}$ denotes the matrix set having the same structure as G_{2R-1} shown in Definition 8.2.

For the optimization problem (8.37), the characteristics of the optimal solution are given in the following lemma.

LEMMA 8.6 *Suppose that Equation* (8.35) *holds and* $\hat{\mathbf{O}}_{2R-1,0:s/2-2}$ *is of full column rank. The optimal solution to Problem* (8.37), *denoted by* $\hat{G}$, *satisfies that*

$$\hat{G} = \left(I_{2R-1} \otimes Q^{-1}\right) G_{2R-1} \left(I_{2R+1} \otimes Q\right), \tag{8.38}$$

where $Q \in \mathbb{R}^{n \times n}$ *is a non-singular matrix.*

Lemma 8.6 implies that the matrices A_l, A, A_r can be determined by solving the optimization problem (8.37) up to a similarity transformation, i.e.

$$\hat{A}_l = Q^{-1}A_l Q, \ \ \hat{A} = Q^{-1}AQ, \ \ \hat{A}_r = Q^{-1}A_r Q.$$

Based on Equation (8.35), the estimates $\hat{C}$ and $\hat{B}$ can be directly obtained from $\hat{\mathbf{O}}_{2R+1,s/2}$ and $\hat{\mathbf{C}}_{2R+1,s/2}$, respectively, satisfying that

$$\hat{C} = CQ, \ \ \hat{B} = Q^{-1}B.$$

For the sake of easy reference, the developed local network identification algorithm is summarized in Algorithm 8.1.

Algorithm 8.1: Local identification of 1D homogeneous systems. ©2019 IEEE. Reprinted, with permission, from Yu et al. (2018b).

Input : The input and output data $\{\mathbf{u}_i(k), \mathbf{y}_i(k)\}_{k=1:N_t}$
Output: The estimates of system matrices C, A, A_l, A_r, B

1. Form the data equation as shown in Equation (8.17).
2. Estimate the convolution matrix $\mathcal{T}_U$ by solving Equation (8.22).
3. Estimate $\mathbf{O}_{2R+1,s/2}$ and $\mathbf{C}_{2R+1,s/2}$ by solving Equation (8.28) or (8.33).
4. Extract the estimates of C and B from the matrices $\mathbf{O}_{2R+1,s/2}$ and $\mathbf{C}_{2R+1,s/2}$, respectively.
5. Estimate A_l, A, A_r from the least-squares problem (8.37).

8.3.1.6 Numerical Simulation Example

To show the effectiveness of the developed identification method – Algorithm 8.1 – we consider the local identification of a line-connected network that consists of 40 identical systems. The system matrices (A, A_l, A_r, B, C) with $A, A_l, A_r \in \mathbb{R}^{3\times 3}$, $B \in \mathbb{R}^{3\times 2}$ and $C \in \mathbb{R}^{2\times 3}$ are generated randomly, with their entries following the standard Gaussian distribution such that Assumption 8.1 holds with probability one.

The data equation of the concerned network system is constructed as shown in Equation (8.17); the corresponding dimension parameters are set to $R = 5$ and $s = 8$. The system input is generated as white noise and the data length is set to 800. The identification performance is evaluated against the signal-to-noise ratio (SNR) which is defined as

$$\text{SNR} = 10\log\left(\frac{\text{var}(y_i(k) - e_i(k))}{\text{var}(e_i(k))}\right).$$

In the numerical simulations, the identification performance will be shown with the SNR ranging from 0 dB to 95 dB. The *impulse-response fitting* criterion will be adopted to evaluate the identification performance. The normalized fitting error of the sequence $\{CA^iB\}$ is defined by

$$\frac{1}{T}\sum_{k=1}^{T}\frac{\sum_{i=0}^{10}\|CA^iB - \hat{C}_k\hat{A}_k^i\hat{B}_k\|_F}{\sum_{i=0}^{10}\|CA^iB\|_F},$$

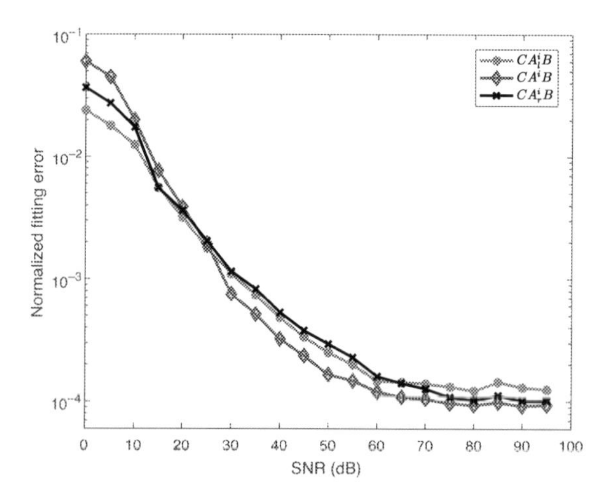

Figure 8.3 Normalized fitting errors under different noise levels with $\lambda = 10^{-3}$.

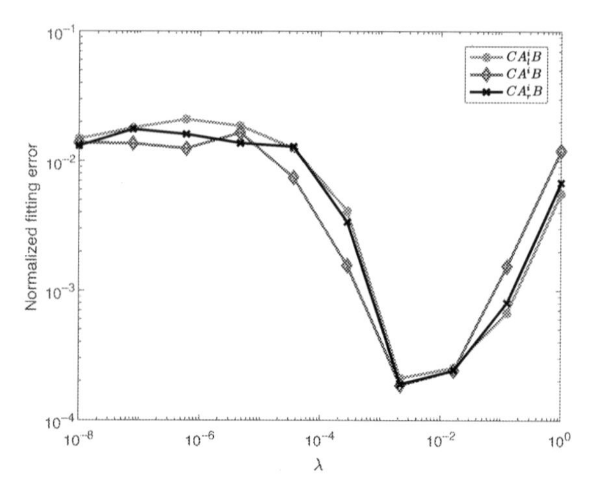

Figure 8.4 Normalized fitting errors against the regularization parameter λ with SNR = 40 dB.

where T is the number of Monte Carlo trials and $\{\hat{C}_k, \hat{A}_k, \hat{B}_k\}$ are the estimates of $\{C, A, B\}$ for the kth trial, respectively. Analogously, the normalized fitting error for the impulse-response sequences CA_l^iB and CA_r^iB are defined. To show the average performance, the normalized fitting errors are calculated by averaging the results of 100 Monte Carlo trials.

Figure 8.3 provides the identification performance of the proposed method by setting $\lambda = 10^{-3}$. It can be seen that the normalized fitting errors obtained by the proposed method decrease as the SNR increases. The normalized fitting error reaches 10^{-4} when the SNR value is about 50 dB.

Figure 8.4 shows the impulse-response fidelity with respect to the parameter λ. In the simulation, the SNR is set to SNR = 40 dB. It can be observed from Figure 8.4 that the identification performance is highly relevant to the value of λ, and better performance can be achieved by choosing λ around 10^{-2}. It indicates that the parameter λ needs to be chosen according to some prior knowledge of the concerned identification problem.

8.3.2 Intersection-Based Subspace Identification

8.3.2.1 Problem Formulation

The case of *heterogeneous* block-triangular systems defined in Equation (4.18) is now considered. With reference to Figure 8.1, and motivated by the approach outlined in Section 8.3.1, we consider the identification of a (small) cluster of systems. Contrary, however, to the discussion in that section, the goal is not to derive the system matrices from (the Markov parameters of) that cluster but to estimate its joint state vector. Furthermore, the heterogeneous case is treated. For that case we consider the estimation of the local state sequence of the local system Σ_{i+1}. This is solved in two parts. First we consider the cluster of $R-1$ subsystems to the right of Σ_{i+1}, including Σ_{i+1} as one cluster. Second, we consider the shifted cluster of R local systems $\{\Sigma_j\}_{j=i-R+2}^{i+1}$. For the first case, the locally lifted state-space model again reads like Equation (8.12):

$$\begin{aligned} \mathbf{x}_i(k+1) &= \mathbf{A}_i\mathbf{x}_i(k) + \mathbf{B}_i\mathbf{u}_i(k) + \mathbf{E}_i\mathbf{v}_i(k), \\ \mathbf{y}_i(k) &= \mathbf{C}_i\mathbf{x}_i(k) + \mathbf{e}_i(k), \end{aligned} \tag{8.39}$$

but with the system matrices $\mathbf{A}_i \in \mathbb{R}^{nR\times nR}, \mathbf{B}_i \in \mathbb{R}^{nR\times mR}, \mathbf{C}_i \in \mathbb{R}^{pR\times nR}$ and $\mathbf{E}_i \in \mathbb{R}^{nR\times 2n}$ given as:

$$\begin{aligned} \mathbf{A}_i &= \begin{bmatrix} A_{i+1} & A_{i+1,r} & 0 & \cdots & 0 \\ A_{i+2,\ell} & A_{i+2} & A_{i+2,r} & & 0 \\ \vdots & \ddots & \ddots & \ddots & \vdots \\ 0 & & & \cdots & A_{i+R} \end{bmatrix}, \\ \mathbf{B}_i &= \operatorname{diag}(B_j \mid j = i+1 : i+R), \\ \mathbf{C}_i &= \operatorname{diag}(C_j \mid j = i+1 : i+R), \\ \mathbf{E}_i &= \begin{bmatrix} A_{i+1,\ell} & 0 \\ 0 & 0 \\ \vdots & \vdots \\ 0 & A_{i+R,r} \end{bmatrix} \quad \mathbf{v}_i(k) = \begin{bmatrix} x_i(k) \\ x_{i+R+1}(k) \end{bmatrix}. \end{aligned} \tag{8.40}$$

Here, we assume that the state dimension does not vary with the spatial index i. This could easily be generalized, but the for the sake of clarity of notation this generalization is not presented. The topology of interest in the identification of the system matrices of the local system Σ_i are the zero block entries in the matrix $\mathbf{E}_{i,R}$.

In such identification problems we are interested in identifying a state-space model up to similarity transformation. Therefore, for the considered class of spatially varying or heterogeneous systems, the notion of similar equivalence is stated in the following definition.

DEFINITION 8.4 (Similar equivalent heterogeneous systems) The quintuple of system matrices $\{C_i, A_{i,\ell}, A_i, A_{i,r}, B_i\}$ for $i = 1 : N$ defining the system matrices $\mathcal{A}, \mathcal{B}, \mathcal{C}$ in Equation (4.18) is similarly equivalent to the quintuple of system matrices $\{C_i^Q, A_{i,\ell}^Q, A_i^Q, A_{i,r}^Q, B_i^Q\}$ for $i = 1 : N$ if there exists a series of non-singular matrices $\{Q_i\}$ for $i = 1 : N$ such that:

$$
\begin{aligned}
&A_i^Q = Q_i^{-1} A_i Q_i, \quad A_{i,\ell}^Q = Q_i^{-1} A_{i,\ell} Q_{i-1}, \quad A_{i,r}^Q = Q_i^{-1} A_{i,r} Q_{i+1},\\
&C_i^Q = C_i Q_i, \quad B_i^Q = Q_i^{-1} B_i
\end{aligned}
$$

This similarity transformation indicates that for the spatially varying (heterogeneous) case, considering the identification of Σ_i there is an additional degree of freedom in the neighbouring states $x_{i-1}(k), x_{i+1}(k)$ given as:

$$
\begin{aligned}
x_{i-1}^{\Gamma}(k) &= \Gamma_\ell x_{i-1}(k) + v_{i-1}(k),\\
x_{i+1}^{\Gamma}(k) &= \Gamma_r x_{i+1}(k) + v_{i+1}(k),
\end{aligned} \tag{8.41}
$$

where $\Gamma_\ell, \Gamma_r \in \mathbb{R}^{n \times n}$ are non-singular transformation matrices and $v_{i-1}(k), v_{i+1}(k)$ are zero-mean ergodic stochastic processes that in the limit for N_t going to ∞ will become uncorrelated with the inputs $u_j(k)$ and the states $x_j(k)$ for all $j = 1 : N$. The introduction of these noise terms then gives rise to the following state-space model of interest to the identification of Σ_i:

$$
\begin{aligned}
x_i(k+1) &= A_i x_i(k) + B_i u_i(k) + \underbrace{\begin{bmatrix} A_{i,\ell}\Gamma_\ell^{-1} & A_{i,r}\Gamma_r^{-1} \end{bmatrix}}_{A_{i,lr}} \left(\underbrace{\begin{bmatrix} x_{i-1}^{\Gamma}(k) \\ x_{i+1}^{\Gamma}(k) \end{bmatrix}}_{x_{i,lr}(k)} - \underbrace{\begin{bmatrix} v_{i-1}(k) \\ v_{i+1}(k) \end{bmatrix}}_{v_{i,lr}(k)} \right),\\
y_i(k) &= C_i x_i(k) + e_i(k).
\end{aligned} \tag{8.42}
$$

Therefore, when one is able to consistently estimate $x_{i,\ell r}(k)$ the estimation of the local system matrices can be done by an errors-in-variables (EIV) type subspace identification method. Before stating the problem, the meaning of a consistent estimation in a subspace framework must be defined.

DEFINITION 8.5 (Consistency for subspace identification) Let the true system matrices of the state-space model in Equation (8.42) be denoted by $\{A_i^\star, \begin{bmatrix} B_i^\star & A_{i,\ell r}^\star \end{bmatrix}, C_i^\star\}$, and define the transfer function as:

$$
H^*(z) = C_i^\star (zI - A_i^\star)^{-1} \begin{bmatrix} B_i^\star & A_{i,\ell r}^\star \end{bmatrix}, \quad z \in \mathbb{C}.
$$

Then the triplet of estimates $\{\hat{A}_i, \begin{bmatrix} \hat{B}_i & \hat{A}_{i,\ell r} \end{bmatrix}, \hat{C}_i\}$ is a consistent estimate of the true system matrices if the following equality holds:

$$
\hat{C}_i (zI - \hat{A}_i)^{-1} \begin{bmatrix} \hat{B}_i & \hat{A}_{i,\ell r} \end{bmatrix} = H^\star(z),
$$

when the number of input-output measurements N_t used to estimate this triplet goes to ∞.

This definition includes the ambiguity due to the transformation matrices Γ_ℓ, Γ_r.

The identification problem(s) of interest, stated in Problem 8.3, make(s) use of the following assumptions.

ASSUMPTION 8.2 (For addressing Problem 8.3)

1. *The global triplet of system matrices* $(\mathcal{A}, \mathcal{B}, \mathcal{C})$ *of the model in Equation* (4.18) *is minimal.*
2. *The input vector* $u(k)$ *to the model in Equation* (4.18) *is ergodic and persistently exciting of any finite order.*
3. *The measurement noise* $e(k)$ *of the model in Equation* (4.18) *is a zero-mean white-noise signal with covariance matrix* $\sigma^2 I \quad (\sigma \in \mathbb{R})$ *and is uncorrelated with* $u(k)$.

PROBLEM 8.3 Assuming Assumption 8.2 holds, and using local input-output measurements of systems $\{\Sigma_j\}_{j=i-R}^{i+R}$ for the number of measurements $N_t \to \infty$, find the estimates of $x_{i+1}^{\Gamma}(k)$ and $x_{i-1}^{\Gamma}(k)$ in Equation (8.42) such that a consistent estimate of the system matrices in that state-space model becomes possible (with an EIV subspace identification method).

8.3.2.2 General Algorithmic Strategy

To estimate the local system matrices in Equation (8.42) the goal of the algorithmic solution is to estimate the neighbouring states $x_{i-1}(k)$ and $x_{i+1}(k)$ satisfying Equation (8.41). the key to obtaining these estimates is the estimation of the lifted state of the 'blind' identification problem for the lifted state-space model (8.40) with unknown inputs present in the vector $\mathbf{v}_i(k)$. The additional information that will be used to tackle this 'blind' identification problem is the zero block entries in the matrix $\mathbf{E}_i$ in Equation (8.40).

The Algorithmic outline is based on (Yu and Verhaegen, 2018).

8.3.2.3 Data Equation for the Lifted Cluster of Systems $\{\Sigma_j\}_{j=i+1}^{i+R}$

The state-space equation of the local cluster consisting of systems $\{\Sigma_j\}_{j=i+1}^{i+R}$ can be written as

$$\begin{aligned} \mathbf{x}_i(k+1) &= \mathbf{A}_i \mathbf{x}_i(k) + \mathbf{B}_i \mathbf{u}_i(k) + \mathbf{E}_i \mathbf{v}_i(k), \\ \mathbf{y}_i(k) &= \mathbf{C}_i \mathbf{x}_i(k) + \mathbf{e}_i(k), \end{aligned} \tag{8.43}$$

where the system matrices are defined in Equation (8.40). As analyzed in Lemma 8.1, in the presence of unknown inputs the system matrices are not identifiable even if the system states are available. In order to overcome this difficulty, a rank-constrained subspace identification method is provided by taking into account the two-layer Toeplitz structure of the associated convolution matrix, which can yield reliable identification results. Here, the additional information for addressing this 'blind' identification problem is the zero block structure in the matrix $\mathbf{E}_i$.

Due to the special property of the line-connected network model, the unknown inputs only influence the boundary block rows of the state equation (8.43). In order to remove the influence of the unknown inputs, the first and last block rows of the state evolution equation are eliminated. The resulting state-space model has time-varying dimensions, which can be represented as

$$\begin{aligned} \mathbf{x}_{i,k}(k+t+1) &= \mathbf{A}_{i,t}\mathbf{x}_{i,k}(k+t) + \mathbf{B}_{i,t}\mathbf{u}_{i,k}(k+t), \\ \mathbf{y}_{i,k}(k+t) &= \mathbf{C}_{i,t}\mathbf{x}_{i,k}(k+t) + \mathbf{e}_{i,k}(k+t), \end{aligned} \tag{8.44}$$

where k is a positive reference time index which can be chosen arbitrarily by the user, $t = 0, 1, \ldots, \lfloor \frac{R}{2} \rfloor$ and the state vector is defined as

$$\mathbf{x}_{i,k}(k+t) = \left[x_{i+t+1}^T(k+t) \ \cdots \ x_{i+R-t}^T(k+t) \right]^T,$$

whose dimension depends on the reference time index k and the real time index $k+t$. The other time-varying vectors $\mathbf{u}_{i,k}(k+t)$, $\mathbf{y}_{i,k}(k+t)$ and $\mathbf{e}_{i,k}(k+t)$ are defined similarly to $\mathbf{x}_{i,k}(k+t)$, the time-varying matrix $\mathbf{A}_{i,t}$ for $0 \leq t \leq \lfloor \frac{R}{2} \rfloor$ is defined as

$$\mathbf{A}_{i,t} = \underbrace{\begin{bmatrix} A_{i+t+2,l} & A_{i+t+2} & A_{i+t+2,r} & & \\ & \ddots & \ddots & \ddots & \\ & & A_{i+R-1-t,l} & A_{i+R-1-t} & A_{i+R-1-t,r} \end{bmatrix}}_{\text{it contains } R-2t \text{ block columns}}, \tag{8.45}$$

and the matrices $\mathbf{B}_{i,t}$ and $\mathbf{C}_{i,t}$ are defined as

$$\begin{aligned} \mathbf{B}_{i,t} &= \operatorname{diag}\left(B_{i+t+2}, \ldots, B_{i+R-1-t}\right), \\ \mathbf{C}_{i,t} &= \operatorname{diag}\left(C_{i+t+1}, \ldots, C_{i+R-t}\right). \end{aligned} \tag{8.46}$$

According to the system (8.44), it can be seen that the dimensions of $\mathbf{x}_{i,k}(k+t)$ and $\mathbf{y}_{i,k}(k+t)$ decrease as the time index t increases, and the state $\mathbf{x}_{i,k}(k+t)$ will disappear when $t > \frac{R}{2}$.

Define the state transition matrix $\Psi_i(t_2, t_1)$, for $t_2 \geq t_1$, as

$$\Psi_i(t_2, t_1) = \mathbf{A}_{i,t_2}\mathbf{A}_{i,t_2-1} \cdots \mathbf{A}_{i,t_1}.$$

The data equation of the state-space model (8.44) is then given as

$$\mathbf{Y}_{s,N_t}^i = \mathcal{O}_s^i \mathbf{x}_{N_t}^i + \mathcal{T}_U^i \mathbf{U}_{s-1,N_t}^i + \mathbf{E}_{s,N_t}^i, \tag{8.47}$$

where, using the definition of the operator $\mathcal{H}$ defined in Equation (8.14), $\mathbf{Y}_{s,N_t}^i$ is defined as $\mathcal{H}_{s,N_t}[\mathbf{y}_{i,k}(k)]$. In a similar way we define $\mathbf{U}_{s-1,N_t}^i$ and $\mathbf{E}_{s,N_t}^i$ from the corresponding lower case signals. Finally, the state sequence $\mathbf{x}_{N_t}^i$ is defined as

$$\mathbf{x}_{N_t}^i = \left[\mathbf{x}_{i,k}(k) \ \mathbf{x}_{i,k+1}(k+1) \cdots \ \mathbf{x}_{i,k+N_t-1}(k+N_t-1)\right],$$

and the extended observability matrix $\mathcal{O}_s^i$ and Toeplitz matrix $\mathcal{T}_U^i$ are now defined as:

$$\mathcal{O}_s^i = \begin{bmatrix} \mathbf{C}_{i,0} \\ \mathbf{C}_{i,1}\Psi_i(0,0) \\ \vdots \\ \mathbf{C}_{i,s-1}\Psi_i(s-2,0) \end{bmatrix},$$

$$\mathcal{T}_U^i = \begin{bmatrix} 0 & & \\ \mathbf{C}_{i,1}\mathbf{B}_{i,0} & \ddots & \\ \vdots & \ddots & 0 \\ \mathbf{C}_{i,s-1}\Psi_i(s-2,1)\mathbf{B}_{i,0} & \cdots & \mathbf{C}_{i,s-1}\mathbf{B}_{i,s-2} \end{bmatrix}.$$

8.3.2.4 Estimating the States of the Lifted Cluster of Systems $\{\Sigma_j\}_{j-i+1}^{i+R}$

Based on the data equation (8.47) for the time-varying state-space model (8.44), the state sequence $\mathbf{x}_{N_t}^i$ is to be estimated, which requires the extended observability matrix $\mathcal{O}_s^i$ to be full column rank. To this end, the following assumption is made.

ASSUMPTION 8.3 (Observability) *There exist dimension parameters R and s such that the matrix $\mathcal{O}_s^i$ in Equation (8.47) is of full column rank for all $R+1 \leq i \leq N-R$.*

In view of the structural properties of $\mathcal{O}_s^i$, the value of s should satisfy $s \leq \lfloor \frac{R+1}{2} \rfloor$. In order to ensure $\mathcal{O}_s^i$ has a full column rank, it is necessary that the spatial-dimension parameter R satisfies that $R \geq \frac{4n}{p} - 1$.

Based on Assumption 8.3, the local state information can be represented in terms of the future local output observations:

$$\mathbf{x}_{N_t}^i = \left(\mathcal{O}_s^i\right)^{\dagger}\left(\mathbf{Y}_{s,N_t}^i - \mathbf{E}_{s,N_t}^i - \mathcal{T}_U^i\mathbf{U}_{s-1,N_t}^i\right). \tag{8.48}$$

From this equation, we can obtain the row space property of $\mathbf{X}_{N_t}^i$ as follows

$$\text{Row}\left[\mathbf{x}_{N_t}^i\right] \subseteq \text{Row}\begin{bmatrix} \mathbf{Y}_{s,N_t}^i - \mathbf{E}_{s,N_t}^i \\ \mathbf{U}_{s-1,N_t}^i \end{bmatrix}. \tag{8.49}$$

Using the strategy described above, the row space of $\mathbf{X}_{N_t}^{i-R+1}$ (states stacked by systems $\{\Sigma_j\}_{j=i-R+1}^{i+1}$) satisfies that

$$\text{Row}\left[\mathbf{x}_{N_t}^{i-R+1}\right] \subseteq \text{Row}\begin{bmatrix} \mathbf{Y}_{s,N_t}^{i-R+1} - \mathbf{E}_{s,N_t}^{i-R+1} \\ \mathbf{U}_{s-1,N_t}^{i-R+1} \end{bmatrix}.$$

8.3.2.5 Estimating the Local State of Σ_{i+1}

According to the topological property of the 1D networked system, it is easy to see that the state of Σ_{i+1} equals the intersection of the stacked state $\{\Sigma_j\}_{j=i+1}^{i+R}$ and that of $\{\Sigma_j\}_{j=i-R+2}^{i+1}$. Denote

$$\bar{x}_{N_t}^{i+1} = [x_{i+1}(k) \ \cdots \ x_{i+1}(k+N_t-1)].$$

Then it can be established that

$$\text{Row}\left[\bar{x}_{N_t}^{i+1}\right] \subseteq \text{Row}\begin{bmatrix} \mathbf{U}_{s-1,N_t}^{i} \\ \mathbf{Y}_{s,N_t}^{i} - \mathbf{E}_{s,N_t}^{i} \end{bmatrix} \cap \text{Row}\begin{bmatrix} \mathbf{U}_{s-1,N_t}^{i-R+1} \\ \mathbf{Y}_{s,N_t}^{i-R+1} - \mathbf{E}_{s,N_t}^{i-R+1} \end{bmatrix}. \tag{8.50}$$

It will be shown that the row subspaces equality in (8.50) holds.

LEMMA 8.7 (State estimation by subspace intersection) *Under Assumptions 8.2 and 8.3, the following holds:*

$$\text{Row}\left[\bar{x}_{N_t}^{i+1}\right] = \text{Row}\begin{bmatrix} \mathbf{U}_{s-1,N_t}^{i} \\ \mathbf{Y}_{s,N_t}^{i} - \mathbf{E}_{s,N_t}^{i} \end{bmatrix} \cap \text{Row}\begin{bmatrix} \mathbf{U}_{s-1,N_t}^{i-R+1} \\ \mathbf{Y}_{s,N_t}^{i-R+1} - \mathbf{E}_{s,N_t}^{i-R+1} \end{bmatrix}. \tag{8.51}$$

Proof By Assumption 8.3 and Equation (8.47), it is straightforward that

$$\text{Row}\begin{bmatrix} \mathbf{U}_{s-1,N_t}^{i} \\ \mathbf{Y}_{s,N_t}^{i} - \mathbf{E}_{s,N_t}^{i} \end{bmatrix} = \text{Row}\begin{bmatrix} \mathbf{U}_{s-1,N_t}^{i} \\ \mathbf{x}_{N_t}^{i} \end{bmatrix}, \tag{8.52}$$

and

$$\text{Row}\begin{bmatrix} \mathbf{U}_{s-1,N_t}^{i-R+1} \\ \mathbf{Y}_{s,N_t}^{i-R+1} - \mathbf{E}_{s,N_t}^{i-R+1} \end{bmatrix} = \text{Row}\begin{bmatrix} \mathbf{U}_{s-1,N_t}^{i-R+1} \\ \mathbf{x}_{N_t}^{i-R+1} \end{bmatrix}. \tag{8.53}$$

Under Assumption 8.2 and by lemma 10.4 in Verhaegen and Verdult (2007), it can be established that the following augmented matrix has full row rank:

$$\begin{bmatrix} \mathbf{U}_{s-1,N_t}^{i-R+1} \\ \mathbf{U}_{s-1,N_t}^{i} \\ \bar{x}_{N_t}^{i-R+2} \\ \vdots \\ \bar{x}_{N_t}^{i+R} \end{bmatrix}. \tag{8.54}$$

Then, by combining Equations (8.52) and (8.53) with the facts that

$$\mathbf{x}_{N_t}^{i} = \begin{bmatrix} \bar{x}_{N_t}^{i+1} \\ \vdots \\ \bar{x}_{N_t}^{i+R} \end{bmatrix} \quad \text{and} \quad \mathbf{x}_{N_t}^{i-R+1} = \begin{bmatrix} \bar{x}_{N_t}^{i-R+2} \\ \vdots \\ \bar{x}_{N_t}^{i+1} \end{bmatrix},$$

the result of the lemma is straightforward. □

Next, the numerical computation of the subspace intersection in Equation (8.51) will be investigated and the influence of the measurement noise specifically handled. According to the numerical approach for the subspace intersection in De Moor (1988), the orthogonal complement for the column subspace of the following matrix needs to be computed:

$$\left[\begin{array}{c} \mathbf{U}^{i}_{s-1,N_t} \\ \mathbf{Y}^{i}_{s,N_t} - \mathbf{E}^{i}_{s,N_t} \\ \hline \mathbf{U}^{i-R+1}_{s-1,N_t} \\ \mathbf{Y}^{i-R+1}_{s,N_t} - \mathbf{E}^{i-R+1}_{s,N_t} \end{array}\right]. \tag{8.55}$$

Due to the unknown measurement noise, it is impossible to obtain an accurate SVD decomposition of the above matrix. To deal with this problem, the variance of the measurement noise, denoted by σ^2, needs to be estimated. This can be done by exploiting the rank deficiency of the matrix $\left[\begin{array}{cc} \mathbf{U}^{i,T}_{s-1,N_t} & \mathbf{Y}^{i,T}_{s,N_t} - \mathbf{E}^{i,T}_{s,N_t} \end{array}\right]^T$ under Assumption 8.2. Given the noise variance σ^2, the following matrix can be exactly known

$$\Delta = \lim_{N_t \to \infty} \frac{1}{N_t} \left[\begin{array}{c} 0 \\ \mathbf{E}^{i}_{s,N_t} \\ 0 \\ \mathbf{E}^{i-R+1}_{s,N_t} \end{array}\right] \left[\begin{array}{c} 0 \\ \mathbf{E}^{i}_{s,N_t} \\ 0 \\ \mathbf{E}^{i-R+1}_{s,N_t} \end{array}\right]^T.$$

It will be shown that an estimate of $\bar{x}^{i+1}_{N_t}$ satisfying Equation (8.41) can be obtained.

THEOREM 8.3 *Denote*

$$\mathbf{R} = \lim_{N_t \to \infty} \frac{1}{N_t} \left[\begin{array}{c} \mathbf{U}^{i}_{s-1,N_t} \\ \mathbf{Y}^{i}_{s,N_t} \\ \mathbf{U}^{i-R+1}_{s-1,N_t} \\ \mathbf{Y}^{i-R+1}_{s,N_t} \end{array}\right] \left[\begin{array}{c} \mathbf{U}^{i}_{s-1,N_t} \\ \mathbf{Y}^{i}_{s,N_t} \\ \mathbf{U}^{i-R+1}_{s-1,N_t} \\ \mathbf{Y}^{i-R+1}_{s,N_t} \end{array}\right]^T - \Delta. \tag{8.56}$$

Let the SVD of $\mathbf{R}$ *be given as*

$$\mathbf{R} = [U_1 \ \ U_2] \left[\begin{array}{cc} S_1 & \\ & S_2 \end{array}\right] \left[\begin{array}{c} V_1^T \\ V_2^T \end{array}\right], \tag{8.57}$$

where S_2 *is a diagonal matrix consisting of the smallest n singular values. Partition the partial orthogonal matrix* U_2 *as*

$$U_2 = \left[U_{21}^T \ U_{22}^T \ U_{23}^T \ U_{24}^T\right]^T,$$

where the dimensions of U_{2j} *for* $j = 1, \ldots, 4$ *are consistent with the block rows of the matrix in (8.55). By Assumption 8.2, the estimate of* $\bar{x}^{i+1}_{N_t}$ *given below satisfies Equation (8.41):*

$$\hat{\bar{x}}^{i+1}_{N_t} = U_{21}^T \mathbf{U}^{i}_{s-1,N_t} + U_{22}^T \mathbf{Y}^{i}_{s,N_t}. \tag{8.58}$$

8.3.2.6 Brief Algorithmic Summary

To sum up, in order to obtain an estimate of the state of a single system, it is necessary to estimate the stacked states of two local clusters with their intersection being exactly the considered single system. After obtaining the state estimates of Σ_{i-1} and Σ_{i+1}, the system matrices of Σ_i in Equation (8.42) can be obtained by identifying an

EIV model. Since the presented algorithm depends only on algebraic operations like SVD decomposition, its computational complexity is cubically proportional to the size $sR(p+m)$.

8.3.2.7 Numerical Simulation Example

To show the performance of the presented identification method, the simulated network consists of 40 line-interconnected systems, where the 20th system is to be identified. The time-varying system matrices are produced by adding random matrices to fixed matrices that are given by

$$A_i = \begin{bmatrix} 0.2728 & -0.2068 \\ 0.1068 & 0.2728 \end{bmatrix}, \qquad A_{i,l} = \begin{bmatrix} -0.1195 & -0.3565 \\ 0.0874 & -0.1048 \end{bmatrix},$$

$$A_{i,r} = \begin{bmatrix} 0.0699 & -0.4278 \\ 0.3842 & 0.1135 \end{bmatrix}, \qquad B_i = \begin{bmatrix} 0.3870 \\ -1.2705 \end{bmatrix},$$

$$C_i = \begin{bmatrix} -0.9075 & -1.3651 \end{bmatrix}, \qquad \text{for } i = 1, \dots, 40.$$

The system input is generated randomly following the standard Gaussian distribution. The values of s and R are set to $s = 10$ and $R = 8$, respectively.

The identification performance for the ith system is measured by the impulse-response-fitting (IRF) criterion which is defined as

$$\text{IRF} = \frac{1}{N} \sum_{k=1}^{N} \frac{\sum_{j=0}^{10} \|\hat{C}_i^k (\hat{A}_i^k)^j \hat{B}_i^k - C_i^* (A_i^*)^j B_j^* \|_F^2}{\sum_{j=0}^{10} \|C_i^* (A_i^*)^j B_i^* \|_F^2}, \tag{8.59}$$

where $N = 200$ denotes the number of Monte Carlo trials, A_i^*, B_i^*, C_i^* are the true system matrices, and $A_i^k, \hat{B}_i^k, \hat{C}_i^k$ are the estimates of the system matrices at the kth trial. In order to show the identification performance with respect to the noise influence, criterion for the signal-to-noise ratio (SNR) is adopted which is defined as

$$\text{SNR (dB)} = 10 \log \frac{\text{var}(y_i(k) - e_i(k))}{\text{var}(e_i(k))}.$$

First, to show the effectiveness of the proposed method, the simulated IRF curve with respect to SNR is plotted in Figure 8.5. The number of input-output data pairs is set to 2000. It can be seen that the IRF values are close to zero when the SNR is large enough, indicating that the local individual systems can be perfectly identified in the absence of measurement noise using the proposed algorithm.

Second, numerical evidence will be provided to show the consistency of the proposed identification method. In the simulation, the SNR is set to SNR = 70 dB. The IRF values with respect to the data lengths are plotted in Figure 8.6, where it can be seen that IRF values decrease as the data length increases, indicating that the estimated poles approach their true values.

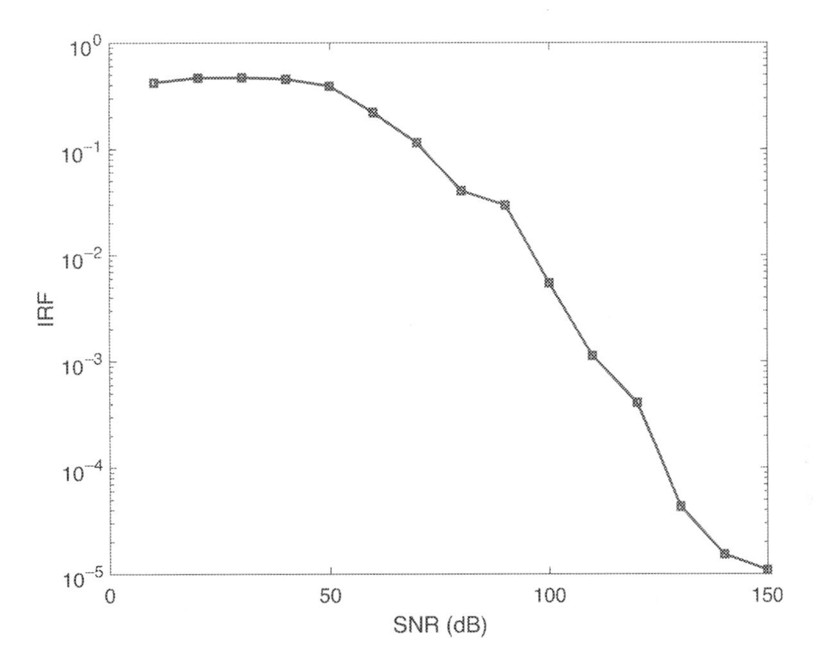

Figure 8.5 IRF values against the SNR for the example defined in Section 8.3.2.7. ©2018 IEEE Reprinted, with permission, from Yu and Verhaegen (2018)

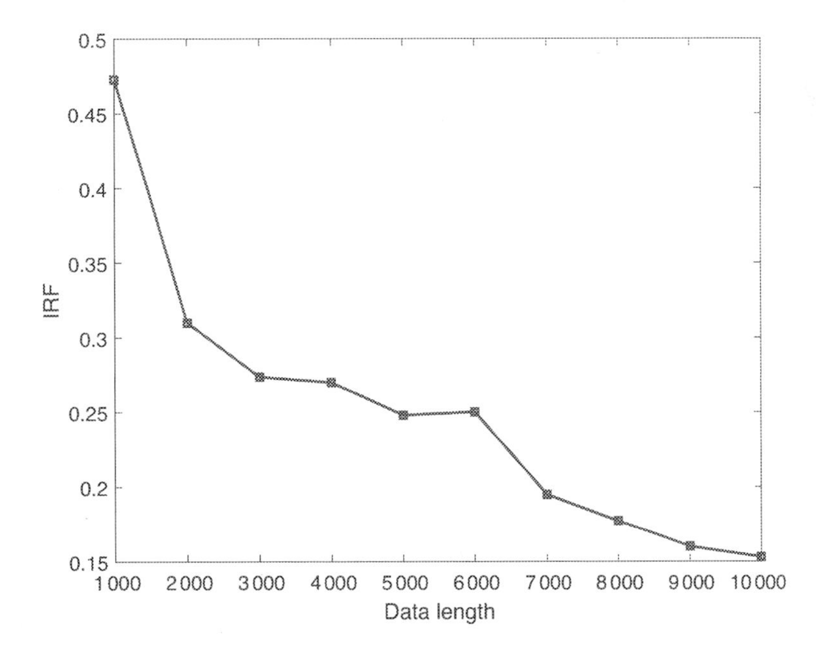

Figure 8.6 IRF of the identified system with respect to the data length for the example defined in Section 8.3.2.7. ©2018 IEEE Reprinted, with permission, from Yu and Verhaegen (2018)

8.4 Networks with Strongly Observable Clusters of Adjacent Local Systems

In Section 8.3.2, it was demonstrated that the identification of a single system boils down to estimating its neighbouring states using local observations. Due to the unmeasurable interconnections among neighbouring systems in a network, the state estimation using local observations needs to handle unknown inputs. To deal with this 'blind' identification problem, the spatial topology of the concerned network is again exploited, as in Section 8.3.2.

The drawback of the solution presented in Section 8.3.2 is the removal of the state evolution equations involving unknown inputs. The consequence is that the resulting time-varying model dimension decreases with time.

To cope with this drawback, another intersection-based subspace method is presented in this section using the assumption of strong observability (Sundaram, 2012) of the cluster of local systems. It is first presented for the class of tri-diagonal systems in Section 8.4.1. Then, in Section 8.4.2, the general applicability is shown in the identification of SSS state-space models. This allows for more general inter system communication between local systems, allowing a transfer of information from one system in the network far down the line beyond only the adjacent system.

8.4.1 Tri-diagonal Systems II

8.4.1.1 Problem Formulation

Following the identification framework provided in Section 8.3.2, the state-space model (8.42) for Σ_i is considered again with the unknown inputs being the states of Σ_{i-1} and Σ_{i+1}. In contrast to the state estimation approach in Section 8.3.2, the unknown inputs in the local cluster model (8.43) are not removed, but compensated by the following observation equations

$$\begin{aligned} y_i(k) &= C_i x_i(k) + e_i(k), \\ y_{i+R+1}(k) &= C_{i+R+1} x_{i+R+1}(k) + e_{i+R+1}(k). \end{aligned}$$

According to these observation equations, the states (or unknown inputs) can be decomposed as

$$\begin{aligned} x_i(k) &= C_i^{\dagger}(y_i(k) - e_i(k)) + P_i \tilde{x}_i(k), \\ x_{i+R+1}(k) &= C_{i+R+1}^{\dagger}(y_{i+R+1}(k) - e_{i+R+1}(k)) + P_{i+R+1}\tilde{x}_{i+R+1}(k), \end{aligned} \tag{8.60}$$

where † denotes the Moore–Penrose inverse operator, the vectors $\tilde{x}_i(k), \tilde{x}_{i+R+1}(k) \in \mathbb{R}^{n-p}$ represent, respectively, the unmeasurable parts of $x_i(k), x_{i+R+1}(k)$, and the coefficient matrices $P_i, P_{i+R+1} \in \mathbb{R}^{n\times(n-p)}$ are chosen such that $[C_i^{\dagger} \quad P_i]$ and $[C_{i+R+1}^{\dagger} \quad P_{i+R+1}]$ are non-singular matrices and $C_i P_i = 0, C_{i+R+1} P_{i+R+1} = 0$.

Using the above partial measurable information, the state-space model (8.43) can be rewritten as

$$\begin{aligned}\mathbf{x}_i(k+1) &= \mathbf{A}_i\mathbf{x}_i(k) + \mathbf{B}_i\mathbf{u}_i(k) + \mathbf{E}_{i1}\left(\tilde{\mathbf{y}}_i(k) - \tilde{\mathbf{e}}_i(k)\right) + \mathbf{E}_{i2}\tilde{\mathbf{v}}_i(k), \\ \mathbf{y}_i(k) &= \mathbf{C}_i\mathbf{x}_i(k) + \mathbf{e}_i(k),\end{aligned} \tag{8.61}$$

where

$$\mathbf{E}_{i1} = \begin{bmatrix} A_{i+1,l}C_i^{\dagger} & 0 \\ 0 & 0 \\ \vdots & \vdots \\ 0 & 0 \\ 0 & A_{i+R,r}C_{i+R+1}^{\dagger} \end{bmatrix}, \quad \mathbf{E}_{i2} = \begin{bmatrix} A_{i+1,l}P_i & 0 \\ 0 & 0 \\ \vdots & \vdots \\ 0 & 0 \\ 0 & A_{i+R,r}P_{i+R+1} \end{bmatrix},$$

$$\tilde{\mathbf{y}}_i(k) = \begin{bmatrix} y_i(k) \\ y_{i+R+1}(k) \end{bmatrix}, \quad \tilde{\mathbf{e}}_i(k) = \begin{bmatrix} e_i(k) \\ e_{i+R+1}(k) \end{bmatrix}, \quad \tilde{\mathbf{v}}_i(k) = \begin{bmatrix} \tilde{x}_i(k) \\ \tilde{x}_{i+R+1}(k) \end{bmatrix}.$$

To estimate the state sequence of this state-space model, we need to deal with the unknown input $\tilde{\mathbf{v}}_i(k)$ (similar to that in (8.43) but with a lower dimension). Instead of removing the state evolution equations corresponding to unknown inputs, the strong observability condition (Sundaram, 2012) will be introduced so that the state sequence can be estimated in the presence of unknown inputs.

To finally state the identification problem to be addressed in this section, we state the data equation for the lifted state-space model (8.61) and define the notion of strong observability.

Data Equation for Lifted Cluster of Systems $\{\Sigma_j\}_{j=i+1}^{i+R}$

In a way analogous that used for the data equation (8.17), the data equation for the state-space model (8.61) can be written as

$$\mathbf{Y}^i_{s,N_t} = \mathcal{O}^i_s\mathbf{x}^i_{N_t} + \mathcal{T}^i_U\mathbf{U}^i_{s-1,N_t} + \mathcal{T}^i_Y\left(\tilde{\mathbf{Y}}^i_{s-1,N_t} - \tilde{\mathbf{E}}^i_{s-1,N_t}\right) + \mathcal{T}^i_V\tilde{\mathbf{V}}^i_{s-1,N_t} + \mathbf{E}^i_{s,N_t}, \tag{8.62}$$

where $\mathbf{Y}^i_{s,N_t} = \mathcal{H}_{s,N_t}[\mathbf{y}_i(k)]$ and $\mathbf{U}^i_{s-1,N_t}, \tilde{\mathbf{Y}}^i_{s-1,N_t}, \tilde{\mathbf{E}}^i_{s-1,N_t}, \tilde{\mathbf{V}}^i_{s-1,N_t}$ are similar Hankel matrices constructed from the corresponding lower case signals, the Toeplitz matrices $\mathcal{T}^i_Y$ and $\mathcal{T}^i_V$ are defined, respectively, as $\mathcal{T}_s(\mathbf{A}_i, \mathbf{E}_{i1}, \mathbf{C}_i, \mathbf{0})$ and $\mathcal{T}_s(\mathbf{A}_i, \mathbf{E}_{i2}, \mathbf{C}_i, \mathbf{0})$.

Using this data equation, we can now define the notion of strong observability.

DEFINITION 8.6 (Strong observability) The state-space model (8.61) is strongly observable if and and only if there exists a dimension parameter $s \geq n$ such that the $\text{rank}\begin{bmatrix}\mathcal{O}^i_s & \mathcal{T}^i_V\end{bmatrix} = n + \text{rank}[\mathcal{T}^i_V]$.

Based on this definition we state the following Assumption.

ASSUMPTION 8.4 (Strong observability condition) *For the state-space model (8.61), there exists a dimension parameter $s \geq n$ such that $[\mathcal{O}^i_s \;\; \mathcal{T}^i_V]$ has full column rank.*

Now we are ready to define the identification problem of interest.

PROBLEM 8.4 Assuming Assumption 8.4 holds and using local input-output measurements of the cluster of local systems $\{\Sigma_j\}_{j=i-R}^{i+R}$ for the number of measurements $N_t \rightarrow \infty$, find estimates of the lifted states $\mathbf{x}_{N_t}^i$ and $\mathbf{x}_{N_t}^{i-R+1}$ such that a consistent estimate of the system matrices (up to similarity transformation) of the local system Σ_i becomes possible.

8.4.1.2 General Algorithmic Strategy

In the spirit of classical intersection-based subspace identification (Moonen et al., 1989), the goal is to find 'past' and 'future' input-output data for a cluster of local systems that allows us to find the state sequence (up to a similarity transformation) of that cluster via intersection. When repeating this subspace intersection step for an adjacent cluster, such that the states of both clusters contain in each of them the state of the system Σ_{i+1} as intersection, the latter can then be found via the solution of a second intersection problem. Crucial in solving the first set of intersection problems is that these clusters of local systems are strongly observable. This is now the additional information exploited to tackle the 'blind' identification problems.

The Algorithmic outline is based on (Yu et al., 2019).

8.4.1.3 Estimating the States of the Lifted Cluster of Systems $\{\Sigma_j\}_{j=i+1}^{i+R}$

In view of the data equation (8.62), it can be established that the sum of state-related term $\mathcal{O}_s^i \mathbf{x}_{N_t}^i$ and the unknown input-related term $\mathcal{T}_V^i \tilde{\mathbf{V}}_{s-1,N_t}^i$ can be expressed as

$$\begin{bmatrix} \mathcal{O}_s^i & \mathcal{T}_V^i \end{bmatrix} \begin{bmatrix} \mathbf{x}_{N_t}^i \\ \tilde{\mathbf{V}}_{s-1,N_t}^i \end{bmatrix} = \begin{bmatrix} I & -\mathcal{T}_U^i & -\mathcal{T}_Y^i \end{bmatrix} \begin{bmatrix} \mathbf{Y}_{s,N_t}^i - \mathbf{E}_{s,N_t}^i \\ \mathbf{U}_{s-1,N_t}^i \\ \tilde{\mathbf{Y}}_{s-1,N_t}^i - \tilde{\mathbf{E}}_{s-1,N_t}^i \end{bmatrix}. \tag{8.63}$$

Then Lemma 8.8 is straightforward.

LEMMA 8.8 *Consider the data equation* (8.63). *If Assumption 8.4 is satisfied, then it has that*

$$Row \begin{bmatrix} \mathbf{x}_{N_t}^i \\ \tilde{\mathbf{V}}_{s-1,N_t}^i \end{bmatrix} \subseteq Row \begin{bmatrix} \mathbf{Y}_{s,N_t}^i - \mathbf{E}_{s,N_t}^i \\ \mathbf{U}_{s-1,N_t}^i \\ \tilde{\mathbf{Y}}_{s-1,N_t}^i - \tilde{\mathbf{E}}_{s-1,N_t}^i \end{bmatrix}. \tag{8.64}$$

Lemma 8.8 implies that the row subspace of the state and unknown input are included in that of the observed data. In order to retrieve the state information, we need to know the exact row subspace of the state and unknown input. Denote $\bar{x}_i = \begin{bmatrix} x_i^T & x_{i+R+1}^T \end{bmatrix}^T$ then its corresponding block Hankel matrix $\bar{\mathbf{X}}_{s-1,N_t}^i$ has the same form as $\tilde{\mathbf{V}}_{s-1,N_t}^i$. It can be derived from the state decomposition equation (8.60) that

$$\text{Row} \begin{bmatrix} \bar{\mathbf{X}}_{s-1,N_t}^i \end{bmatrix} = \text{Row} \begin{bmatrix} \tilde{\mathbf{V}}_{s-1,N_t}^i \\ \tilde{\mathbf{Y}}_{s-1,N_t}^i - \tilde{\mathbf{E}}_{s-1,N_t}^i \end{bmatrix}. \tag{8.65}$$

Then the row subspace of $\mathbf{X}_{N_t}^i$ can be represented in terms of local observations, as shown in the following lemma.

LEMMA 8.9 *If Assumption 8.4 holds, then we have that*

$$Row \begin{bmatrix} \mathbf{x}^i_{N_t} \\ \bar{\mathbf{X}}^i_{s-1,N_t} \\ \mathbf{U}^i_{s-1,N_t} \end{bmatrix} = Row \begin{bmatrix} \mathbf{Y}^i_{s,N_t} - \mathbf{E}^i_{s,N_t} \\ \mathbf{U}^i_{s-1,N_t} \\ \tilde{\mathbf{Y}}^i_{s-1,N_t} - \tilde{\mathbf{E}}^i_{s-1,N_t} \end{bmatrix}. \tag{8.66}$$

Proof It can be derived from the data equation (8.62) that

$$\begin{bmatrix} \mathcal{O}^i_s & \mathcal{T}^i_V & \mathcal{T}^i_Y & \mathcal{T}^i_U \\ 0 & 0 & 0 & I \\ 0 & 0 & I & 0 \end{bmatrix} \begin{bmatrix} \mathbf{x}^i_{N_t} \\ \tilde{\mathbf{V}}^i_{s-1,N_t} \\ \tilde{\mathbf{Y}}^i_{s-1,N_t} - \tilde{\mathbf{E}}^i_{s-1,N_t} \\ \mathbf{U}^i_{s-1,N_t} \end{bmatrix} = \begin{bmatrix} \mathbf{Y}^i_{s,N_t} - \mathbf{E}^i_{s,N_t} \\ \mathbf{U}^i_{s-1,N_t} \\ \tilde{\mathbf{Y}}^i_{s-1,N_t} - \tilde{\mathbf{E}}^i_{s-1,N_t} \end{bmatrix}.$$

By Assumption 8.4, it is can easily verified that the coefficient matrix of the above equation has full column rank; therefore, it can be derived that

$$\text{Row} \begin{bmatrix} \mathbf{x}^i_{N_t} \\ \tilde{\mathbf{V}}^i_{s-1,N_t} \\ \tilde{\mathbf{Y}}^i_{s-1,N_t} - \tilde{\mathbf{E}}^i_{s-1,N_t} \\ \mathbf{U}^i_{s-1,N_t} \end{bmatrix} = \text{Row} \begin{bmatrix} \mathbf{Y}^i_{s,N_t} - \mathbf{E}^i_{s,N_t} \\ \mathbf{U}^i_{s-1,N_t} \\ \tilde{\mathbf{Y}}^i_{s-1,N_t} - \tilde{\mathbf{E}}^i_{s-1,N_t} \end{bmatrix}.$$

By Equation (8.65), the lemma is then proven. □

It can be seen from Lemma 8.9 that the state $\mathbf{x}^i_{N_t}$ of the local cluster can be linearly represented in terms of local observations. Next, following the spirit of the classical subspace identification methods (Moonen et al., 1989), the row subspace of the state sequence $\mathbf{x}^i_{N_t}$ will be estimated by conducting subspace intersection between the past input-output data and future data. Analogous to the future data equation (8.62), the past data equation is written as

$$\mathbf{Y}^{i,p}_{s,N_t} = \mathcal{O}^i_s \mathbf{x}^{i,p}_{N_t} + \mathcal{T}^i_U \mathbf{U}^{i,p}_{s-1,N_t} + \mathcal{T}^i_Y \left(\tilde{\mathbf{Y}}^{i,p}_{s-1,N_t} - \tilde{\mathbf{E}}^{i,p}_{s-1,N_t} \right) + \mathcal{T}^i_V \tilde{\mathbf{V}}^{i,p}_{s-1,N_t} + \mathbf{E}^{i,p}_{s,N_t}, \tag{8.67}$$

where the superscript ^{p} denotes 'past', the block Hankel matrices $\mathbf{U}^{i,p}_{s-1,N_t}$, $\tilde{\mathbf{Y}}^{i,p}_{s-1,N_t}$, $\tilde{\mathbf{E}}^{i,p}_{s-1,N_t}$, $\tilde{\mathbf{V}}^{i,p}_{s-1,N_t}$, $\mathbf{E}^{i,p}_{s,N_t}$ are defined similarly to the Hankel matrix $\mathbf{Y}^{i,p}_{s,N_t} = \mathcal{H}_{s,N_t}[\mathbf{y}_i(k-s+1)]$. The state sequence $\mathbf{x}^{i,p}_{N_t}$ is defined as

$$\mathbf{x}^{i,p}_{N_t} = [\mathbf{x}_i(k-s+1) \ \ \mathbf{x}_i(k-s+2) \ \cdots \ \mathbf{x}_i(k-s+N_t)].$$

According to Lemma 8.9, it is easy to obtain that

$$\text{Row} \begin{bmatrix} \mathbf{x}^{i,p}_{N_t} \\ \bar{\mathbf{X}}^{i,p}_{s-1,N_t} \\ \mathbf{U}^{i,p}_{s-1,N_t} \end{bmatrix} = \text{Row} \begin{bmatrix} \mathbf{Y}^{i,p}_{s,N_t} - \mathbf{E}^{i,p}_{s,N_t} \\ \mathbf{U}^{i,p}_{s-1,N_t} \\ \tilde{\mathbf{Y}}^{i,p}_{s-1,N_t} - \tilde{\mathbf{E}}^{i,p}_{s-1,N_t} \end{bmatrix}. \tag{8.68}$$

Based on Equations (8.52) and (8.68), the row subspace of $\mathbf{x}^i_{N_t}$ will be obtained in the following theorem.

THEOREM 8.4 *Suppose that the following data matrix stacked by the initial state and inputs of Equation* (8.40) *has full row rank*

$$\begin{bmatrix} \mathbf{x}_{N_t}^{i,p} \\ \bar{\mathbf{X}}_{s-1,N_t}^{i,p} \\ \mathbf{U}_{s-1,N_t}^{i,p} \\ \bar{\mathbf{X}}_{s-1,N_t}^{i} \\ \mathbf{U}_{s-1,N_t}^{i} \end{bmatrix}.$$

If Assumption 8.4 holds, then the row subspace of $\mathbf{x}_{N_t}^i$ *can be obtained as*

$$Row\left[\ \mathbf{x}_{N_t}^i\ \right] = Row\begin{bmatrix} \mathbf{Y}_{s,N_t}^{i,p} - \mathbf{E}_{s,N_t}^{i,p} \\ \mathbf{U}_{s-1,N_t}^{i,p} \\ \tilde{\mathbf{Y}}_{s-1,N_t}^{i,p} - \tilde{\mathbf{E}}_{s-1,N_t}^{i,p} \end{bmatrix} \cap Row\begin{bmatrix} \mathbf{Y}_{s,N_t}^{i} - \mathbf{E}_{s,N_t}^{i} \\ \mathbf{U}_{s-1,N_t}^{i} \\ \tilde{\mathbf{Y}}_{s-1,N_t}^{i} - \tilde{\mathbf{E}}_{s-1,N_t}^{i} \end{bmatrix}. \tag{8.69}$$

It is noted that the persistently exciting condition for the (measurable and unmeasurable) inputs to the local network to be identified is necessary for the estimation of the row space of $\mathbf{x}_{N_t}^i$.

Proof By Lemma 8.9, we has that

$$\text{Row}\begin{bmatrix} \mathbf{Y}_{s,N_t}^{i,p} - \mathbf{E}_{s,N_t}^{i,p} \\ \mathbf{U}_{s-1,N_t}^{i,p} \\ \tilde{\mathbf{Y}}_{s-1,N_t}^{i,p} - \tilde{\mathbf{E}}_{s-1,N_t}^{i,p} \end{bmatrix} = \text{Row}\begin{bmatrix} \mathbf{x}_{N_t}^{i,p} \\ \bar{\mathbf{X}}_{s-1,N_t}^{i,p} \\ \mathbf{U}_{s-1,N_t}^{i,p} \end{bmatrix} = \text{Row}\begin{bmatrix} \mathbf{x}_{N_t}^{i,p} \\ \bar{\mathbf{X}}_{s-1,N_t}^{i,p} \\ \mathbf{U}_{s-1,N_t}^{i,p} \\ \mathbf{Y}_{s,N_t}^{i,p} - \mathbf{E}_{s,N_t}^{i,p} \end{bmatrix}.$$

By the full column rank property of the matrix $\mathcal{O}_s$, the future system state $\mathbf{x}_{N_t}^i$ can be linearly represented by the past state $\mathbf{x}_{N_t}^{i,p}$, past input $\{\mathbf{U}_{s-1,N_t}^{i,p}, \bar{\mathbf{X}}_{s-1,N_t}^{i,p}\}$ and past output $\mathbf{Y}_{s,N_t}^{i,p} - \mathbf{E}_{s,N_t}^{i,p}$, i.e.

$$\text{Row}[\mathbf{x}_{N_t}^i] \subseteq \text{Row}\begin{bmatrix} \mathbf{x}_{N_t}^{i,p} \\ \bar{\mathbf{X}}_{s-1,N_t}^{i,p} \\ \mathbf{U}_{s-1,N_t}^{i,p} \\ \mathbf{Y}_{s,N_t}^{i,p} - \mathbf{E}_{s,N_t}^{i,p} \end{bmatrix} = \text{Row}\begin{bmatrix} \mathbf{x}_{N_t}^{i,p} \\ \bar{\mathbf{X}}_{s-1,N_t}^{i,p} \\ \mathbf{U}_{s-1,N_t}^{i,p} \end{bmatrix}.$$

Therefore, we have

$$\text{Row}\begin{bmatrix} \mathbf{Y}_{s,N_t}^{i,p} - \mathbf{E}_{s,N_t}^{i,p} \\ \mathbf{U}_{s-1,N_t}^{i,p} \\ \tilde{\mathbf{Y}}_{s-1,N_t}^{i,p} - \tilde{\mathbf{E}}_{s-1,N_t}^{i,p} \end{bmatrix} = \text{Row}\begin{bmatrix} \mathbf{x}_{N_t}^{i,p} \\ \bar{\mathbf{X}}_{s-1,N_t}^{i,p} \\ \mathbf{U}_{s-1,N_t}^{i,p} \\ \mathbf{x}_{N_t}^{i} \end{bmatrix}.$$

It then follows that

$$\begin{aligned}
&\text{Row}\begin{bmatrix} \mathbf{Y}^{i,p}_{s,N_t} - \mathbf{E}^{i,p}_{s,N_t} \\ \mathbf{U}^{i,p}_{s-1,N_t} \\ \tilde{\mathbf{Y}}^{i,p}_{s-1,N_t} - \tilde{\mathbf{E}}^{i,p}_{s-1,N_t} \end{bmatrix} \cap \begin{bmatrix} \mathbf{Y}^{i}_{s,N_t} - \mathbf{E}^{i}_{s,N_t} \\ \mathbf{U}^{i}_{s-1,N_t} \\ \tilde{\mathbf{Y}}^{i}_{s-1,N_t} - \tilde{\mathbf{E}}^{i}_{s-1,N_t} \end{bmatrix} \\
&= \text{Row}\begin{bmatrix} \mathbf{x}^{i,p}_{N_t} \\ \bar{\mathbf{X}}^{i,p}_{s-1,N_t} \\ \mathbf{U}^{i,p}_{s-1,N_t} \\ \mathbf{x}^{i}_{N_t} \end{bmatrix} \cap \text{Row}\begin{bmatrix} \mathbf{x}^{i}_{N_t} \\ \bar{\mathbf{X}}^{i}_{s-1,N_t} \\ \mathbf{U}^{i}_{s-1,N_t} \end{bmatrix} \\
&= \text{Row}[\mathbf{x}^{i}_{N_t}]
\end{aligned}$$

This completes the proof. □

Theorem 8.4 shows that the stacked state of the local cluster consisting of $\{\Sigma_j\}_{j=i+1}^{i+R}$ can be estimated using the (noisy) local observations. The corresponding state estimation needs to compute the subspace intersection in Equation (8.69). The influence of the measurement noise can be handled using the same technique used in Section 8.3.2. The noise variance σ^2 can be estimated by exploiting the low-rank property of the matrix on the right-hand side of Equation (8.66). In order to compute the subspace intersection in Equation (8.69), it is necessary to obtain the left null subspace of the following matrix

$$\Xi_i = \lim_{N_t \to \infty} \begin{bmatrix} \mathbf{Y}^{i,p}_{s,N_t} - \mathbf{E}^{i,p}_{s,N_t} \\ \mathbf{U}^{i,p}_{s-1,N_t} \\ \tilde{\mathbf{Y}}^{i,p}_{s-1,N_t} - \tilde{\mathbf{E}}^{i,p}_{s-1,N_t} \\ \mathbf{Y}^{i}_{s,N_t} - \mathbf{E}^{i}_{s,N_t} \\ \mathbf{U}^{i}_{s-1,N_t} \\ \tilde{\mathbf{Y}}^{i}_{s-1,N_t} - \tilde{\mathbf{E}}^{i}_{s-1,N_t} \end{bmatrix} \begin{bmatrix} \mathbf{Y}^{i,p}_{s,N_t} - \mathbf{E}^{i,p}_{s,N_t} \\ \mathbf{U}^{i,p}_{s-1,N_t} \\ \tilde{\mathbf{Y}}^{i,p}_{s-1,N_t} - \tilde{\mathbf{E}}^{i,p}_{s-1,N_t} \\ \mathbf{Y}^{i}_{s,N_t} - \mathbf{E}^{i}_{s,N_t} \\ \mathbf{U}^{i}_{s-1,N_t} \\ \tilde{\mathbf{Y}}^{i}_{s-1,N_t} - \tilde{\mathbf{E}}^{i}_{s-1,N_t} \end{bmatrix}^T. \tag{8.70}$$

Given the noise variance σ^2, the matrix Ξ_i can be computed. Let the SVD decomposition of Ξ_i be given as

$$\Xi_i = \begin{bmatrix} U_1^i & U_2^i \end{bmatrix} \begin{bmatrix} \Lambda_1^i & \\ & \Lambda_2^i \end{bmatrix} \begin{bmatrix} V_1^{i,T} \\ V_2^{i,T} \end{bmatrix}, \tag{8.71}$$

where Λ_2^i is a diagonal matrix consisting of the Rn least singular values with Rn being the number of rows of $\mathbf{x}^i_{N_t}$. Partition the matrix U_2^i which is left null space of Ξ_i into two parts of the same size $U_2^i = [U_{21}^{i,T} \;\; U_{22}^{i,T}]^T$. Then, the subspace intersection in Equation (8.69) can be carried out by

$$\text{Row}[\mathbf{x}^i_{N_t}] = \text{Row}\left(U_{21}^{i,T} \begin{bmatrix} \mathbf{Y}^{i,p}_{s,N_t} - \mathbf{E}^{i,p}_{s,N_t} \\ \mathbf{U}^{i,p}_{s-1,N_t} \\ \tilde{\mathbf{Y}}^{i,p}_{s-1,N_t} - \tilde{\mathbf{E}}^{i,p}_{s-1,N_t} \end{bmatrix} \right), \tag{8.72}$$

or

$$\mathbf{x}_{N_t}^i = Q_i U_{21}^{i,T} \begin{bmatrix} \mathbf{Y}_{s,N_t}^{i,p} - \mathbf{E}_{s,N_t}^{i,p} \\ \mathbf{U}_{s-1,N_t}^{i,p} \\ \tilde{\mathbf{Y}}_{s-1,N_t}^{i,p} - \tilde{\mathbf{E}}_{s-1,N_t}^{i,p} \end{bmatrix}, \tag{8.73}$$

where Q_i is an unknown non-singular matrix. Due to the unknown measurement noise, the estimate of $\mathbf{x}_{N_t}^i$ is provided as follows

$$\hat{\mathbf{x}}_{N_t}^i = Q_i U_{21}^{i,T} \begin{bmatrix} \mathbf{Y}_{s,N_t}^{i,p} \\ \mathbf{U}_{s-1,N_t}^{i,p} \\ \tilde{\mathbf{Y}}_{s-1,N_t}^{i,p} \end{bmatrix}. \tag{8.74}$$

Then the estimated state and its true value are related as follows

$$\hat{\mathbf{x}}_{N_t}^i = Q_i^{-1} \mathbf{x}_{N_t}^i + U_{21}^{i,T} \underbrace{\begin{bmatrix} \mathbf{E}_{s,N_t}^{i,p} \\ 0 \\ \tilde{\mathbf{E}}_{s-1,N_t}^{i,p} \end{bmatrix}}_{\Delta_{N_t}^i}. \tag{8.75}$$

Since the true state is determined by the initial state and the system input, it is uncorrelated with the measurement noise, namely $\mathbf{x}_{N_t}^i$ is uncorrelated with $\Delta_{N_t}^i$. In addition, given the coefficient matrix U_{21}^i and the noise variance σ^2, the covariance matrix of the perturbation term $U_{21}^{i,T} \Delta_{N_t}^i$ can be obtained. This is important for the consistent identification of the system Σ_i.

8.4.1.4 Estimating the Local State of Σ_{i+1}

The state sequence of Σ_{i+1}, which is denoted as $\bar{x}_{N_t}^{i+1}$, will be computed by the subspace intersection as done in Section 8.3.2. By Assumption 8.2, it can be verified that the state sequence of the whole network has full row rank. According to the topological property of the 1D networked system, the state sequence can be obtained as follows

$$\text{Row}[\bar{x}_{N_t}^{i+1}] = \text{Row}[\mathbf{x}_{N_t}^i] \cap \text{Row}[\mathbf{x}_{N_t}^{i-R+1}]. \tag{8.76}$$

To compute the above subspace intersection, the noise influence shown in Equation (8.75) needs to be removed. Let the estimates of $\mathbf{x}_{N_t}^i$ and $\mathbf{x}_{N_t}^{i-R+1}$ be given as

$$\begin{aligned} \hat{\mathbf{x}}_{N_t}^i &= Q_i^{-1} \mathbf{x}_{N_t}^i + U_{21}^{i,T} \Delta_{N_t}^i \\ \hat{\mathbf{x}}_{N_t}^{i-R+1} &= Q_{i-R+1}^{-1} \mathbf{x}_{N_t}^{i-R+1} + U_{21}^{i-R+1,T} \Delta_{N_t}^{i-R+1}. \end{aligned} \tag{8.77}$$

Since the covariance matrices of the perturbation terms in the above equation are available, the subspace intersection can be computed by first carrying out the following SVD decomposition:

$$\lim_{N_t\to\infty}\frac{1}{N_t}\begin{bmatrix} \hat{\mathbf{x}}^i_{N_t} - U_{21}^{i,T}\Delta^i_{N_t} \\ \hat{\mathbf{x}}^{i-R+1}_{N_t} - U_{21}^{i-R+1,T}\Delta^{i-R+1}_{N_t}\end{bmatrix}\begin{bmatrix} \hat{\mathbf{x}}^i_{N_t} - U_{21}^{i,T}\Delta^i_{N_t} \\ \hat{\mathbf{x}}^{i-R+1}_{N_t} - U_{21}^{i-R+1,T}\Delta^{i-R+1}_{N_t}\end{bmatrix}^T$$
$$= \begin{bmatrix} U_1 & U_2 \end{bmatrix}\begin{bmatrix} \Lambda_1 & \\ & \Lambda_2 \end{bmatrix}\begin{bmatrix} V_1^T \\ V_2^T \end{bmatrix}, \tag{8.78}$$

where Λ_2 is a diagonal matrix consisting of the n least singular values. Partition U_2 into two block entries of the same size $U_2^T = \begin{bmatrix} U_{21}^T & U_{22}^T \end{bmatrix}$. Then the state sequence $\bar{x}^{i+1}_{N_t}$ corresponding to the system Σ_{i+1} can be estimated as

$$\hat{\bar{x}}^{i+1}_{N_t} = U_{21}^T \hat{\mathbf{x}}^i_{N_t}. \tag{8.79}$$

By Equation (8.77), the above estimate can be written as

$$\begin{aligned} \hat{\bar{x}}^{i+1}_{N_t} &= U_{21}^T \hat{\mathbf{x}}^i_{N_t} \\ &= U_{21}^T Q_i^{-1}\mathbf{x}^i_{N_t} + U_{21}^T U_{21}^{i,T}\Delta^i_{N_t} \end{aligned} \tag{8.80}$$

where $U_{21}^T Q_i^{-1}$ is a non-singular ambiguity matrix, and the error $U_{21}^T U_{21}^{i,T}\Delta^i_{N_t}$ is uncorrelated with $\mathbf{x}^i_{N_t}$. Therefore, the state estimate (8.80) satisfies Equation (8.41), which is essential for the consistent identification of the system Σ_i.

8.4.1.5 Summary of the Algorithm to Estimate the Local State of Σ_{i+1}

Summary

The estimation for the state of Σ_{i+1} is carried out by **three sequential subspace intersections**:

1. first, for the state estimation of $\{\Sigma_j\}_{j=i+1}^{i+R}$,
2. second, for the state estimation of $\{\Sigma_j\}_{j=i-R+2}^{i+1}$,
3. third, for the state estimation of Σ_{i+1}.

The first two subspace intersections are performed in the temporal domain, namely the past and future system observations are used for estimating the current state; the third subspace intersection is performed in the spatial domain, namely the state of Σ_{i+1} is the intersection between the stacked states of $\{\Sigma_j\}_{j=i+1}^{i+R}$ and $\{\Sigma_j\}_{j=i-R+2}^{i+1}$. As a result, both the temporal dynamic property and the spatial topological property are utilized for the state estimation of Σ_{i+1}.

Computational Complexity

The computational complexity of the three subspace intersections are analyzed as follows. For the state estimation of $\{\Sigma_j\}_{j=i+1}^{i+R}$, it is required to compute the subspace intersection in Equation (8.69), which boils down to making the SVD decomposition of the matrix in Equation (8.70). The computation of the covariance matrix Equation (8.70) takes about $\mathcal{O}(s^2 N_t)$ multiplication and addition flops, while the SVD decomposition of the matrix in (8.70) scales with $\mathcal{O}(s^3)$. The computational burden involved in the state estimation of $\{\Sigma_j\}_{j=i-R+2}^{i+1}$ is the same as that of $\{\Sigma_j\}_{j=i+1}^{i+R}$. The state estimation of Σ_{i+1} boils down to carrying out the SVD decomposition in (8.78) which scales with $\mathcal{O}(R^2 N_t) + \mathcal{O}(R^3)$. To sum up the total computational complexity for estimating the state of Σ_{i+1} scales with $\mathcal{O}(R^2(R+N_t))+\mathcal{O}(s^2(s+N_t))$. It indicating that, when $s, R \ll N_t$, the computational complexity is mainly caused by the computation of the associated covariance matrix which is linearly proportional to the data length N_t.

8.4.1.6 Theoretical Comparison with Method of Section 8.3.2

As shown in Theorem 8.4, the state estimation for the system Σ_{i+1} mainly depends on the strong observability condition in Assumption 8.4. For the concerned 1D network, it will be shown that this condition relies on the size of the chosen local cluster.

The definition of the generic full rank for a structured matrix is provided below.

DEFINITION 8.7 A structured matrix has a *generic* full-rank if the full rank property holds for almost every choice of its non-zero entries.

LEMMA 8.10 *The matrix* $[\mathcal{O}_s^i \;\; \mathcal{T}_V^i]$ *is of (generic) full rank if*

1) *The matrix products* $C_{i+1}A_{i+1,l}P_i$ *and* $C_{i+R}A_{i+R,r}P_{i+R+1}$ *have full column rank, where* P_i *and* P_{i+R+1} *are defined in (8.60);*
2) $R \geq \frac{2(n-p)}{p} + 1$.

Proof To prove the lemma, the following two statements are to be verified:

i) $\text{rank}[\mathcal{O}_s^i \;\; \mathcal{T}_V^i] = \text{rank}[\mathcal{O}_s^i] + \text{rank}[\mathcal{T}_V^i]$ and $\mathcal{O}_s^i$ has full column rank;
ii) the convolution matrix $\mathcal{T}_V^i$ has full column rank.

The first statement can be verified by showing that the matrix quadruple $(\mathbf{A}_i, \mathbf{E}_{i2}, \mathbf{C}_i, 0)$ is strongly observable, namely the following matrix pencil has full column rank for all $z \in \mathbb{C}$:

$$P(z) = \begin{bmatrix} \mathbf{A}_i - zI & \mathbf{E}_{i2} \\ \mathbf{C}_i & 0 \end{bmatrix}. \tag{8.81}$$

By theorem C.6 in Sundaram (2012), the matrix pencil $P(z)$ has a generic full column rank if condition 2) (of Lemma 8.10) holds.

The second statement can be verified as follows. When the condition 1) is satisfied, i.e. $A_{i+1,l}P_i$ and $A_{i+R,r}P_{i+R+1}$ have full column rank, it is easy to see that the lower-triangular block matrix $\mathcal{T}_V^i$ has full column rank. The lemma is then proven. □

REMARK 8.3 To show the necessity of condition 1) in Lemma 8.10, two extreme cases will be discussed. First, when the observability matrix C_i has full column rank, the matrix $\mathcal{T}_V^i$ will disappear so that the full column rank condition of $[\mathcal{O}_s^i \;\; \mathcal{T}_V^i]$ is degraded to be the full column rank condition of $\mathcal{O}_s$. On the other hand, if the extra information of the unknown inputs is not used, the augmented matrices $[\mathbf{C}_i \mathbf{E}_{i2}]$ and $\mathcal{T}_V^i$ cannot have full rank.

Condition 2) provides the size condition for the local network such that the strong observability condition can be satisfied. When $p = n$ or the matrix C_i is square and non-singular, it is easy to see that the system state can be obtained from a single system, namely the size of local network only requires to be one. For the local identification method in Section 8.3.2, the required smallest size of the local cluster is $\frac{4n}{p} - 1$, which is much larger than $\frac{2(n-p)}{p} + 1$ for the presented method by using the extra information about the unknown inputs.

8.4.1.7 Numerical Simulation Example

In order to demonstrate the effectiveness of the proposed identification method and to compare with the performance of the method in Section 8.3.2, the local identification of a line-interconnected network consisting of 40 systems is considered and the central system in the line network will be identified. The system matrices are produced by adding random matrices to fixed matrices which are given as

$$\begin{aligned} A_i &= \begin{bmatrix} 0.2728 & -0.2068 \\ 0.1068 & 0.2728 \end{bmatrix}, & A_{i,i-1} &= \begin{bmatrix} -0.1195 & -0.3565 \\ 0.0874 & -0.1048 \end{bmatrix}, \\ A_{i,i+1} &= \begin{bmatrix} 0.0699 & -0.4278 \\ 0.3842 & 0.1135 \end{bmatrix}, & B_i &= \begin{bmatrix} 0.3870 \\ -1.2705 \end{bmatrix}, \\ C_i &= \begin{bmatrix} -0.9075 & -1.3651 \end{bmatrix}, & &\text{for } i = 1, \ldots, 40. \end{aligned} \tag{8.82}$$

The system input in the simulation is generated randomly following the standard Gaussian distribution.

To identify the central (or the 20th) system in the network, we choose the local cluster from 12th system to the 28th system. For comparison purposes, the developed methods in this section and Section 8.3.2 are simulated. The dimension parameter s defined in data equation (8.62) is set to $s = 10$. The impulse-response-fitting (IRF) criterion is used for evaluating the identification performance. For the ith system, the IRF criterion is defined as

$$\text{IRF} = \frac{1}{K}\sum_{k=1}^{K} \frac{\sum_{j=1}^{10} \|\hat{C}_i^k(\hat{A}_i^k)^j \hat{B}_i^k - C_i^*(A_i^*)^j B_j^*\|_F^2}{\sum_{j=1}^{10} \|C_i^*(A_i^*)^j B_i^*\|_F^2}, \tag{8.83}$$

where K is the number of Monte Carlo trials which is set to 200; A_i^*, B_i^*, C_i^* are true system matrices, while $A_i^k, \hat{B}_i^k, \hat{C}_i^k$ are the estimates at the kth Monte Carlo trial.

For notational simplicity, the proposed identification method in this section is called subspace identification by compensating unknown inputs (SIC) method, while the identification method provided in Section 8.3.2 is called subspace identification by removing unknown inputs (SIR) method.

First, the identification performance of the SIC method will be demonstrated. By setting the data length to 2000, the simulated IRF curve against the SNR is shown in Figure 8.7. It can be seen that the IRF values are close to zero when the SNR is large enough, indicating that the local system identification can be perfectly addressed by the SIC and SIR methods in the absence of measurement noise. In addition, it can be found that the SIC method has a better performance than the SIR method, which is

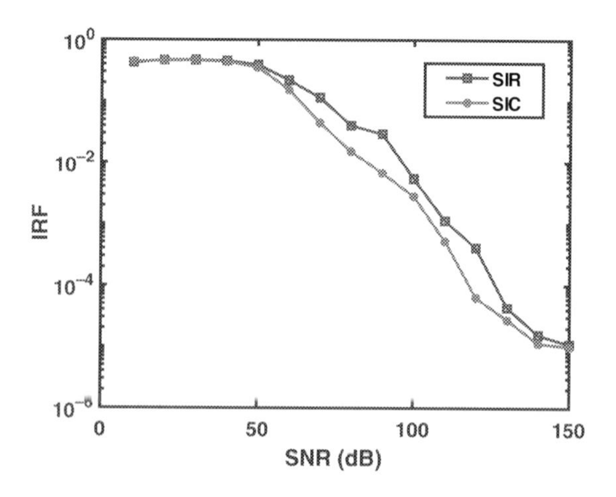

Figure 8.7 IRFs of the SIR and SIC methods with respect to the SNR for the example in Section 8.4.1.7. ©2019 IEEE Reprinted, with permission, from Yu et al. (2019)

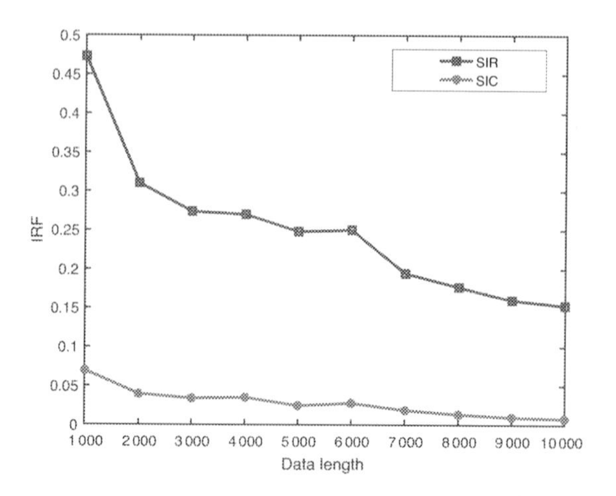

Figure 8.8 IRFs of the SIR and SIC methods with respect to the data length for the example defined in Section 8.4.1.7.

caused by exploiting the extra information about the unknown inputs embedded in the system measurements.

Second, numerical evidence will be provided to show the consistent identification of the SIC method. Two criteria will be adopted to demonstrate the consistent identification: one is that the IRF curve decays as the data length increases, and the other is that the pole estimation errors decrease along with the increase of the data length.

Figure 8.8 shows the IRF curve of the SIC method against the length of input-output data, where the SNR value is set to 70 dB. It can be observed that the IRF values decrease as the data length increases. Figure 8.9 shows the scattering plots of the pole estimates at the data lengths 2 000, 4 000, and 8 000. It can be seen that the pole estimation errors become much smaller as the data length increases. Both Figures 8.8 and 8.9 provide experimental evidence for the consistent identification of

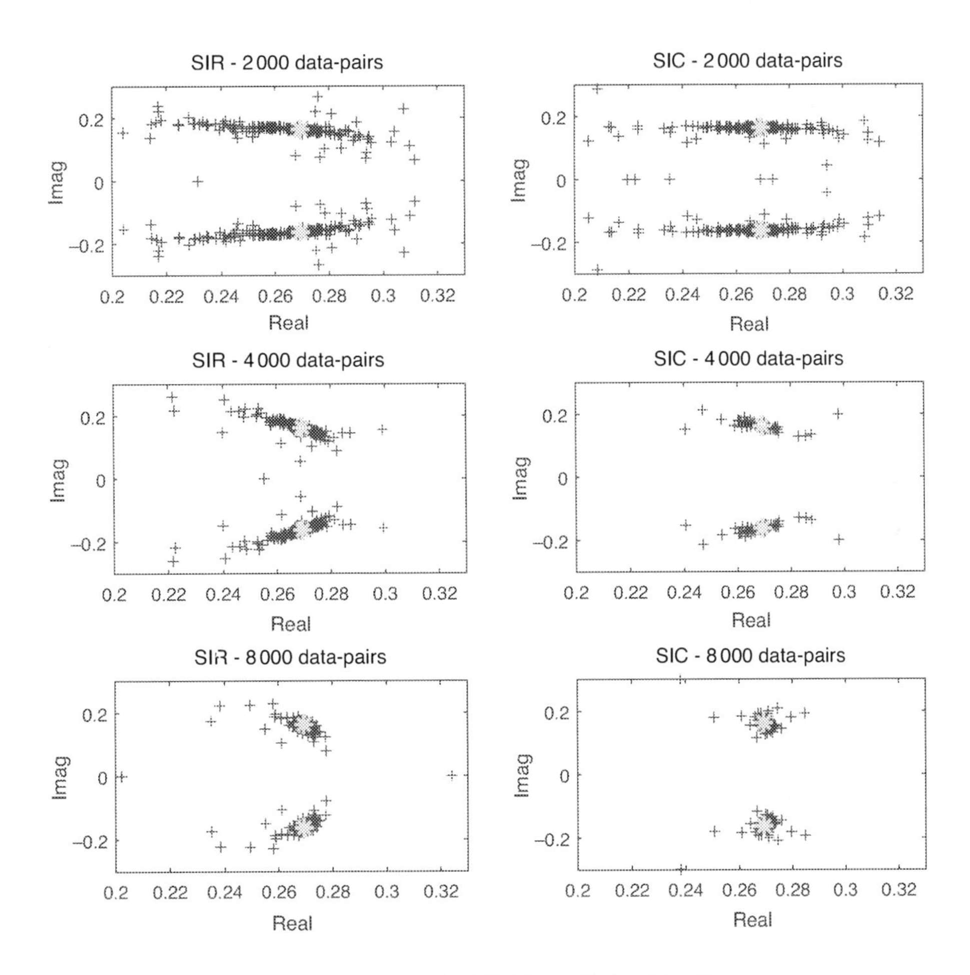

Figure 8.9 Estimated poles of 20th system by 200 Monte Carlo trials. The stars represent true poles, while the crosses denote estimated poles for the example defined in Section 8.4.1.7.

the SIC method and also the better performance of the SIC method compared to the SIR method.

8.4.2 The Identification of SSS Systems

8.4.2.1 Preliminaries and Problem Formulation

We again consider the string of N subsystems as depicted in Figure 8.1, where now subsystem Σ_i has the following structure:

$$\Sigma_i : \begin{bmatrix} x_i(k+1) \\ y_i(k) \\ v_{i-1}^p(k) \\ v_{i+1}^m(k) \end{bmatrix} = \begin{bmatrix} A_i & B_i & B_i^p & B_i^m \\ C_i & D_i & H_i^p & H_i^m \\ C_i^p & V_i^p & W_i^p & 0 \\ C_i^m & V_i^m & 0 & W_i^m \end{bmatrix} \begin{bmatrix} x_i(k) \\ u_i(k) \\ v_i^p(k) \\ v_i^m(k) \end{bmatrix}, \tag{8.84}$$

where $x_i(k) \in \mathbb{R}^n$ is the local state, $v_i^m \in \mathbb{R}^\ell$ and $v_i^p \in \mathbb{R}^\ell$ are interconnections to other subsystems, and $y_i \in \mathbb{R}^p$ and $u_i \in \mathbb{R}^m$ are measured outputs and inputs, respectively. As the heterogeneous case is considered, the dimensions of the local state, input and output dimension are allowed to vary as well. However, for the sake of brevity, and to avoid complicating the notation, we consider them as constant in this section.

If these N subsystems are lifted in one global state-space model, assuming zero boundary conditions, i.e. $v_1^m = 0, v_N^p = 0$ and resolving the interconnection variables $v_\bullet^p(k)$ and $v_\bullet^m(k)$ (where $\bullet$ indicates the spatial coordinate), then the global (or lifted) state-space model with noise added to the output becomes:

$$\Sigma : \begin{bmatrix} x(k+1) \\ y(k) \end{bmatrix} = \begin{bmatrix} \mathcal{A} & \mathcal{B} \\ \mathcal{C} & \mathcal{D} \end{bmatrix} \begin{bmatrix} x(k) \\ u(k) \end{bmatrix} + \begin{bmatrix} 0 \\ e(k) \end{bmatrix}, \tag{8.85}$$

where $x(k) \in \mathbb{R}^{Nn}, u(k) \in \mathbb{R}^{Nm}, y(k) \in \mathbb{R}^{Np}$ are the the lifted state, input and output vectors, respectively, denoted explicitly for the state vector $x(k)$ as $x(k) = \begin{bmatrix} x_1(k)^T & \dots & x_N(k)^T \end{bmatrix}^T$. Apart from Remark 8.5, the noise term will be neglected in the exposure of this section.

By the lifting process, Rice and Verhaegen (2009) showed that the system matrices A, B, C, D become highly structured and belong to the class of SSS matrices (Dewilde and van der Veen, 1998). A matrix within this class is determined by its so-called generator of seven sub-matrices, denoted for the $\mathcal{A}$-matrix as:

$$\mathcal{A} = SSS_{i=1}^N \left(B_i^m, W_i^m, C_i^m, A_i, B_i^p, W_i^p, C_i^p \right).$$

Here, $A_i \in \mathbb{R}^{n \times n}, B_i^p \in \mathbb{R}^{n \times \ell_p}, W_i^p \in \mathbb{R}^{\ell_p \times \ell_p}, C_i^p \in \mathbb{R}^{\ell_p \times n}$ and $B_i^m \in \mathbb{R}^{n \times \ell_m}, W_i^m \in \mathbb{R}^{\ell_m \times \ell_m}, C_i^m \in \mathbb{R}^{\ell_m \times n}$. The structural parameters that determine the size of these matrices are

$$n, \ell_p, \ell_m.$$

Though the methods described in this section can deal with these structure parameters being spatially varying, for the sake of brevity in notation, we restrict the description to constant dimensions only.

An illustration, for $N = 4$, of the SSS matrix structure of the quadruple of system matrices (8.85) is shown in Figure 8.10. Similarly to the matrix $\mathcal{A}$ above, the other system matrices could be denoted as,

$$\begin{aligned} \mathcal{B} &= SSS_{i=1}^N (B_i^m, W_i^m, V_i^m, B_i, B_i^p, W_i^p, V_i^p), \\ \mathcal{C} &= SSS_{i=1}^N (H_i^m, W_i^m, C_i^m, C_i, H_i^p, W_i^p, C_i^p), \\ \mathcal{D} &= SSS_{i=1}^N (H_i^m, W_i^m, V_i^m, D_i, H_i^p, W_i^p, V_i^p). \end{aligned}$$

From Figure 8.10, we observe that the sub-matrices in the dashed boxes are of low rank matrices. For example, the following factorization holds for the 2×2 block matrix in the upper right corner:

$$
\left(\begin{array}{cccc|cccc}
A_1 & B_1^p C_2^p & B_1^p W_2^p C_3^p & B_1^p W_2^p W_3^p C_4^p & B_1 & B_1^p V_2^p & B_1^p W_2^p V_3^p & B_1^p W_2^p W_3^p V_4^p \\
B_2^m C_1^m & A_2 & B_2^p C_3^p & B_2^p W_3^p C_4^p & B_2^m V_1^m & B_2 & B_2^p V_3^p & B_2^p W_3^p V_4^p \\
B_3^m W_2^m C_1^m & B_3^m C_2^m & A_3 & B_3^p C_4^p & B_3^m W_2^m V_1^m & B_3^m V_2^m & B_3 & B_3^p V_4^p \\
B_4^m W_3^m W_2^m C_1^m & B_4^m W_3^m C_2^m & B_4^m C_3^m & A_4 & B_4^m W_3^m W_2^m V_1^m & B_4^m W_3^m V_2^m & B_4^m V_3^m & B_4 \\
\hline
C_1 & H_1^p C_2^p & H_1^p W_2^p C_3^p & H_1^p W_2^p W_3^p C_4^p & D_1 & H_1^p V_2^p & H_1^p W_2^p V_3^p & H_1^p W_2^p W_3^p V_4^p \\
H_2^m C_1^m & C_2 & H_2^p C_3^p & H_2^p W_3^p C_4^p & H_2^m V_1^m & D_2 & H_2^p V_3^p & H_2^p W_3^p V_4^p \\
H_3^m W_2^m C_1^m & H_3^m C_2^m & C_3 & H_3^p C_4^p & H_3^m W_2^m V_1^m & H_3^m V_2^m & D_3 & H_3^p V_4^p \\
H_4^m W_3^m W_2^m C_1^m & H_4^m W_3^m C_2^m & H_4^m C_3^m & C_4 & H_4^m W_3^m W_2^m V_1^m & H_4^m W_3^m V_2^m & H_4^m V_3^m & D_4
\end{array}\right)
$$

Figure 8.10 Illustration of the quadruple of SSS matrices $(\mathcal{A} = SSS_{i=1}^{4}(B_i^m, W_i^m, C_i^m, A_i, B_i^p, W_i^p, C_i^p), \mathcal{B}, \mathcal{C}, \mathcal{D})$.

$$
\begin{aligned}
\mathcal{H}_A &= \begin{bmatrix} B_1^p W_2^p C_3^p & B_1^p W_2^p W_3^p C_4^p \\ B_2^p C_3^p & B_2^p W_3^p C_4^p \end{bmatrix} \\
&= \begin{bmatrix} B_1^p W_2^p \\ B_2^p \end{bmatrix} \begin{bmatrix} C_3^p & W_3^p C_4^p \end{bmatrix}.
\end{aligned} \tag{8.86}
$$

The property that SSS matrices are characterized by their generators only is a direct consequence of the above (often low) rank decomposition of off-diagonal block matrices in SSS matrices. This rank equals ℓ_p and can vary with the spatial index s.

The lifted system (8.85) is a 'simple' LTI system with highly structured matrices that are dense but data sparse (only governed by their generators). Though the SSS algebra is a closed algebra under operations of $+$, $\times$,$(.)^{-1}$, the ambiguity in the similarity transformation by which existing subspace identification methods determine an LTI state-space model from input-output data completely destroys these matrix structures. This is because such a similarity transformation would (in general) not be SSS. For that reason, a dedicated structure preserving subspace identification is developed in this section.

Acceptable similarity transformations on the local system level (8.84), which preserve the input-output behaviour, as well as the SSS structure of the system matrices, has to have the following form:

$$
T_s = \begin{bmatrix} T_{s,x} & 0 & 0 & 0 \\ 0 & I & 0 & 0 \\ 0 & 0 & T_{s-1,p} & 0 \\ 0 & 0 & 0 & T_{s+1,m} \end{bmatrix}, \tag{8.87}
$$

$$
T_s^{-1} = \begin{bmatrix} T_{s,x}^{-1} & 0 & 0 & 0 \\ 0 & I & 0 & 0 \\ 0 & 0 & T_{s,p}^{-1} & 0 \\ 0 & 0 & 0 & T_{s,m}^{-1} \end{bmatrix}. \tag{8.88}
$$

In the identification solution use will again be made of a cluster of R local systems $\{\Sigma_j\}_{j=i}^{i+R-1}$.

For this class of systems the following identification problem is considered.

PROBLEM 8.5 Given the lifted input-output data vector sequences $\{u(k), y(k)\}_{k=1}^{N_t}$ associated to system (8.85), the problem of interest is to find the generators of the SSS system matrices, denoted as follows:

$$A_i, B_i, C_i, D_i, B_i^m, B_i^p, C_i^m, C_i^p, H_i^m, H_i^p, V_i^m, V_i^p, W_i^m, W_i^p,$$

for all $i \in \{1, \ldots, N\}$ up to a local similarity transformation (8.87), under Assumption 8.5.

ASSUMPTION 8.5 (Identifying SSS systems)

1. *The lifted input signal $u(k)$ in system* (8.85) *is persistently exciting at any finite order.*
2. *The lifted system matrices $[\mathcal{A}, \mathcal{B}, \mathcal{C}]$ as well as the system matrices of any considered cluster of local systems in the network defined by the quadruple $[\overline{A}, \overline{B}, \overline{C}, \overline{D}]$ as defined later on in Equation* (8.89) *are minimal.*
3. *For the sake of simplicity in notation, we assume the dimensions of the SSS generators are constant.*
4. *Each cluster of local adjacent systems considered in the identification of the state of such a cluster is strongly observable (Sundaram, 2012), and,*
5. *The noise $e(k)$ is zero, except in Remark 8.5.*

8.4.2.2 General Algorithmic Strategy

The identification scheme proposed can be subdivided into three steps: (1) The identification of the state (up to a similarity transformation) of the cluster $\{\Sigma_j\}_{j=i}^{i+R-1}$ for all $i = 1 : N - R + 1$, (2) using the cluster states to find the state sequence (up to similarity transformation) of each individual system in the network and finally (3) using all estimated state quantities to estimate the system matrices having an SSS structure with a computational complexity linear in the number of subsystems. The latter we indicate as *linear computational complexity*.

REMARK 8.4 Only in the last step have we stipulated the linear complexity constraint. This is because the calculations necessary here are global, while for the other two steps the calculation could be performed locally and in parallel. The linear computational complexity and/or parallel computability is crucial for dealing with large scale systems.

These three steps of the identification methods are described in the following three subsections.

8.4.2.3 Estimating the State Sequence of a Cluster of Systems

Consider a cluster $\{\Sigma_j\}_{j=i}^{i+R-1}$ of R adjacent local systems. If the state and direct measurable input vector of this cluster be denoted by $\overline{x}(k)$ and $\overline{u}(k)$, respectively, then we can re-order the global state and input vector of the state-space model (8.85) in the following partitioned manner:

$$\begin{bmatrix} \bar{x}(k) \\ \mathbf{x}(k) \end{bmatrix} \quad \text{and} \quad \begin{bmatrix} \bar{u}(k) \\ \mathbf{u}(k) \end{bmatrix}.$$

Here, $\bar{x}(k) \in \mathbb{R}^{Rn}, \bar{u}(k) \in \mathbb{R}^{Rm}$ and $\mathbf{x}(k) \in \mathbb{R}^{Mn}, \mathbf{u}(k) \in \mathbb{R}^{Mm}$, for $M = N - R$. Based on this partitioning, the state-space model of that cluster as well as its environment can be denoted as,

$$\begin{bmatrix} \bar{x}(k+1) \\ \bar{y}(k) \end{bmatrix} = \begin{bmatrix} \bar{A} & \bar{B} \\ \bar{C} & \bar{D} \end{bmatrix} \begin{bmatrix} \bar{x}(k) \\ \bar{u}(k) \end{bmatrix} + \begin{bmatrix} \mathbf{A} & \mathbf{B} \\ \mathbf{C} & \mathbf{D} \end{bmatrix} \begin{bmatrix} \mathbf{x}(k) \\ \mathbf{u}(k) \end{bmatrix}. \tag{8.89}$$

The matrices with a $\bar{(.)}$ have the following SSS structure,

$$\begin{aligned}
\bar{A} &= SSS_{j=i}^{i+R-1}(B_i^m, W_i^m, C_i^m, A_i, B_i^p, W_i^p, C_i^p), \\
\bar{B} &= SSS_{j=i}^{i+R-1}(B_i^m, W_i^m, V_i^m, B_i, B_i^p, W_i^p, V_i^p), \\
\bar{C} &= SSS_{j=i}^{i+R-1}(H_i^m, W_i^m, C_i^m, C_i, H_i^p, W_i^p, C_i^p), \\
\bar{D} &= SSS_{j=i}^{i+R-1}(H_i^m, W_i^m, V_i^m, D_i, H_i^p, W_i^p, V_i^p).
\end{aligned}$$

The matrices in bold in Equation (8.89) are the off-diagonal sub-matrices of the SSS system matrices of the global system. They have, in general, a low rank property that is illustrated in the following example.

Example 8.4 (Low-rank property of off-diagonal blocks of SSS matrices) Consider the global system for $N = 4$, as shown in Figure 8.10. When the cluster of $R = 2$ local networks of the local system Σ_1 is considered, the associated matrices $\tilde{A}, \tilde{B}, \tilde{C}, \tilde{D}$ in Equation (8.89) allow the following decomposition:

$$\left[\begin{array}{c|c} \mathbf{A} & \mathbf{B} \\ \hline \mathbf{C} & \mathbf{D} \end{array}\right] = \left[\begin{array}{cc|cc} B_1^p W_2^p C_3^p & B_1^p W_2^p W_3^p C_4^p & B_1^p W_2^p V_3^p & B_1^p W_2^p W_3^p V_4^p \\ B_2^p C_3^p & B_2^p W_3^p C_4^p & B_2^p V_3^p & B_2^p W_3^p V_4^p \\ \hline H_1^p W_2^p C_3^p & H_1^p W_2^p W_3^p C_4^p & H_1^p W_2^p V_3^p & H_1^p W_2^p W_3^p V_4^p \\ H_2^p C_3^p & H_2^p W_3^p C_4^p & H_2^p V_3^p & H_2^p W_3^p V_4^p \end{array}\right]$$

$$= \begin{bmatrix} B_1^p W_2^p \\ B_1^p \\ H_1^p W_2^p \\ H_1^p \end{bmatrix} \begin{bmatrix} C_3^p & W_3^p C_4^p & V_3^p & W_3^p V_4^p \end{bmatrix}$$

This factorization is assumed to be of rank ℓ_p. Careful inspection in the spatial invariant case will reveal that the above factorization can be looked upon as the factorization of a Hankel matrix in an observability and controllability matrix.

The matrix in Equation (8.89) consisting of the matrices $\tilde{A}, \tilde{B}, \tilde{C}, \tilde{D}$ is, in general, for a cluster of systems with an excitation both from the left and the right a low rank matrix of rank $(\ell_p + \ell_m)$. On the boundaries, this low rank property (with a reduced rank) is preserved. This property will be exploited in the procedure outlined next.

For notational convenience we define,

$$\mathcal{W}_{i,j}^{x} = \begin{cases} W_i^x W_{i+1}^x \dots W_{j-1}^x W_j^x & \text{if } i \le j, \\ W_i^x W_{i-1}^x \dots W_{j+1}^x W_j^x & \text{if } i > j, \end{cases} \tag{8.90}$$

and let,

$$P_a = \begin{bmatrix} B_i^m & B_i^p \mathcal{W}_{i+1,i+M}^p \\ B_{i+1}^m W_i^m & \vdots \\ B_{i+2}^m \mathcal{W}_{i+1,i}^m & B_{i+M-2}^p \mathcal{W}_{i+M-1,i+M}^p \\ \vdots & B_{i+M-1}^p W_{i+M}^p \\ B_{i+M}^m \mathcal{W}_{i+M-1,i}^m & B_{i+M}^p \end{bmatrix},$$

$$P_c = \begin{bmatrix} H_i^m & H_i^p \mathcal{W}_{i+1,i+M}^p \\ H_{i+1}^m W_i^m & \vdots \\ H_{i+2}^m \mathcal{W}_{i+1,i}^m & H_{i+M-2}^p \mathcal{W}_{i+M-1,i+M}^p \\ \vdots & H_{i+M-1}^p W_{i+M}^p \\ H_{i+M}^m \mathcal{W}_{i+M-1,i}^m & H_{i+M}^p \end{bmatrix},$$

$$B_x = \begin{bmatrix} \mathcal{W}_{i-1,2}^m C_1^m & \dots & C_{i-1}^m & 0 & \dots & 0 \\ 0 & \dots & 0 & C_{i+M+1}^m & \dots & \mathcal{W}_{i+M+1,N-1}^m V_N^m \end{bmatrix},$$

$$B_u = \begin{bmatrix} \mathcal{W}_{i-1,2}^m V_1^m & \dots & V_{i-1}^m & 0 & \dots & 0 \\ 0 & \dots & 0 & V_{i+M+1}^m & \dots & \mathcal{W}_{i+M+1,N-1}^m V_N^m \end{bmatrix},$$

then by construction,

$$\begin{bmatrix} \mathbf{A} & \mathbf{B} \\ \mathbf{C} & \mathbf{D} \end{bmatrix} = \begin{bmatrix} P_a \\ P_c \end{bmatrix} \begin{bmatrix} B_x & B_u \end{bmatrix}. \tag{8.91}$$

Substituting Equation (8.91) into Equation (8.89) yields,

$$\begin{bmatrix} \bar{x}(k+1) \\ \bar{y}(k) \end{bmatrix} = \begin{bmatrix} \bar{A} & \bar{B} \\ \bar{C} & \bar{D} \end{bmatrix} \begin{bmatrix} \bar{x}(k) \\ \bar{u}(k) \end{bmatrix} + \begin{bmatrix} P_a \\ P_c \end{bmatrix} v(k), \tag{8.92}$$

where

$$v(k) = \begin{bmatrix} B_x & B_u \end{bmatrix} \begin{bmatrix} \mathbf{x}(k) \\ \mathbf{u}(k) \end{bmatrix}. \tag{8.93}$$

This shows that due to the SSS structure of the system matrices of the global system (8.85), the influence of the environment on a cluster of R local system is represented by the term $\begin{bmatrix} P_a \\ P_c \end{bmatrix} v(k)$ in Equation (8.92) with $v(k) \in \mathbb{R}^{\ell_p+\ell_m}$. By increasing the size of the local cluster R, we can make the size pR of the local output of the cluster (much) larger than $\ell_p + \ell_m$. It should be remarked that, due to the SSS structure, that latter rank does not increase. This is a welcome property for making the cluster of systems strongly observable.

The signal $v(k)$ in Equation (8.93) acts as an unknown input to the cluster $\{\Sigma_j\}_{j=i}^{i+R-1}$ of R local systems. In Lemma 8.11, which is slight variation of Theorem 8.4, we show under which conditions the state of the cluster can be estimated (up to similarity transformation), despite the presence of this unknown input. In that lemma the following two data equations will be used,

$$\bar{Y}_p = O\bar{X}_p + T_{\bar{u}}\bar{U}_p + T_v V_p, \tag{8.94}$$

$$\bar{Y}_f = O\bar{X}_f + T_{\bar{u}}\bar{U}_f + T_v V_f, \tag{8.95}$$

where the extended observability matrix O is defined as $O = \mathcal{O}_s(\bar{A}, \bar{C})$, the state sequence $\overline{X}_p$ and $\overline{X}_f$ are defined as:

$$\bar{X}_p = \begin{bmatrix} \bar{x}(k-s) & \bar{x}(k-s+1) & \cdots & x(k-s+N_t-1) \end{bmatrix},$$

$$\bar{X}_f = \begin{bmatrix} \bar{x}(k) & \bar{x}(k+1) & \cdots & x(k+N_t-1) \end{bmatrix},$$

and the block-Toeplitz matrices $T_{\bar{u}}, T_v$ are defined as

$$T_{\bar{u}} = \mathcal{T}_s(\bar{A}, \bar{B}, \bar{C}, \bar{D}), \quad T_v = \mathcal{T}_s(\bar{A}, P_a, \bar{C}, P_c). \tag{8.96}$$

The past and future block Hankel matrices are, respectively, defined as

$$\bar{Y}_p = \mathcal{H}_{s,N_t}[\bar{y}(k-h)], \qquad \bar{Y}_f = \mathcal{H}_{s,N_t}[\bar{y}(k)],$$

$$\bar{U}_p = \mathcal{H}_{s,N_t}[\bar{u}(k-h)], \qquad \bar{U}_f = \mathcal{H}_{s,N_t}[\bar{u}(k)],$$

$$V_p = \mathcal{H}_{s,N_t}[v(k-h)], \qquad V_f = \mathcal{H}_{s,N_t}[v(k)].$$

LEMMA 8.11 *Let the system $(\bar{A}, P_a, \bar{C}, P_c)$ be strongly observable (Sundaram, 2012), i.e.* $\mathbf{rank}\begin{bmatrix} O & T_v \end{bmatrix} = n + \mathbf{rank}[T_v]$ *with the matrix T_v in* (8.96) *full column rank. Let the compound matrix*

$$\begin{bmatrix} X_p^T & U_p^T & V_p^T & U_f^T & V_f^T \end{bmatrix}^T,$$

have full row rank, then the row subspace of the state sequence $\bar{X}_f$ can be determined by the following intersection:

$$\bar{X}_f = \mathbf{row}\begin{bmatrix} \bar{U}_p \\ \bar{Y}_p \end{bmatrix} \cap \begin{bmatrix} \bar{U}_f \\ \bar{Y}_f \end{bmatrix}. \tag{8.97}$$

Proof In the data equation (8.94), the sum of the state-related term $O\bar{X}_f$ and the unknown input-related term $T_v V_f$ can be expressed as

$$\begin{bmatrix} O & T_v \end{bmatrix}\begin{bmatrix} \bar{X}_f \\ V_f \end{bmatrix} = \begin{bmatrix} I & -T_{\bar{u}} \end{bmatrix}\begin{bmatrix} \bar{Y}_f \\ \bar{U}_f \end{bmatrix}. \tag{8.98}$$

By the full column rank of $\begin{bmatrix} O & T_v \end{bmatrix}$, Equation (8.98) implies that

$$\mathbf{row}\begin{bmatrix} \bar{X}_p \\ V_p \end{bmatrix} \subseteq \mathbf{row}\begin{bmatrix} \bar{Y}_p \\ \bar{U}_p \end{bmatrix}.$$

And with a similar reasoning,

$$\textbf{row}\begin{bmatrix}\bar{X}_f\\V_f\end{bmatrix}\subseteq\textbf{row}\begin{bmatrix}\bar{Y}_f\\\bar{U}_f\end{bmatrix}. \tag{8.99}$$

By the state evolution equation (8.92), we have that

$$\textbf{row}\begin{bmatrix}\bar{X}_f\end{bmatrix}\subseteq\textbf{row}\begin{bmatrix}\bar{X}_p\\\bar{U}_p\\V_p\end{bmatrix}=\textbf{row}\begin{bmatrix}\bar{Y}_p\\\bar{U}_p\end{bmatrix}, \tag{8.100}$$

where the last equality follows from

$$\begin{bmatrix}O & T_{\bar{u}} & T_v\\0 & I & 0\end{bmatrix}\begin{bmatrix}\bar{X}_p\\\bar{U}_p\\V_p\end{bmatrix}=\begin{bmatrix}\bar{Y}_p\\\bar{U}_p\end{bmatrix},$$

and the full column rank property of the compound matrix constructed from the system matrices on the left-hand side. By Equations (8.99) and (8.100), it is easy to verify that

$$\textbf{row}\begin{bmatrix}\bar{X}_f\end{bmatrix}\subseteq\textbf{row}\begin{bmatrix}\bar{Y}_p\\\bar{U}_p\end{bmatrix}\cap\begin{bmatrix}\bar{Y}_f\\\bar{U}_f\end{bmatrix}=\textbf{row}\begin{bmatrix}\bar{X}_p\\\bar{U}_p\\V_p\end{bmatrix}\cap\begin{bmatrix}\bar{X}_f\\\bar{U}_f\\V_f\end{bmatrix}. \tag{8.101}$$

By the assumption that the augmented matrix $\begin{bmatrix}X_p^T & U_p^T & V_p^T & U_f^T & V_f^T\end{bmatrix}^T$ has full row rank. Equation (8.101) then shows that

$$\textbf{row}\begin{bmatrix}\bar{X}_f\end{bmatrix}=\textbf{row}\begin{bmatrix}\bar{X}_p\\\bar{U}_p\\V_p\end{bmatrix}\cap\begin{bmatrix}\bar{X}_f\\\bar{U}_f\\V_f\end{bmatrix}, \tag{8.102}$$

which completes the proof of the lemma. □

According to Lemma 8.11, the state sequence of a cluster of local systems can be estimated by solving a (row-)subspace intersection problem between matrices constructed from cluster input and output only.

This intersection problem is conceptually solved in the following manner. First compute the left null space of the matrix $\begin{bmatrix}\bar{Y}_p^T & \bar{U}_p^T & \bar{Y}_f^T & \bar{U}_f^T\end{bmatrix}^T$, i.e.

$$\begin{bmatrix}N_1 & N_2\end{bmatrix}=N_L\left(\begin{bmatrix}\bar{Y}_p\\\bar{U}_p\\\bar{Y}_f\\\bar{U}_f\end{bmatrix}\right), \tag{8.103}$$

with N_1, N_2 being of equal dimension. A reliable way to solve this null space determination problem is via the singular value decomposition. Knowing the null space, the solution to the subspace intersection then is given as,

$$\hat{X}_f=N_1\begin{bmatrix}\bar{Y}_p\\\bar{U}_p\end{bmatrix}. \tag{8.104}$$

REMARK 8.5 When the noise $e(k)$ in Equation (8.85) is present, special modifications to Lemma 8.11 and the derived methodology need to be undertaken. For example, when this noise is zero-mean and white with covariance matrix equal to a multiple of the identity matrix, a similar approach to that presented in the paragraph after the proof of Theorem 8.4 to cope with this noise in a consistent manner could also be derived for the current methodology. This approach would first estimate the variance σ^2 of $e(k)$ from data, and then use this estimate in the estimation of the intersections.

8.4.2.4 Estimating the State of a Local Subsystem Σ_i

The previous subsection gives the conditions and the computational details for estimating the state of a cluster of R local systems. Using this information, we first consider the generic case of estimating the state by the intersection of the states of two lifted clusters. This introduces some constraints for finding the states of the systems at the boundaries. Therefore, we first address the generic case, then estimate the states of the system at the boundary.

In these procedures, use will be made of state sequences. However, for the sake of brevity the state sequence will simply be denoted without the argument. For example, consider the following notation for the state of system Σ_i:

$$x_i = \begin{bmatrix} x_i(k) & x_i(k+1) & \cdots & x_i(k+\overline{N}-1) \end{bmatrix},$$

where k and $\overline{N}$ follow from the case considered. Further, for outlining these approaches use will be made of Figure 8.11.

LEMMA 8.12 *Let assumptions 1–2 of Assumption 8.5 hold. Consider the two clusters* $\mathcal{C}_1 = \{\Sigma_j\}_{j=i-R+1}^{i}$ *and* $\mathcal{C}_2 = \{\Sigma_j\}_i^{j=i+R-1}$ *for* $R \leq i \leq N - R + 1$. *Let the state sequence of the corresponding clusters be denoted by* $x_{\mathcal{C}_1}$ *and* $x_{\mathcal{C}_2}$, *respectively, then the row subspace of the state sequence of the subsystem* Σ_i *can be estimated as,*

$$\mathbf{row}\left[x_i\right] = \mathbf{row}\left[x_{\mathcal{C}_1}\right] \cap \mathbf{row}\left[x_{\mathcal{C}_2}\right]. \tag{8.105}$$

Proof Note that we have,

$$\mathbf{row}\left[x_{\mathcal{C}_1}\right] = \mathbf{row}\begin{bmatrix} x_{i-R+1} \\ \vdots \\ x_i \end{bmatrix}, \quad \mathbf{row}\left[x_{\mathcal{C}_2}\right] = \mathbf{row}\begin{bmatrix} x_i \\ \vdots \\ x_{i+R-1} \end{bmatrix}.$$

Due to assumptions 1–2, the matrix $\begin{bmatrix} x_{i-R+1}^T & \cdots & x_{i-1}^T & x_{i+1}^T & \cdots & x_{i+R-1}^T \end{bmatrix}^T$ has full row rank (Verhaegen and Verdult, 2007). Therefor, it follows that

$$\mathbf{row}\begin{bmatrix} x_{i-R+1} \\ \vdots \\ x_i \end{bmatrix} \cap \mathbf{row}\begin{bmatrix} x_i \\ \vdots \\ x_{i+R-1} \end{bmatrix} = \mathbf{row}\left[x_i\right],$$

which concludes the proof. □

Figure 8.11 Illustration of the state estimation of a single system, using the stacked states of two local networks, $C_1 = \{\Sigma_j\}_{j=i-R+1}^{i}$ and $C_2 = \{\Sigma_j\}_{j=i}^{i+R-1}$ for $R = 3$ and $i = 3$.

In Lemma 8.12, the individual states sequence x_i (up to a similarity transformation) is known for $R \leq i \leq N - R + 1$. To find the states of the boundary system, a number of possibilities may exist depending on the situation at hand. One possible scenario is discussed in Example 8.6. Another scenario is discussed next.

The description of this scenario is restricted to finding the state of Σ_1 in Figure 8.11.

Knowing the state of clusters $\{\Sigma_j\}_{j=1}^{R}$ and $\{\Sigma_j\}_{j=2}^{R+1}$, we can, via intersection calculations, find the state sequence,

$$\begin{bmatrix} x_2 \\ \vdots \\ x_R \end{bmatrix}.$$

From the 'last' cluster $\{\Sigma_j\}_{j=N-R+1}^{N}$ we find the state sequence,

$$\begin{bmatrix} x_{N-R+1} \\ \vdots \\ x_N \end{bmatrix},$$

and from Lemma 8.12 we have all the state sequences in between and are able to define the following 'external' input to system Σ_1:

$$v_E = \begin{bmatrix} \begin{bmatrix} x_2 \\ \vdots \\ x_R \end{bmatrix} \\ x_{R+1} \\ \vdots \\ x_{N-R} \\ \begin{bmatrix} x_{N-R+1} \\ \vdots \\ x_N \end{bmatrix} \end{bmatrix},$$

with all state sequences up to similarity equivalence. If we now collect the input sequences in the 'extended' input vector as,

$$u_E = \begin{bmatrix} u_1 \\ u_2 \\ \vdots \\ u_N \end{bmatrix},$$

we can write the local state-space model Σ_1 as:

$$\begin{aligned} x_1(k+1) &= A_1 x_1(k) + B_u u_E(k) + B_v v_E(k), \\ y_1(k) &= C_1 x_1(k) + D_v v_E(k). \end{aligned} \tag{8.106}$$

This is an ordinary LTI system and finding the state (and system matrices) can be done using a variety of subspace identification methods.

Example 8.5 (Identification of the local states) The 1D mass-spring-damper system of Example 4.7 is considered here as a test case. In the continuous-time case, this results for N masses in an SSS state-space with the following description for the local system Σ_i given, similar to Equation (8.84), as:

$$\Sigma_i : \begin{bmatrix} \dot{x}_i(t) \\ y_i(t) \\ v^p_{i-1}(t) \\ v^m_{i+1}(t) \end{bmatrix} = \begin{bmatrix} A_i & B_i & B^p_i & B^m_i \\ C_i & D_i & H^p_i & H^m_i \\ C^p_i & V^p_i & W^p_i & 0 \\ C^m_i & V^m_i & 0 & W^m_i \end{bmatrix} \begin{bmatrix} x_i(t) \\ u_i(t) \\ v^p_i(t) \\ v^m_i(t) \end{bmatrix}, \tag{8.107}$$

where t is now the continuous-time parameter. Considering the spatially heterogeneous case where the masses m_i, and spring and damper constants c_i and k_i are allowed to vary randomly with the spatial index i, the non-zero matrices in Equation (8.107) have the following form:

$$A_i = \begin{bmatrix} -\frac{(c_i+c_{i+1})}{m_i} & -\frac{(k_i+k_{i+1})}{m_i} \\ 1 & 0 \end{bmatrix},$$

$$B^p_i = \begin{bmatrix} \frac{c_{i+1}}{m_i} & \frac{k_{i+1}}{m_i} \\ 0 & 0 \end{bmatrix}, B^m_i = \begin{bmatrix} \frac{c_{i+1}}{m_{i+1}} & \frac{k_{i+1}}{m_{i+1}} \\ 0 & 0 \end{bmatrix}, B_i = \begin{bmatrix} b_i \\ 0 \end{bmatrix},$$

$$C_i = \begin{bmatrix} -\frac{(c_i+c_{i+1})}{m_i} & -\frac{(k_i+k_{i+1})}{m_i} \\ 0 & 1 \end{bmatrix}; H^p_i = \begin{bmatrix} \frac{c_{i+1}}{m_i} & \frac{k_{i+1}}{m_i} \\ 0 & 0 \end{bmatrix}; H^m_i = \begin{bmatrix} \frac{c_{i+1}}{m_{i+1}} & \frac{k_{i+1}}{m_{i+1}} \\ 0 & 0 \end{bmatrix}$$

$$C^p_i = \begin{bmatrix} 1 & 0 \\ 0 & 1 \end{bmatrix}$$

$$C^m_i = \begin{bmatrix} 1 & 0 \\ 0 & 1 \end{bmatrix}$$

The random nature of the variation of the parameters is determined via the standard normal distribution.

Considering the masses at the boundaries to be clamped, then the discretization using Tustin (see Lemma 5.1) again has an SSS structure. In the current examples we simulate for $N = 25$ and a sample period $\Delta T = 1$ (sec) these mass-spring-damper systems using uncorrelated zero-mean white-noise signals $u_i(k)$ for $k = 1 : N_t$

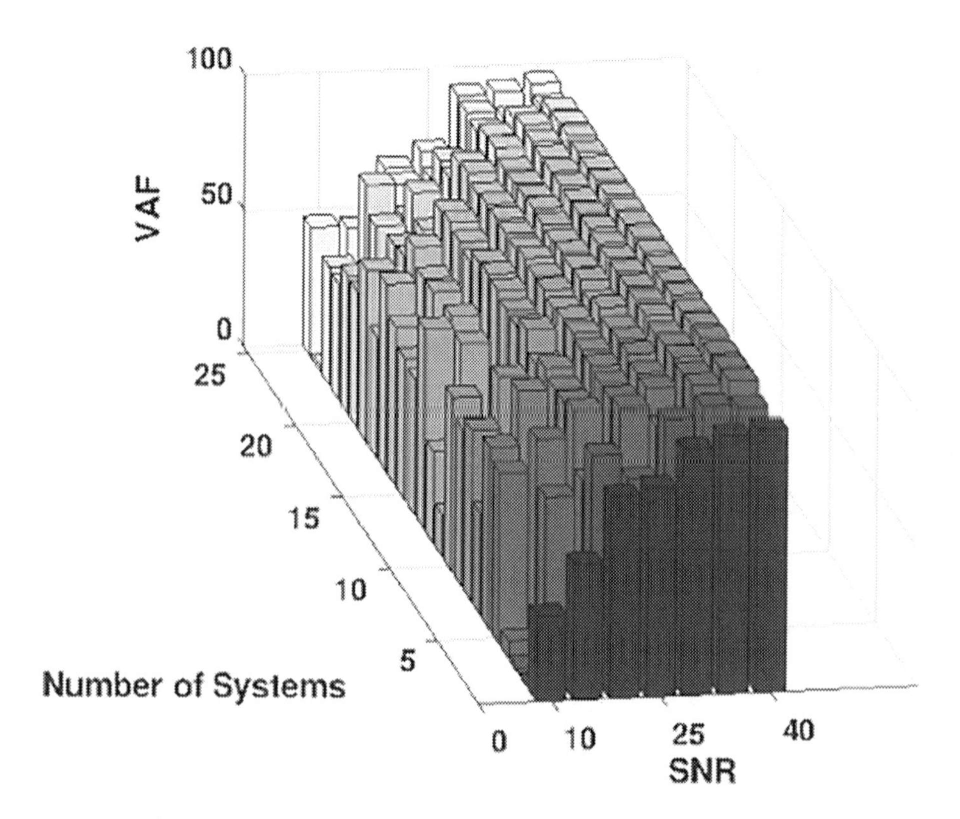

Figure 8.12 The application of Lemma 8.11 to the mass-spring-damper system of Example 8.5 for decreasing SNR values due to the additive zero-mean white noise to the individual output signals. The number of local systems is taken equal to 25 and the SNR reduces from 40 db to 10 db.

(with $N_t = 2500$). As it turns out that the systems on the boundaries are strongly observable, we can estimate the local states of the systems Σ_1 and Σ_N on the boundaries using the methodology highlighted by Lemma 8.11.

As the clustering of two adjecent systems turned out to be be strongly observable, we use Lemma 8.11 to estimate the joined state of systems Σ_i and Σ_{i+1} for $i = 1 : N - 1$. Then the local states of all systems are found by intersection between these clustered states. The results[2] for one particular example are shown in Figure 8.12. In this figure, the resilience of additive noise to the output is investigated in such a way that a particular SNR results. The VAF in this graph is computed as follows. When $\hat{x}_i(k)$ is the estimated state and $x_i(k)$ is the true state of the ith system, accounting for the fact that both are in a different state basis, we define the following VAF metric for the ith system:

[2] The results presented in this example have been made possible by the work of Engineer Wiegert Krijgsman, during the time he was affiliated with the Delft Center for Systems and Control in 2019.

$$VAF(i) = \left(1 - \frac{\sum_{k=1}^{end} \|\hat{x}_i(k) - x_i^p(k)\|}{\sum_{k=1}^{end} \|\hat{x}_i(k)\|}\right) \cdot 100\%$$

Here, $x_i^p(k) = T_i x_i(k)$, where the matrix T is least-squares solution to the following problem:

$$\min_{T_i} \sum_{k=1}^{end} \|\hat{x}_i(k) - T x_i(s+k)\|^2.$$

This figure shows that the results of Lemma 8.11 hold in the noise-free case as well as for high SNR (SNR $\geq$ 30). Figure 8.12 also indicates that, for increasing SNRs, the accuracy of the states of the boundary systems start to decay first, while the systems in the middle remain more accurate. This is probably due to the fact that the latter states are obtained via intersection themselves.

8.4.2.5 Challenges for Identifying the SSS-Structured System Matrices

When all the individual state sequences of the local systems (up to a similarity transformation) are estimated, we now address the estimation of the system matrices preserving the SSS structure. Let, for that purpose, the following notation be introduced:

$$\begin{aligned}
x_i^+ &= \begin{bmatrix} x_i(2) & \cdots & x_i(N_t) \end{bmatrix}, \\
x_i &= \begin{bmatrix} x_i(1) & \cdots & x_i(N_t - 1) \end{bmatrix}, \\
y_i &= \begin{bmatrix} y_i(1) & \cdots & y_i(N_t - 1) \end{bmatrix}, \\
u_i &= \begin{bmatrix} u_i(1) & \cdots & u_i(N_t - 1) \end{bmatrix}.
\end{aligned}$$

By ordering these estimated state sequences and the input sequences as,

$$\begin{aligned}
\mathcal{X} &= \begin{bmatrix} x_1^T & u_1^T & x_2^T & u_2^T & \ldots & x_N^T & u_N^T \end{bmatrix}^T, \\
\mathcal{Y} &= \begin{bmatrix} (x_1^+)^T & y_1^T & (x_2^+)^T & y_2^T & \ldots & (x_N^+)^T & y_N^T \end{bmatrix}^T,
\end{aligned}$$

and defining the matrices $J_i, K_i^m, K_i^p, L_i^m, L_i^p$, as

$$J_i = \begin{bmatrix} A_i & B_i \\ C_i & D_i \end{bmatrix}, \quad K_i^m = \begin{bmatrix} B_i^m \\ H_i^m \end{bmatrix}, K_i^m = \begin{bmatrix} B_i^p \\ H_i^p \end{bmatrix},$$

$$L_i^m = \begin{bmatrix} C_i^m & V_i^m \end{bmatrix}, \quad K_i^m = \begin{bmatrix} B_i^p \\ H_i^p \end{bmatrix}, \quad L_i^p = \begin{bmatrix} C_i^p & V_i^p \end{bmatrix},$$

the identification of the structured SSS system matrices of the global system boils down to estimating the generator matrices of $SSS(K_i^m, W_i^m, L_i^m, J_i, K_i^p, W_i^p, L_i^p)$, and gives rise to the following structured optimization problem,

$$\min_{\substack{J_i, \\ K_i^m, W_i^m, L_i^m, \\ K_i^p, W_i^p, L_i^p}} \|\mathcal{Y} - SSS(K_i^m, W_i^m, L_i^m, J_i, K_i^p, W_i^p, L_i^p)\mathcal{X}\|_F^2. \tag{8.108}$$

The two challenges of the above structured optimization problem are:

1. the problem is multi-linear in the elements of the generators of the SSS matrices.
2. the size of the problem determined, for example, by the data matrices becomes large when the number of systems in the network, denoted by N, becomes large.

The solution to the second challenge is overcome by exploiting the multi-linear structure of the problem as outlined in the next section.

8.4.3 Identifying the SSS-Structured System Matrices by Multilinear Regression

The SSS-structure preserving methodology will be presented, for the sake of brevity, for the following generic case:

$$\min_{\substack{D_i,\\ U_i, W_i, V_i,\\ P_i, R_i, Q_i}} \left\| \begin{bmatrix} y_1 \\ \vdots \\ y_N \end{bmatrix} - SSS(U_i, W_i, V_i, D_i, P_i, R_i, Q_i) \begin{bmatrix} x_1 \\ \vdots \\ x_N \end{bmatrix} \right\|_F^2, \tag{8.109}$$

with the notation $SSS(U_i, W_i, V_i, D_i, P_i, R_i, Q_i)$ made explicit as,

$$SSS(U_i, W_i, V_i, D_i, P_i, R_i, Q_i) = \begin{bmatrix} D_1 & U_1 V_2 & \dots & U_1 W_2 \dots V_N \\ P_2 Q_1 & D_2 & \dots & U_2 W_3 \dots V_N \\ \vdots & & \ddots & \vdots \\ P_N \dots Q_1 & & \dots & D_N \end{bmatrix}.$$

This problem can be denoted more compactly as,

$$\min_{\substack{D_i,\\ U_i, W_i, V_i,\\ P_i, R_i, Q_i}} \|\mathcal{Y} - SSS(U_i, W_i, V_i, D_i, P_i, R_i, Q_i)\mathcal{X}\|_F^2. \tag{8.110}$$

The main idea of the algorithm is to optimize over a single element of the generator $U_i, W_i, V_i, D_i, P_i, R_i, Q_i$ of the SSS-matrix at a time, while keeping the others constant and continue this process sequentially until all elements of the generator have been estimated and then repeat the whole cycle. By exploiting the special structure of the problem, a sequence of linear-least squares problems that can each be solved with low complexity is obtained.

In the following subsection, each of these linear least-squares problems is discussed. A single iteration t is described and the estimates obtained from iteration $t-1$ are denoted with a $\hat{(\cdot)}$.

8.4.3.1 Minimizing over D_i

The first step is to minimize over the diagonal entries in the SSS-matrix (8.110). For this purpose, define the matrix $\hat{E}_{t-1}$ as,

$$\hat{E}_{t-1} = \mathcal{Y} - SSS(\hat{U}_i, \hat{W}_i, \hat{V}_i, 0, \hat{P}_i, \hat{R}_i, \hat{Q}_i)\mathcal{X}.$$

Then updating the estimates D_i is done via the (parallel) solution of the following least-squares problems,

$$\hat{D}_i^+ = \arg\min_{D_i} \| \hat{E}_{t-1} - SSS(0,0,0,D_i,0,0,0)\mathcal{X}\|_F^2.$$

The construction of the matrix $\hat{E}_{t-1}$ can, due to the SSS structure, be done with linear complexity in N. Furthermore, the N parallel least-squares problem, we show that the overall computational complexity remains linear in N for this step.

8.4.3.2 Minimizing over U_i, W_i, V_i

With the updated matrices D_i^+ we update the measurement matrix $\mathcal{Y}$ as follows,

$$\hat{Z}_{t-1} = \mathcal{Y} - SSS(\hat{P}_i, \hat{R}_i, \hat{Q}_i, \hat{D}_i^+, 0,0,0)X.$$

Using the estimates of W_s and V_s we compute the sequence of quantities,

$$\Psi_{k-1} = W_k \Psi_k + V_k x_k, \tag{8.111}$$

for $k = N : -1 : 1$ and $\Psi_N = 0$. Then the *updating of the matrices* U_i is done via the following, again parallel, least-squares problems,

$$\hat{U}_s^+ = \arg\min_{U_s} \left\| \hat{Z}_{t-1} - \textbf{diag}\,(U_1, \ldots, U_{N-1}) \begin{bmatrix} \Psi_1 \\ \vdots \\ \Psi_{N-1} \end{bmatrix} \right\|_F^2.$$

The combination of SSS structure and the parallel nature of the least-squares again leads to linear computation complexity for this update. Next, we discuss the update of the matrices W_s. Here, use will be made of the following Lemma. For solving the least-squares problem of the form (8.112) the following Lemma, stated without proof, is useful.

LEMMA 8.13 *The solution to following least squares problem:*

$$\min_X \|C - AXB\|_F^2, \tag{8.112}$$

with full column rank A and full row rank B, with $A^\dagger$ denoting the (left) pseudo-inverse of A, given as $(A^T A)^{-1} A^T$, and $B^\dagger$ denoting the (right) pseudo-inverse $B^T (BB^T)^{-1}$, is given as:

$$\hat{X} = A^\dagger C B^\dagger. \tag{8.113}$$

The *updating of the matrices* W_i starts by updating W_2, etc. For that reason we assume the updates of the matrices W_2 up to W_{i-1} have been performed and present the least-squares problem to update the estimate of the matrix W_i. If we denote the top $i-1$-block rows of the matrix $\hat{Z}_{t-1}$ by $\hat{Z}_{t-1}^{i-1}$, then the update of the matrix W_i follows from the following least-squares problem,

$$
\min_{W_i} \left\| \hat{Z}_{t-1}^{i-1} - \left[\underbrace{\begin{matrix} \hat{U}_1^+ \hat{V}_2 & \hat{U}_1^+ \hat{W}_2^+ \hat{V}_3 & \dots & \hat{U}_1^+ \hat{W}_2^+ \cdots \hat{V}_i \\ 0 & \hat{U}_2^+ \hat{V}_3 & & \hat{U}_2^+ \hat{W}_3^+ \dots \hat{V}_i \\ \vdots & & \ddots & \vdots \\ 0 & & \dots & \hat{U}_{i-1}^+ \hat{V}_i \end{matrix}}_{F_i} \right| \cdots \right.
$$

$$
\left| \begin{matrix} 0 & \dots & 0 \\ 0 & \dots & 0 \\ \vdots & & \vdots \\ 0 & \dots & 0 \end{matrix} \right] \begin{bmatrix} x_2 \\ x_3 \\ \vdots \\ x_i \\ \hline x_{i+1} \\ \vdots \\ x_N \end{bmatrix}
$$

$$
\left. - \underbrace{\begin{bmatrix} \hat{U}_1^+ \hat{W}_2^+ \dots \hat{W}_{i-1}^+ \\ \hat{U}_2^+ \hat{W}_3^+ \dots \hat{W}_{i-1}^+ \\ \vdots \\ \hat{U}_{i-1}^+ \end{bmatrix}}_{G_i} W_i \begin{bmatrix} 0 \\ \vdots \\ 0 \\ \hline (\hat{V}_{i+1})^T \\ (\hat{W}_{i+1} \hat{V}_{i+2})^T \\ \vdots \\ (\hat{W}_{i+1} \cdots V_N)^T \end{bmatrix}^T \begin{bmatrix} x_2 \\ x_3 \\ \vdots \\ x_i \\ \hline x_{i+1} \\ \vdots \\ x_N \end{bmatrix} \right\|_F^2 \tag{8.114}
$$

Using the sequence defined in Equation (8.111), the definition of the matrices F_i, G_i in Equation (8.114) and the corresponding part of the signal matrices, we can write the above least-squares problem compactly as,

$$
\min_{W_i} \| \hat{Z}_{t-1}^{i-1} - F_i X_i - G_i W_i \Psi_i \|_F^2. \tag{8.115}
$$

Using Lemma 8.13, the explicit formula for the update is given by,

$$
W_i^+ = (G_i^T G_i)^{-1} G_i^T \left(\hat{Z}_{t-1}^{i-1} - F_i X_i \right) \Psi_i^\dagger. \tag{8.116}
$$

In order to subsequently update W_{i+1} efficiently, we denote the corresponding least-squares problem like that in (8.115) as,

$$
\min_{W_{i+1}} \| \hat{Z}_{t-1}^{i} - F_{i+1} X_{i+1} - G_{i+1} W_{i+1} \Psi_{i+1} \|_F^2. \tag{8.117}
$$

The following relationship exists between the data matrices in the least-squares problems (8.115) and (8.117):

$$G_{i+1} = \begin{bmatrix} G_i W_i \\ U_i \end{bmatrix}, \quad F_{i+1} = \begin{bmatrix} F_i & G_{i+1} V_{i+1} \\ 0 & \end{bmatrix},$$

$$X_{i+1} = \begin{bmatrix} X_i \\ x_{i+1} \end{bmatrix}, \quad Z_{i+1} = \begin{bmatrix} Z_i \\ z_i \end{bmatrix}.$$

An efficient way of updating the least-squares is by updating QR factorizations. For that purpose, consider that for solving (8.115) the following QR factorization is performed:

$$G_i = \begin{bmatrix} Q_{G,i} & \star \end{bmatrix} \begin{bmatrix} R_{G,i} \\ 0 \end{bmatrix}. \tag{8.118}$$

Then problem (8.115) is equivalent to,

$$\min_{W_i} \left\| Q_{G,i}^T \hat{Z}_{t-1}^{i-1} - Q_{G,i}^T F_i X_i - \begin{bmatrix} R_{G,i} \\ 0 \end{bmatrix} W_i \Psi_{i+1} \right\|_F^2,$$

and is denoted as,

$$\min_{W_i} \left\| \begin{bmatrix} \hat{Z}^1 \\ \hat{Z}^2 \end{bmatrix} - \begin{bmatrix} F^1 \\ F^2 \end{bmatrix} - \begin{bmatrix} R_{G,i} \\ 0 \end{bmatrix} W_i \Psi_{i+1} \right\|_F^2,$$

which has as solution (see Lemma 8.13),

$$\hat{W}_i^+ = R_{G,i}^{-1} \left(\hat{Z}^1 - F^1 \right) \Psi_{i+1}^\dagger.$$

To compute $\hat{W}_{i+1}^+$ we need to update the QR factorization of the matrix G_{i+1}. This can be done efficiently using compact Householder transformations working on trapezoidal matrices (Golub and Van Loan, 1996).

The *update of the* V_i matrices is also done sequentially. We start with the update of V_N, proceed with the update of V_{N-1} and end with the update of V_2. Meanwhile, the update of the new sequence Ψ_k according to the recursion (8.111) is computed. Here, we consider the update of a single matrix V_i. Suppose that the updates of the matrices $V_N, V_{N-1}, \ldots, V_{i+1}$ have been performed and that the sequence Ψ_i has been computed using W_i^+, V_i^+ for $k = N, N-1, \ldots, i+1$. Analogous to the decomposition in (8.114), the minimization problem can be expressed in terms of an isolated V_i term as follows,

$$\min_{V_i} \left\| Z_{t-1}^{i-1} - \begin{bmatrix} F_{i-1} X_{i-1} \\ 0 \end{bmatrix} - G_i V_i x_i - G_i W_i \Psi_i \right\|_F^2, \tag{8.119}$$

where F_{i-1}, X_{i-1} and G_i, are defined as in Equation (8.114). Using the same QR factorization which was used in the W_i update (8.118), the above is equivalent to,

$$\min_{V_i} \left\| Q_{G,i}^T Z_{t-1}^{i-1} - Q_{G,i}^T \begin{bmatrix} F_{i-1} X_{i-1} \\ 0 \end{bmatrix} - R_{G,i} V_i x_i - R_{G,i} W_i \Psi_i \right\|_F^2$$

Note that $Q_{G,i}^T Z_{t-1}^{i-1}$ was formed during the update of W_i. Having computed V_i, we then compute Ψ_{i-1} and proceed to the update of V_{i-1}.

8.4.3.3 Minimizing over P_i, R_i, Q_i

The updates of the matrices P_i, R_i, Q_i are very much related to the updates of the matrices U_i, W_i, V_i; the main difference is that matrices now form a lower triangular matrix. Due to space constraints the derivation is not given, but summarized. During every iteration the following steps need to be performed:

1. Update the P_i matrices similarly to the update of U_i.
2. Update the R_i matrices similarly to the update of W_i, but starting at $i = N - 1$ and descending to $i = 2$.
3. Update the Q_i matrices similarly to the update of V_i, but starting at $i = 2$, and ascending to $i = N - 1$. Meanwhile, the update of the entries in the sequence Φ_i are performed, according to the recursion,

$$\Phi_{k+1} = R_k \Phi_i + Q_k x_k \quad k = 2, \ldots, \tag{8.120}$$

for $\Phi_1 = 0$.

8.4.3.4 Initialization

As stated before, the previous problem is a non-convex problem. In this section we will develop a means of providing the above algorithm with an initial estimate. The method of providing the initial estimate will make use of the spatial decay in the SSS matrix and is illustrated again for a generic case. We will approximate the SSS matrix by a block tridiagonal matrix with block-bandwidth r,

$$\min_{\substack{A_{i,j} \\ i=1,\ldots,N \\ j=-r,\ldots,r}} \left\| Y - \begin{bmatrix} A_{1,0} & A_{1,1} & \cdots & A_{1,r} & 0 & \cdots & 0 \\ A_{2,-1} & A_{2,0} & & & & & \\ \vdots & & \ddots & \ddots & & & \\ 0 & \cdots & 0 & A_{i,-r} \cdots & A_{i,0} & \cdots A_{i,r} & 0 \\ \vdots & & & & \ddots & \ddots & \\ 0 & & & 0 & A_{N,-r} & \cdots & A_{N,0} \end{bmatrix} X \right\|_F^2.$$

As $r \ll N$, the above can be solved with linear complexity in N. Now suppose that the $A_{i,j}$ terms have been determined. Then we use the $A_{i,0}$ terms to form the generators of D_i. Let the $U_p \Sigma_p V_p$ be the singular value decomposition of the matrix $\begin{bmatrix} A_{i,0} & \cdots & A_{i,r} \end{bmatrix}$, then the first ℓ columns of U_p will be used to construct $\hat{P}_i$. Similarly, let $U_m \Sigma_m V_m = \begin{bmatrix} A_{i,-r} & \cdots & A_{i,-1} \end{bmatrix}$, then the first ℓ column of U_m will be used to form $\hat{U}_i$. Having determined $\hat{U}_i$ and $\hat{P}_i$, and provided that $\begin{bmatrix} \hat{U}_i & \hat{P}_i \end{bmatrix}$ is full column rank, the estimate of the sequence terms $\hat{\Phi}_i$ and $\hat{\Psi}_i$ can be computed as follows:

$$\min_{\Phi_i, \Psi_i} \| Y_i - \hat{D}_i x_i - U_i \Phi_i - P_i \Psi_i \|.$$

Based on the recursive equation (8.120), the terms $\hat{R}_i$ and $\hat{Q}_i$ can be determined from the following series of least-squares problem,

$$\min_{R_i, Q_i} \|\Phi_{k+1} - R_k \Phi_k - Q_k x_k\|_F^2,$$

and similarly $\hat{W}_i$ and $\hat{V}_k$ from,

$$\min_{W_i, V_i} \|\Psi_{k-1} - \hat{W}_k \Psi_k - \hat{V}_k x_k\|_F^2. \tag{8.121}$$

Example 8.6 (Identification of the SSS matrices) To test the algorithm in this subsection, we generated an SSS matrix A through its generator. In this case, the entries of the generator matrices are zero-mean random numbers from a standard normal distribution with the local state dimension equal to five and the dimension of the interconnection variables equal to two. Special care is taken to ensure that the resulting SSS matrix is stable. Then a random white-noise sequence $x(k) = \begin{bmatrix} x_1(k)^T & x_2(k)^T & \cdots & x_N(k)^T \end{bmatrix}^T$ was generated of size such that the product $Ax(k)$ exists. This product defines the sequence $y(k)$. The sequences $x(k)$ and $y(k)$ are then used by the iterative procedure outlined in Section 8.4.3, starting with (other) random initializations for the generator matrices. The results[3] for an increasing number N of local systems are summarized in Figure 8.13. In this figure,

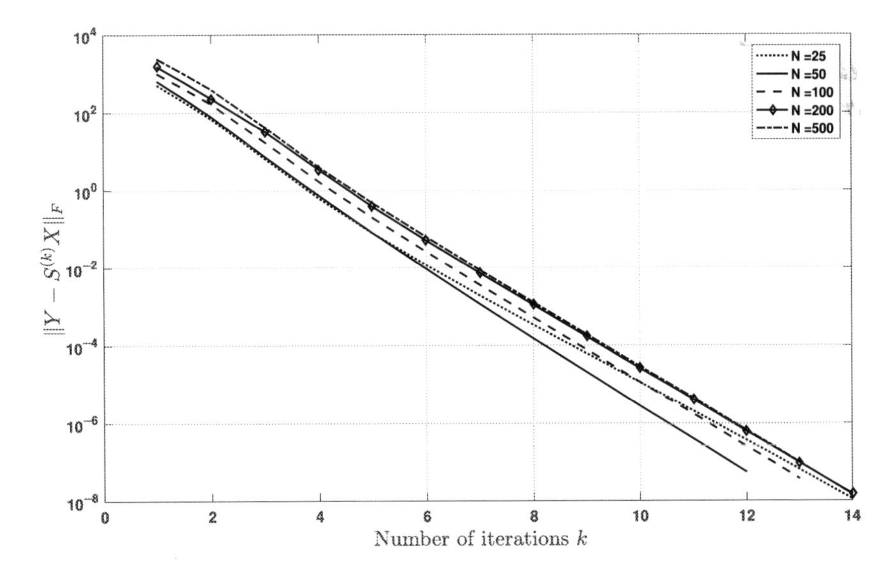

Figure 8.13 The convergence of the multilinear regression solution to estimate an SSS matrix as described in Section 8.4.3 for the SSS matrix and data generated in Example 8.6 for the number of local systems N equal to 25, 50, 100, 200, and 500 and the local state order equal to five with the dimension of the local interaction variables equal to two.

[3] The results presented in this example have been made possible by the work of Engineer Wiegert Krijgsman, during the time he was affiliated with the Delft Center for Systems and Control in 2019.

$S^{(k)}$ is the estimated SSS matrix based on the estimated components of its generator at iteration k. The convergence curves in Figure 8.13 show that independently of the size of the network, linear convergence results are obtained (in a logarithmic plot). The convergence is also monotonic.

8.5 Conclusions

In this chapter we have presented a number of subspace identification methods to identify local system matrices characterizing a single local dynamic systems in a global network using only local input-output measurements. These algorithms use partial topological information as defined in Definition 4.2. The subspace methods not only identify the size of the local system matrices (including the local state dimension) but also their values.

The main class of systems taken as the driver for development were the block tri-diagonal systems. But further extensions to other classes of systems are possible based on the methods derived. These include SSS systems as shown in Section 8.4.2. Further extensions are possible. These extensions could, for example, be based on properties such as strong observability analysed for networks in Kibangou et al. (2016).

9 Estimating Structured State-Space Models

9.1 Introduction

This chapter focuses on the estimation of parameters in parametrized state-space models. Such parametrized models may result from first principle information about dynamic (networks of) systems, like that presented in Example 4.1. We start by briefly reviewing classical prediction error methods and expectation maximization that can be used for estimating parameters in these models and that allow us to include regularization to enforce sparsity in the vector of parameters. Then a constrained subspace method for the identification of structured state-space models (COSMOS) is presented. This new parameter estimation method tries to combine ideas from subspace identification with prediction error methods. It leads to a difference of convex programming problems and has been demonstrated, in general, to have improved convergence properties compared to the considered alternatives. It can also be used for a parametrized network of systems, as will be illustrated in Section 9.6.

By stacking all the unknowns of the system matrices of the parametrized (structured) network model (8.10) into a parameter vector θ, it can be abstractly written as,

$$\begin{aligned} x(k+1) &= A(\theta)x(k) + B(\theta)u(k), \\ y(k) &= C(\theta)x(k) + e(k), \end{aligned} \tag{9.1}$$

where $x(k) \in \mathbb{R}^n, u(k) \in \mathbb{R}^m, y(k), e(k) \in \mathbb{R}^p$ are, respectively, the system state, input, output and measurement noise. The parameter vector $\theta \in \mathbb{R}^d$ contains the entries to be estimated of the system matrices operating in a network. Then, the corresponding network identification becomes the identification of a parametrized state-space model. This will be the point of investigation in this chapter.

Although extensions to innovation models are possible, for the sake of brevity in this chapter we mostly restrict ourselves to state-space models of the form (9.1). It is noted that the parametrized (structured) state-space model is not only restricted to the 1D network system investigated in Chapter 8, but can also model networks of any interconnection patterns. In this chapter, it assumed that the system matrices (9.1) are affine in θ and given as

$$A(\theta) = A_0 + \sum_{j=1}^{d} A_i \theta(i), \quad B(\theta) = B_0 + \sum_{j=1}^{d} B_i \theta(i),$$

$$C(\theta) = C_0 + \sum_{j=1}^{d} C_i \theta(i), \tag{9.2}$$

where the coefficient matrices $\{A_i, B_i, C_i\}_{i=0}^{d}$ define a known matrix basis. The above affinely parametrized structure can be used to represent various interconnection patterns of networked systems, such as circular patterns (Massioni, 2015) and general topologies (Yu and Verhaegen, 2017).

The organization of this chapter is as follows. The parametrization of the state-space model and the identifiability of the model parameters are briefly reviewed in Section 9.2. Then, Section 9.3 reviews the use of the Maximum Likelihood approach for estimating θ, while Section 9.4 reviews the use of the well-known expectation maximization method for that purpose. The new method, based on difference of convex programming, is presented in Section 9.5. Here, a number of illustrative examples are included to illustrate the improved convergence properties. Finally, Section 9.6 presents some extensions that apply the COSMOS method to the identification of parametrized networks.

9.2 Parametrizing the Model

The output prediction of Equation (9.1) can be parametrized as

$$x(k+1;\theta) = A(\theta)x(k;\theta) + B(\theta)u(k),$$
$$\hat{y}(k;\theta) = C(\theta)x(k;\theta), \tag{9.3}$$

where the parameter vector $\theta \in \mathbb{R}^d$ contains the entries to be estimated of the system matrices, the initial state conditions and information about the measurement noise $e(k)$ to define the maximum likelihood cost function. Finding such a generic parametrization is far from trivial.

One key element in selecting a parametrization is the parameter identifiability. The topic of identifiability of parameters has been discussed in e.g. van den Hof (2002) where it is called structural identifiability. This is different from the notion of network identifiability considered in Gevers et al. (2019). For estimating parameters, structural identifiability is of interest. This topic is defined next.

DEFINITION 9.1 (Structural Identifiability of (9.3)) The parametrization of the state-space model (9.3) is (locally) structurally identifiable (in θ_m) for a given input $u(k)$ and initial condition $x(0;\theta_m)$, if for all θ_1, θ_2 (in a neighbourhood of θ_m) the following holds:

$$\forall\, k : \hat{y}(k;\theta_1) = \hat{y}(k;\theta_2) \Rightarrow \theta_1 = \theta_2.$$

REMARK 9.1 Special parametrizations are under study such that it results in network identifiability. One such example is the parametrization of a special case of an innovation model in Yue et al. (2018c). This model considers the output matrix of the innovation model in Equation (2.14) to be of the form:

$$C = \begin{bmatrix} I & 0 \end{bmatrix},$$

and considers the covariance matrix of the innovation signal $e(k)$ in Equation (2.14) to be a scalar multiple of the identity matrix. Further, the pair (A, B) is parametrized such that a DSF derived from this parametrized state-space model is (network) identifiable in the sense considered by Gevers et al. (2019).

9.3 Maximum Likelihood Method

A standard and widely used framework in system identification, in general, and also used in the identification of NDS is the maximum likelihood method. This method is briefly reviewed in this subsection, based on the approach in Wills et al. (2018).

Let the joint input-output data $\{u(k), y(k)\}_{k=1}^{N_t}$ be denoted by Y^{N_t} and let $p_\theta(Z, Y^{N_t})$ denote the joint probability density function of both the observed variables Y^{N_t} and the so-called latent variables $Z = \{x(k)\}_{k=1}^{N_t+1}$, then the *joint log likelihood function* reads,

$$L_\theta(Z, Y^{N_t}) = \log p_\theta(Z, Y^{N_t}). \tag{9.4}$$

The aim of maximum likelihood estimation is to estimate both θ and the state sequence Z that maximizes this joint log likelihood function. It is, however, common practice to work with the negative of this function denoted as,

$$V(Z, \theta, Y^{N_t}) = -\log p_\theta(Z, Y^{N_t}). \tag{9.5}$$

With this notation, the maximum likelihood estimate is formulated abstractly and stated as,

$$\begin{bmatrix} \hat{\theta}^{\mathrm{ML}} \\ \hat{Z} \end{bmatrix} = \arg\min_{\theta, Z} V(Z, \theta, Y^{N_t}). \tag{9.6}$$

For the special output error state-space model (9.3), the perturbations $e(k)$ are zero-mean, white and have identity covariance matrix, then the negative log likelihood function (NLLF) takes the following form (Ljung (1999), chapter 7, section 4] (neglecting constants that do not affect the optimization):

$$\begin{aligned} V(Z, \theta, Y^{N_t}) &= \frac{1}{N_t} \sum_{k=1}^{N_t} \|y(k) - \hat{y}(k; \theta)\|_F^2, \\ \text{such that } \; x(k+1; \theta) &= A(\theta)x(k; \theta) + B(\theta)u(k), \\ \hat{y}(k; \theta) &= C(\theta)x(k; \theta) \quad \text{for } k = 1 : N_t. \end{aligned} \tag{9.7}$$

The unknown state sequence in constraint (9.7) only depends on the initial state vector $x(1)$ in the following way:

$$x(k;\theta) = A(\theta)^{k-1}x(1) + \sum_{j=1}^{k-1} A(\theta)^{j-1}B(\theta)u(k-j). \tag{9.8}$$

The cost function $V\left(Z,\theta,Y^{N_t}\right)$ does not depend on the full state sequence but only on the initial condition $x(1)$. With the expression for the state in Equation (9.8), the output $\hat{y}(k;\theta)$ becomes,

$$\hat{y}(k;\theta) = C(\theta)A(\theta)^{k-1}x(1) + \sum_{j=1}^{k-1} C(\theta)A(\theta)^{j-1}B(\theta)u(k-j).$$

By this expression, $V\left(x(1),\theta,Y^{N_t}\right)$ reduces to,

$$V\left(x(1),\theta,Y^{N_t}\right)$$
$$= \frac{1}{N_t}\sum_{k=1}^{N_t}\left\| y(k) - C(\theta)A(\theta)^{k-1}x(1) + \sum_{j=1}^{k-1} C(\theta)A(\theta)^{j-1}B(\theta)u(k-j)\right\|_F^2$$

Having an estimate of, or knowing, θ shows that this cost function is a linear least-squares cost function in $x(1)$. However, the dependency on θ is much more complex and in the case of the affine parametrization in system (9.2) becomes a high-order polynomial in the unknown parameter vector θ. This may result in this cost function having several local minima, causing the performance of gradient based optimization methods (Verhaegen and Verdult, 2007) to highly depend on the initial estimates of the parameter vector θ. When no, or only poor information on this initial estimate is available, the chances of these maximum likelihood optimization methods or related prediction error methods finding the global optimum is very small for parameter vectors of realistic size (Ljung and Parrilo, 2003).

REMARK 9.2 (Regularization) Problems with the identifiability of the parameter vector θ may be relaxed by the use of regularization. Such regularizations have been discussed in the identification of Sparse VAR(X) models in Section 6.1. It would correspond to modifying the cost function as:

$$V\left(x(1),\theta,Y^{N_t}\right) + \gamma g(\theta),$$

with γ an additional tuning parameter to be selected and the function $g(.)$ a function that may induce sparsity (or zeros) in θ (see Wen et al. (2018)). In a statistical setting, the term $\gamma g(\theta)$ may be replaced by the following Gaussian prior to inducing sparsity,

$$\frac{1}{(2\pi)^{\frac{d}{2}}}\frac{1}{\sqrt{\det(\Lambda)}}\exp^{-\frac{1}{2}\theta^T\Lambda^{-1}\theta},$$

with $\Lambda = \mathrm{diag}(\Lambda)$ an additional (hyper-)parameter to be estimated.

9.4 Expectation Maximization

In the context of expectation maximization (EM), a number of parameters are usually added to the parameter vector θ with respect to the maximum likelihood criterium in Equation (9.6). For that purpose, we augment the parameter vector θ with a parameter vector η, containing the unknowns of the distribution of $x(1)$ and the unknowns of the noise $e(k)$ in Equation (9.1), as defined later on explicitly on 4 on page 195. This augmented parameter vector β is then defined as:

$$\beta = \begin{bmatrix} \eta \\ \theta \end{bmatrix}. \tag{9.9}$$

The aim of expectation maximization is to separate the problem of jointly estimating this parameter vector β simultanously with the latent variables Z in Equation (9.6) (Dempster et al., 1977).

To induce a separation, use is made of the conditional probability density function of the latent variable conditioned on the measured data Y^{N_t}. As $x(1)$ is the only latent variable in the case of the identification of the parametrized output error state-space models, this conditional density function is denoted as:

$$p_\beta\big(x(1)|Y^{N_t}\big) = \frac{p_\beta\big(x(1), Y^{N_t}\big)}{p_\beta(Y^{N_t})}, \tag{9.10}$$

and the log likelihood function $\log p_\beta(Y^{N_t})$ becomes,

$$\log p_\beta\big(Y^{N_t}\big) = \log p_\beta\big(x(1), Y^{N_t}\big) - \log p_\beta\big(x(1)|Y^{N_t}\big).$$

Let $\hat{\beta}^{(\ell)}$ be an estimate of β (at some iteration ℓ of an optimization method), then a separation between the estimation of $x(1)$ and β results when we take the conditional expectation as follows:

$$\begin{aligned} E_{\hat{\beta}^{(\ell)}}\Big[\log p_\beta\big(Y^{N_t}\big)\Big] &= \int \log p_\beta\big(Y^{N_t}\big) p_{\hat{\beta}^{(\ell)}}\big(x(1)|Y^{N_t}\big)dx(1) \\ &= \int \log p_\beta\big(x(1), Y^{N_t}\big) p_{\hat{\beta}^{(\ell)}}\big(x(1)|Y^{N_t}\big)dx(1) \\ &\quad - \int \log p_\beta\big(x(1)|Y^{N_t}\big) p_{\hat{\beta}^{(\ell)}}\big(x(1)|Y^{N_t}\big)dx(1) \\ &= E_{\hat{\beta}^{(\ell)}}\Big[\log p_\beta\big(x(1), Y^{N_t}\big)|Y^{N_t}\Big] \\ &\quad - E_{\hat{\beta}^{(\ell)}}\Big[\log p_\beta\big(x(1)|Y^{N_t}\big)|Y^{N_t}\Big] \\ &= \mathcal{Q}\big(\beta, \hat{\beta}^{(\ell)}\big) - \mathcal{V}\big(\beta, \hat{\beta}^{(\ell)}\big), \end{aligned} \tag{9.11}$$

where the last line in this array of equations introduces the functions $\mathcal{Q}(.,.)$ and $\mathcal{V}(.,.)$.

It can be shown, using the definition of $\mathcal{V}(.)$, that this negative logarithm is a convex function and using Jensen's inequality (Rudin, 1986) that,

$$\mathcal{V}(\hat{\beta}^{(\ell)}, \hat{\beta}^{(\ell)}) - \mathcal{V}(\beta, \hat{\beta}^{(\ell)}) \geq 0.$$

Therefore, it is sufficient to ensure that $\mathcal{Q}(\beta, \hat{\beta}^{(\ell)})$ is larger or equal to $\mathcal{Q}(\hat{\beta}^{(\ell)}, \hat{\beta}^{(\ell)})$ in order to make the log likelihood function at β larger than that at $\hat{\beta}^{(\ell)}$. Hence maximizing the log likelihood function boils down to first, the calculation of the conditional expectation $\mathcal{Q}(\beta, \hat{\beta}^{(\ell)})$ for a given estimate $\hat{\beta}^{(\ell)}$ and then a maximization of this expectation with respect to β. This is conceptually summarized in Algorithm 9.1. The evaluation of the function $\mathcal{Q}(\beta, \hat{\beta}^{(\ell)})$ usually requires a Kalman filter step (or

Algorithm 9.1: Conceptual summary of the EM algorithm to estimate the parameter vector β in a mathematical model that defines $p_\beta(Z, Y^{N_t})$ where $Z = x(1)$ is the latent variable and Y^{N_t} are the measured variables.

Input : $\ell = 1$, `tol` and initialize $\hat{\beta}^{(1)}$ such that $L_{\beta^{(1)}}(Y^{N_t})$ is finite

Output: $\hat{\beta}^{(\ell+1)}$

`/* Expectation (E) Step: */`

1 Compute,

2 $\mathcal{Q}(\beta, \hat{\beta}^{(\ell)}) = E_{\hat{\beta}^{(\ell)}}\left[\log p_\beta \left(x(1), Y^{N_t}\right) | Y^{N_t}\right]$

`/* Maximization (M) Step: */`

3 $\hat{\beta}^{(\ell+1)} = \arg\max_\beta \mathcal{Q}(\beta, \hat{\beta}^{(\ell)})$

`/* Check stopping criterion: */`

4 $\left| \log p_{\hat{\beta}^{(\ell+1)}}(Y^{N_t}) - \log p_{\hat{\beta}^{(\ell)}}(Y^{N_t}) \right| \leq$ `tol`

particle filter or related state reconstructor). For the output-error state-space model (8.10) this has been done in Wills et al. (2018) and is summarized next. For this purpose, let the initial state $x(1)$ be Gaussian distributed with mean μ and covariance matrix P_1 and let $e(k)$ be zero-mean Gaussian distributed white noise with covariance matrix R, then $\mathcal{Q}(\beta, \hat{\beta}^{(\ell)})$ can be given as in Wills et al. (2018),

$$\begin{aligned} \mathcal{Q}(\beta, \hat{\beta}^{(\ell)}) = & - \log \det P_1 - N_t \log \det R \\ & - \operatorname{Tr}\left(P_1^{-1} \left((\hat{x}_{1|N_t} - \mu)(\hat{x}_{1|N_t} - \mu)^T + P_{1|N_t} \right) \right) \\ & - \operatorname{Tr}\left(R^{-1} \sum_{k=1}^{N_t} \epsilon(k)\epsilon(k)^T \right) - \operatorname{Tr}\left(R^{-1} \sum_{k=1}^{N_t} C(\beta) P_k C(\beta))^T \right), \end{aligned} \tag{9.12}$$

where the following notation has been used,

$$\hat{x}_{1|N_t} := E_{\hat{\beta}^{(\ell)}}\left[x(1) | Y^{N_t} \right], \tag{9.13}$$

$$P_{1|N} := E_{\hat{\beta}^{(\ell)}}\left[(x(1) - \hat{x}_{1|N_t})(x(1) - \hat{x}_{1|N_t})^T | Y^{N_t} \right], \tag{9.14}$$

$$\epsilon(k) = y(k) - \hat{y}(k|k-1), \tag{9.15}$$

$$\hat{y}(k|k-1) = E_{\hat{\beta}^{(\ell)}}\left[y(k)|Y^{N_t-1}\right], \tag{9.16}$$

$$P_k = E_{\hat{\beta}^{(\ell)}}\left[(x(k) - \hat{x}(k|k-1))(x(k) - \hat{x}(k|k-1))^T|Y^{N_t}\right]. \tag{9.17}$$

For the case of a linear output error model, the quantities in (9.13)–(9.14) can be computed with a Kalman smoother (Kailath et al., 2000), while the other quantities (9.15)–(9.17) only require a standard recursive Kalman filter (Kailath et al., 2000).

The computation of the function $\mathcal{Q}(\beta, \hat{\beta}^{(\ell)})$ has introduced some other parameters beside θ. These additional parameters are collected in the vector η and comprise the parameters (μ, P_0, R).

The M-step to optimize $\mathcal{Q}(\beta, \hat{\beta}^{(\ell)})$ can be done in two parts. The first part with respect to the elements of η that yield an analytic solution and second with respect to θ that generally has to be done numerically by e.g. a gradient search.

The optimization of Equation (9.12) with respect to μ of η simply results in:

$$\mu = \hat{x}_{1|N_t}.$$

Substituting this expression in Equation (9.12), the maximization with respect to P_0 yields,

$$P_1 = P_{1|N_t}.$$

By an analogous argument we obtain,

$$R = \frac{1}{N_t}\sum_{k=1}^{N_t}\left(\epsilon(k)\epsilon(k)^T + CP_kC^T\right).$$

Substituting the above estimates for (μ, P_1, R) in Equation (9.12) yields the reduced form of the Q-function of the EM algorithm given as,

$$Q^{\text{red}}\left(\theta, \hat{\beta}^{(\ell)}\right) = -\log\det\left(\frac{1}{N_t}\sum_{k=1}^{N_t}\left(\epsilon(k)\epsilon(k)^T + CP_kC^T\right)\right). \tag{9.18}$$

In general, no analytic solution exists for optimizing this cost function with respect to θ and one has to make use of gradient search methods to numerically perform the optimization. See e.g. Wills et al. (2018).

REMARK 9.3 In the M-step of the EM algorithm, additional regularization can be included. For example, by using the Gaussian prior to inducing sparsity as stated in Remark 9.2, the iterate of the M-step could conceptually be denoted as:

$$\hat{\beta}^{(\ell+1)} = \arg\max_{\beta} Q^{\text{red}}\left(\theta, \hat{\beta}^{(\ell)}\right) + \log\left(\frac{1}{(2\pi)^{\frac{d}{2}}}\frac{1}{\sqrt{\det(\Lambda)}}\exp^{-\frac{1}{2}\theta^T\Lambda^{-1}\theta}\right).$$

As this regularization further introduces the hyperparameters λ, a separate E-step is required prior to solving the above maximization problem. This approach is followed in Yue et al. (2018c).

9.5 Difference of Convex Programming

To deal with the identification problem for the structured state-space model (9.1) with affine parametrization, a subspace inspired identification framework will be presented that poses constraints on a linear regression problem with the constraints being formulated in terms of Markov parameters and the system parameters to be determined. Due to the non-convex property of the rank-constrained optimization problem, it is transformed into a difference-of-convex optimization problem that is then handled by the sequential convex programming strategy.

The standard assumptions for the identification problem are given below.

ASSUMPTION 9.1 (For addressing the rank-constrained optimization problem (9.39))

1. *The system represented by the triplet* $(A(\theta), B(\theta), C(\theta))$ *is minimal.*
2. *The input vector* $u(k)$ *is ergodic and persistently exciting of any order.*
3. *The measurement noise* $e(k)$ *is white and uncorrelated with* $u(k)$.

PROBLEM 9.1 According to the affinely parametrized system structure $\{A(\theta), B(\theta), C(\theta)\}$, and assuming Assumptions 9.1 hold, *the problem of interest* is to identify the system parameters in θ from the input-output observations $\{u(k), y(k)\}$.

REMARK 9.4 Due to the possible non-unique realization of the state-space model, we are going to find one realization which can well describe the input-output mapping.

9.5.1 General Algorithmic Strategy

Using the available input-output data, we start with postulating the linear regression problem to estimate the Markov parameters of system (9.1) in a subspace identification context. Then two key ideas from subspace identification are used to formulate a rank-constrained least-squares problem to address Problem 9.1. First, the fact that the Hankel matrix constructed from the Markov parameters is the product of the (extended) observability and controllability matrix. Second, the latter matrices bear a special shift structure from which the system matrix $A(\theta)$ can be estimated. Both pieces of information are cast into a rank constraint to the least-squares cost function. One way to solve this rank-constrained optimization problem via difference of convex programming.

The Algorithmic outline is based on (Yu et al., 2020).

9.5.2 Identification of Markov Parameters

When the state-space model is fully parametrized, its Markov parameters can be obtained by solving a structured least-squares problem (Peternell et al., 1996). This identification approach will be reviewed and the associated consistency will be analyzed. As the block-Hankel matrix constructed by Markov parameters can be expressed as the product of an extended observability matrix and an extended controllability matrix, it has a low rank which equals the system order. By exploiting this property, we propose a rank-constrained optimization solution, which is instrumental in developing the COSMOS method.

Analogous to the data equation (8.47) for a single system (therefore removing the top index i), the corresponding data equation for the state-space model (9.1) can be written as

$$\mathbf{Y}_{s,N_t} = \mathcal{O}_s \mathbf{x}_{N_t} + \mathcal{T}_U \mathbf{U}_{s-1,N_t} + \mathbf{E}_{s,N_t}, \tag{9.19}$$

where $\mathbf{Y}_{s,N_t} = \mathcal{H}_{s,N_t}[y(k)]$ and $\mathbf{U}_{s-1,N_t}, \mathbf{E}_{s,N_t}$ are defined similarly. The state sequence is defined as

$$\mathbf{x}_{N_t} = [x(k) \;\; x(k+1) \;\; \cdots \;\; x(k+N_t-1)] .$$

For the data equation (9.19), due to the unknown state-related term $\mathcal{O}_s \mathbf{x}_{N_t}$, the finite-length Markov parameters embedded in the block Toeplitz matrix $\mathcal{T}_U$ cannot be directly estimated from the data equation. Inspired by the blind subspace identification approach (Scobee et al., 2015), if the basis of the unknown state sequence $\mathbf{x}_{N_t}$ can be estimated beforehand, the finite-length Markov parameters can be estimated by the least-squares method. In the following we will first remove the row space of x_{N_t} from the data equation (9.20), prior to the estimation of the Markov parameters.

In view of the spirit of the N4SID method (Van Overschee and De Moor, 1994) and the PO-MOESP method (Verhaegen and Verdult, 2007), it can be derived that the space of the current state is subsumed in the past system observations or the future system observations. To show this, the data equation of the past system observations is given as :

$$\mathbf{Y}^p_{s,N_t} = \mathcal{O}_s \mathbf{x}^p_{N_t} + \mathcal{T}_U \mathbf{U}^p_{s-1,N_t} + \mathbf{E}^p_{s,N_t}, \tag{9.20}$$

where the superscript ^{p} denotes 'past', the block-Hankel matrix $\mathbf{Y}^p_{s,N_t}$ equals $\mathcal{H}_{s,N_t}[y(k-s+1)]$ and $\mathbf{U}^p_{s-1,N_t}, \mathbf{E}^p_{s,N_t}$ have the same form as $\mathbf{Y}^p_{s,N_t}$, and the state sequence $\mathbf{x}^p_{N_t}$ is defined as

$$\mathbf{x}^p_{N_t} = [x(k-s+1) \;\; x(k-s+2) \;\; \cdots \;\; x(k-s+N_t)] .$$

When the observability matrix $\mathcal{O}_s$ in Equation (9.20) has full column rank, the state sequence $\mathbf{x}^p_{N_t}$ can be represented as:

$$\mathbf{x}^p_{N_t} = \mathcal{O}_s^\dagger \mathbf{Y}^p_{s,N_t} - \mathcal{O}_s^\dagger \mathcal{T}_U \mathbf{U}^p_{s-1,N_t} - \mathcal{O}_s^\dagger \mathbf{E}^p_{s,N_t}. \tag{9.21}$$

In addition, the state sequence $\mathbf{x}_{N_t}$ can be represented as

$$\mathbf{x}_{N_t} = A^{s-1} \mathbf{x}^p_{N_t} + \mathcal{C}_U \mathbf{U}^p_{s-1,N_t}, \tag{9.22}$$

where $\mathcal{C}_U = \begin{bmatrix} CA^{s-2}B & \cdots & CB \end{bmatrix}$. Substituting the expression for $\mathbf{x}^p_{N_t}$ in Equation (9.21) into Equation (9.22) yields

$$\mathbf{x}_{N_t} = A^{s-1} \mathcal{O}_s^\dagger \mathbf{Y}^p_{s,N_t} + [\mathcal{C}_U - A^{s-1} \mathcal{O}_s^\dagger \mathcal{T}_U] \mathbf{U}^p_{s-1,N_t} - A^{s-1} \mathcal{O}_s^\dagger E^p_{s,N_t}. \tag{9.23}$$

It can be seen from this equation that the current state sequence $\mathbf{x}_{N_t}$ can be represented as a linear combination of the past system observations. As a result, the data equation (9.20) can be rewritten as

$$\begin{aligned}\mathbf{Y}_{s,N_t} = & \underbrace{\mathcal{O}_s A^{s-1} \mathcal{O}_s^{\dagger}}_{\Phi_s} \mathbf{Y}^p_{s,N_t} + \underbrace{\mathcal{O}_s \left[\mathcal{C}_U - A^{s-1} \mathcal{O}_s^{\dagger} \mathcal{T}_U \right]}_{\Psi_s} \mathbf{U}^p_{s-1,N_t} + \mathcal{T}_U \mathbf{U}_{s-1,N_t} \\ & - \underbrace{\mathcal{O}_s A^{s-1} \mathcal{O}_s^{\dagger}}_{\Phi_s} \mathbf{E}^p_{s,N_t} + \mathbf{E}_{s,N_t}.\end{aligned} \tag{9.24}$$

Then, the least-squares solution of the block Toeplitz matrix $\mathcal{T}_U$ constructed by Markov parameters is formulated as (Peternell et al., 1996):

$$\min_{\Phi_s, \Psi_s, \mathcal{T}_U} \left\| \mathbf{Y}_{s,N_t} - \Phi_s \mathbf{Y}^p_{s,N_t} - \Psi_s \mathbf{U}^p_{s-1,N_t} - \mathcal{T}_U \mathbf{U}_{s-1,N_t} \right\|_F^2. \tag{9.25}$$

According to the equivalence between the least-squares optimization problem and the oblique projection operator (Van Overschee and De Moor, 2011), the Toeplitz matrix $\mathcal{T}_U$ in the above optimization problem can be estimated by the oblique projection technique. This will not be elaborated on. Next, the properties of the least-squares solution (9.25) for the Toeplitz matrix $\mathcal{T}_U$ will be discussed.

LEMMA 9.1 *Let Assumption 9.1 hold and the value of s be larger than the observability index of (A, C). Define*

$$\lim_{N_t \to \infty} \frac{1}{N_t} \left\{ \begin{bmatrix} \mathbf{Y}^p_{s,N_t} \\ \mathbf{U}^p_{s-1,N_t} \\ \hline \mathbf{U}_{s-1,N_t} \end{bmatrix} \begin{bmatrix} \mathbf{Y}^p_{s,N_t} \\ \mathbf{U}^p_{s-1,N_t} \\ \hline \mathbf{U}_{s-1,N_t} \end{bmatrix}^T \right\} = \begin{bmatrix} R_{11} & R_{12} \\ R_{21} & R_{22} \end{bmatrix}, \tag{9.26}$$

and

$$\lim_{N_t \to \infty} \frac{1}{N_t} \mathbf{E}_{s,N_t} \mathbf{E}^T_{s,N_t} = R_{ee}. \tag{9.27}$$

Denote by $\mathcal{T}_U^{N_t}$ the least-squares estimate of $\mathcal{T}_U$ in (9.25) that relies on data length N_t. As the value of N_t tends to infinity, it can be derived that

$$\lim_{N_t \to \infty} \mathcal{T}_U^{N_t} = \mathcal{T}_U^* - [\Phi_s^* R_{ee} \;\; 0] \left[R_{11} - R_{12} R_{22}^{-1} R_{21} \right]^{-1} R_{12} R_{22}^{-1}, \tag{9.28}$$

where $\mathcal{T}_U^*$ and Φ_s^* represent the true values of $\mathcal{T}_U$ and Φ_s, respectively.

Proof First, by Assumption 9.1, it can be derived straightforwardly that the following matrix is of full rank (Verhaegen and Verdult, 2007), i.e.

$$\lim_{N_t \to \infty} \frac{1}{N_t} \left\{ \begin{bmatrix} \mathbf{Y}^p_{s,N_t} \\ \mathbf{U}^p_{s-1,N_t} \\ \hline \mathbf{U}_{s-1,N_t} \end{bmatrix} \begin{bmatrix} \mathbf{Y}^p_{s,N_t} \\ \mathbf{U}^p_{s-1,N_t} \\ \hline \mathbf{U}_{s-1,N_t} \end{bmatrix}^T \right\} = \begin{bmatrix} R_{11} & R_{12} \\ R_{21} & R_{22} \end{bmatrix} > 0.$$

As $e(k)$ is a white noise, it can be obtained that

$$\lim_{N_t\to\infty}\frac{1}{N_t}\left\{\mathbf{E}_{s,N_t}\left[\mathbf{E}^{p,T}_{s,N_t}\ \ \mathbf{Y}^{p,T}_{s,N_t}\ \ \mathbf{U}^{p,T}_{s-1,N_t}\ \ \mathbf{U}^{T}_{s-1,N_t}\right]\right\}=0. \tag{9.29}$$

Then, it can be derived from the first-order optimality condition (9.25) that

$$\begin{bmatrix}\Phi_s^{N_t} & \Psi_s^{N_t} & \mathcal{T}_U^{N_t}\end{bmatrix}\begin{bmatrix}\mathbf{Y}^p_{s,N_t}\\ \mathbf{U}^p_{s-1,N_t}\\ \mathbf{U}_{s-1,N_t}\end{bmatrix}\begin{bmatrix}\mathbf{Y}^p_{s,N_t}\\ \mathbf{U}^p_{s-1,N_t}\\ \mathbf{U}_{s-1,N_t}\end{bmatrix}^T=\mathbf{Y}_{s,N_t}\begin{bmatrix}\mathbf{Y}^p_{s,N_t}\\ \mathbf{U}^p_{s-1,N_t}\\ \mathbf{U}_{s-1,N_t}\end{bmatrix}^T. \tag{9.30}$$

Let $\mathcal{T}_U^*,\Phi_s^*,\Psi_s^*$ be, respectively, the true values of $\mathcal{T}_U,\Phi_s,\Psi_s$ satisfying that

$$\mathbf{Y}_{s,N_t}=\Phi_s^*\mathbf{Y}^p_{s,N_t}+\Psi_s^*\mathbf{U}^p_{s-1,N_t}+\mathcal{T}_U^*\mathbf{U}^p_{s-1,N_t}-\Phi_s^*\mathbf{E}^p_{s,N_t}+\mathbf{E}_{s,N_t}.$$

Substituting this expression into Equation (9.30) yields that

$$\begin{aligned}&\begin{bmatrix}\Phi_s^{N_t} & \Psi_s^{N_t} & \mathcal{T}_U^{N_t}\end{bmatrix}\begin{bmatrix}\mathbf{Y}^p_{s,N_t}\\ \mathbf{U}^p_{s-1,N_t}\\ \mathbf{U}_{s-1,N_t}\end{bmatrix}\begin{bmatrix}\mathbf{Y}^p_{s,N_t}\\ \mathbf{U}^p_{s-1,N_t}\\ \mathbf{U}_{s-1,N_t}\end{bmatrix}^T\\ &=\begin{bmatrix}\Phi_s^* & \Psi_s^* & \mathcal{T}_u^*\end{bmatrix}\begin{bmatrix}\mathbf{Y}^p_{s,N_t}\\ \mathbf{U}^p_{s-1,N_t}\\ \mathbf{U}_{s-1,N_t}\end{bmatrix}\begin{bmatrix}\mathbf{Y}^p_{s,N_t}\\ \mathbf{U}^p_{s-1,N_t}\\ \mathbf{U}_{s-1,N_t}\end{bmatrix}^T\\ &\quad+\left[-\Phi_s^*\mathbf{E}^p_{s,N_t}+\mathbf{E}_{s,N_t}\right]\begin{bmatrix}\mathbf{Y}^p_{s,N_t}\\ \mathbf{U}^p_{s-1,N_t}\\ \mathbf{U}_{s-1,N_t}\end{bmatrix}^T.\end{aligned} \tag{9.31}$$

According to the equalities in Equation (9.29) and taking limits on both hand sides of the above equation as $N_t\to\infty$, we can obtain that

$$\begin{aligned}\begin{bmatrix}\Phi_s^\infty & \Psi_s^\infty & \mathcal{T}_U^\infty\end{bmatrix}\begin{bmatrix}R_{11} & R_{12}\\ R_{21} & R_{22}\end{bmatrix}&=\begin{bmatrix}\Phi_s^* & \Psi_s^* & \mathcal{T}_U^*\end{bmatrix}\begin{bmatrix}R_{11} & R_{12}\\ R_{21} & R_{22}\end{bmatrix}\\ &\quad+\begin{bmatrix}-\Phi_s^*R_{ee} & 0 & 0\end{bmatrix}.\end{aligned}$$

It can then be derived that

$$\begin{aligned}\begin{bmatrix}\Phi_s^\infty & \Psi_s^\infty & \mathcal{T}_U^\infty\end{bmatrix}&=\begin{bmatrix}\Phi_s^* & \Psi_s^* & \mathcal{T}_U^*\end{bmatrix}\\ &\quad+\begin{bmatrix}-\Phi_s^*R_{ee} & 0 \mid 0\end{bmatrix}\begin{bmatrix}R_{11} & R_{12}\\ R_{21} & R_{22}\end{bmatrix}^{-1}.\end{aligned} \tag{9.32}$$

Since the matrix $\begin{bmatrix}R_{11} & R_{12}\\ R_{21} & R_{22}\end{bmatrix}$ is positive definite, the estimate of $\mathcal{T}_U$ can be expressed as (Golub and Van Loan, 1996):

$$\mathcal{T}_U^\infty=\mathcal{T}_U^*-[\Phi_s^*R_{ee}\ \ 0]\left[R_{11}-R_{12}R_{22}^{-1}R_{21}\right]^{-1}R_{12}R_{22}^{-1}.$$

This completes the proof of the lemma. □

REMARK 9.5 According to the least-squares solution of $\mathcal{T}_U$ given in Equation (9.28), several special cases shall be discussed and analyzed.

First, by using expression $\Phi_s^* = \mathcal{O}_s A^{s-1} \mathcal{O}_s^\dagger$, it can be derived from Equation (9.28) that $\Phi_s^* \to 0$ as $s \to \infty$. As a result, the last block row of $\mathcal{T}_U^{N_t}$ approaches the last block row of $\mathcal{T}_U^*$; namely the Markov parameters can be consistently estimated, as studied in Peternell et al. (1996).

Second, when the input signal $u(k)$ is a white noise, it is easy to obtain that $\mathbf{U}_{s-1,N_t}$ and $\begin{bmatrix} \mathbf{Y}_{s,N_t}^{p,T} & \mathbf{U}_{s-1,N_t}^{p,T} \end{bmatrix}^T$ are uncorrelated, i.e., $R_{21} = R_{12}^T = 0$. Then, it can be deduced from equation (9.28) that the Markov parameters can be consistently estimated, as studied in Peternell et al. (1996).

Third, when the measurement noise is absent, i.e. $e(k) = 0$, it can be seen from the Lemma 9.1 that the matrix $\mathcal{T}_U$ can be recovered exactly. In other words, the finite-length Markov parameters can be perfectly estimated using noise-free measurements.

REMARK 9.6 In the least-squares solution (9.25), the term $\Phi_s \mathbf{Y}_{s,N_t}^p + \Psi_s \mathbf{U}_{s-1,N_t}^p$ can be regarded as the output predictor and the term $\mathcal{T}_U \mathbf{U}_{s-1,N_t}$ is the associated prediction error. As a result, the provided least-squares solution is inherently a combination of the prediction-error method and the subspace identification approach.

According to the previous analysis, it can be seen that the least-squares solution (9.25) may yield erroneous Markov parameters under the influence of measurement noise. To mitigate the estimation error, the structural properties embedded in the Markov parameters will be explored and a structured least-squares method will be developed.

For the identification of the concerned grey-box model (9.1), it is always assumed that the system order is available. For the sake of notational simplicity, we denote

$$\mathcal{T}_s[M_1, \ldots, M_{s-1}] = \begin{bmatrix} 0 & & \\ M_1 & \ddots & \\ \vdots & \ddots & 0 \\ M_{s-1} & \cdots & M_1 \end{bmatrix},$$
$$\mathcal{H}_{l_1,l_2}[M_1, \ldots, M_{s-1}] = \begin{bmatrix} M_1 & M_2 & \cdots & M_{l_2} \\ M_2 & M_3 & \cdots & M_{l_2+1} \\ \vdots & \vdots & \ddots & \vdots \\ M_{l_1} & M_{l_1+1} & \cdots & M_{s-1} \end{bmatrix}, \tag{9.33}$$

where the dimensions l_1, l_2 and s are chosen such that $l_1 + l_2 = s - 1$ and $l_1, l_2 \geq n$. By exploiting low-rank property of the block-Hankel matrix, a rank-constrained convex optimization problem is given as follows

$$\begin{aligned} \min_{\Phi_s, \Psi_s, \mathcal{T}_U, M_i} \quad & \left\| \mathbf{Y}_{s,N_t} - \Phi_s \mathbf{Y}_{s,N_t}^p - \Psi_s \mathbf{U}_{s-1,N_t}^p - \mathcal{T}_U \mathbf{U}_{s-1,N_t} \right\|_F^2, \\ \text{such that} \quad & \mathcal{T}_U = \mathcal{T}_s \left[M_0 \ \cdots \ M_{s-1} \right], \\ & H_U = \mathcal{H}_{l_1,l_2} \left[M_0 \ \cdots \ M_{s-1} \right], \\ & \text{rank}(H_U) = n. \end{aligned} \tag{9.34}$$

Since the rank-constrained optimization problem cannot be solved within a polynomial time scale, the heuristic optimization approaches such as the nuclear norm method (Fazel et al., 2012) are often adopted. In view of the characteristics of the nuclear norm optimization method, the Markov parameters together with the system order can be simultaneously estimated without any prior knowledge. In Section 9.5.3, a difference-of-convex approach will be proposed to deal with the rank-constrained optimization problem, which can yield much better performance than the nuclear norm optimization method.

9.5.3 A Rank-Constrained Optimization Problem

Here, we aim to identify system parameters such that the parametrized state-space model can be realized from the input and output data. Since the Markov parameters are parametrized by the system parameters, the rank-constrained optimization approach (9.34) will be adopted for the system-parameter estimation.

Besides the low-rank property of the block-Hankel matrix H_U, the corresponding matrix fraction factors, such as the extended observability matrix and the extended controllability matrix, have shifting features (Verhaegen and Verdult, 2007). It is well known that the shifting structure is crucial for the system matrix identification in the classical subspace identification methods, which will be exploited here. To sum up, the structural properties of the block-Hankel matrix include the low-rank property and the shifting structures of its factorization factors, which will be treated as the structural constraints to the optimization problem in (9.34). Due to the non-convex nature of the optimization problem, the iterative optimization method (Yu et al., 2018a) will be adopted.

As analyzed above, the block-Hankel matrix H_U has a *structured* and *low-rank* factorization. The *low-rank* property is explicitly shown as follows:

$$H_U = O_{l_1} C_{l_2}, \tag{9.35}$$

where $O_{l_1} = \begin{bmatrix} C^T & (CA)^T & \cdots & (CA^{l_1-1})^T \end{bmatrix}^T$ and $C_{l_2} = \begin{bmatrix} B & AB & \cdots & A^{l_2-1}B \end{bmatrix}$. When the considered system model is minimal, it can be established that $\text{rank}\,(H_U) = n$.

The *structural* property of the block-Hankel matrix refers to the shifting property of the block entries of the extended observability/controllability matrix. Based on the shifting structure, the parametrized system matrices can be explicitly shown in the optimization problem. For example, the matrices C and B correspond to the first block entries of O_{l_1} and C_{l_2}, respectively; thus, their structural constraints can be straightforwardly written as

$$C = O_{l_1}(1 : p, :), \quad B = C_{l_2}(: , 1 : m). \tag{9.36}$$

The parametrized system matrix A can be expressed by the shifting property of the block entries of O_{l_1} or C_{l_2}. Let

$$\bar{O}_{l_1} = \begin{bmatrix} CA \\ \vdots \\ CA^{l_1+1} \end{bmatrix} \text{ and } \bar{C}_{l_2} = \begin{bmatrix} AB & \cdots & A^{l_2+1}B \end{bmatrix}.$$

Then the system matrix A can be explicitly written as

$$\bar{O}_{l_1} = O_{l_1} A, \quad \bar{C}_{l_2} = AC_{l_2}. \tag{9.37}$$

To sum up, the two-layer structural properties of the block-Hankel matrix H_U are explicitly described in Equations (9.35), (9.36) and (9.37). However, it can be found that the parameters in Equations (9.35) and (9.37) are coupled together, leading the corresponding optimization problem to be challenging. To deal with the bilinear equality constraints, the bilinear equalities are equivalently formulated as a compact rank constraint.

LEMMA 9.2 *The bilinear equalities in Equations* (9.35) *and* (9.37) *hold if and only if there exists a matrix variable* $\bar{A}$ *such that the following rank equality holds:*

$$rank \begin{bmatrix} H_U & O_{l_1} & \bar{O}_{l_1} \\ C_{l_2} & I & A \\ \bar{C}_{l_2} & A & \bar{A} \end{bmatrix} = n. \tag{9.38}$$

Proof Sufficiency: due to the identity block entry, it is easy to see that the second block row (column) has rank n. As a consequence, all the other block rows (columns) can be linearly represented in terms of the second block column (row), and the bilinear equalities in Equations (9.35) and (9.37) can be obtained.

Necessity: when the bilinear equalities hold and the matrix $\bar{A}$ is set to $\bar{A} = A^2$, it is not difficult to prove that all the block rows (columns) can be linearly expressed in terms of the second block row (column). Hence, the rank of the matrix is n. □

Given the system order n, the two-layer structural properties of the block-Hankel matrix is now embedded in the rank constraint (9.38). As a result, the structured system identification problem can be formulated as a rank-constrained optimization problem:

$$\begin{aligned} \min_{\Theta} \quad & \left\| \mathbf{Y}_{s,N_t} - \Phi_s \mathbf{Y}^p_{s,N_t} - \Psi_s \mathbf{U}^p_{s-1,N_t} - \mathcal{T}_U \mathbf{U}_{s-1,N_t} \right\|_F^2, \\ \text{such that} \quad & \mathcal{T}_U = \mathcal{T}_s \begin{bmatrix} M_0 & \cdots & M_{s-1} \end{bmatrix}, \\ & H_U = \mathcal{H}_{l_1,l_2} \begin{bmatrix} M_0 & \cdots & M_{s-1} \end{bmatrix}, \\ & \text{rank} \begin{bmatrix} H_U & O_{l_1} & \bar{O}_{l_1} \\ C_{l_2} & I & A(\theta) \\ \bar{C}_{l_2} & A(\theta) & \bar{A} \end{bmatrix} = n, \\ & C(\theta) = O_{l_1}(1:p,:), \quad B(\theta) = C_{l_2}(:,1:m), \end{aligned} \tag{9.39}$$

where $\Theta = \{\Phi_s, \Psi_s, \mathcal{T}_U, M_i, \theta, O_{l_1}, \bar{O}_{l_1}, C_{l_2}, \bar{C}_{l_2}, \bar{A}\}$ represents the variable set. Apart from the rank constraint, the above optimization problem is convex.

REMARK 9.7 In contrast to the two-step identification method of Yu et al. (2018a), which estimates the Markov parameters and, subsequently, the parametrized system matrices, the identification approach presented here provides an integrated framework where the system parameters, as well as the Markov parameters, can be simultaneously identified from the input and output observations.

9.5.4 A Difference of Convex Programming Solution

To deal with the challenging rank-constrained optimization problem (9.39), a difference of convex programming solution will be provided. For the sake of notational simplicity, let

$$\begin{aligned} f(\Theta) &= \left\| \mathbf{Y}_{s,N_t} - \Phi_s \mathbf{Y}^p_{s,N_t} - \Psi_s \mathbf{U}^p_{s-1,N_t} - \mathcal{T}_U \mathbf{U}_{s-1,N_t} \right\|_F^2, \\ \text{such that} \quad \mathcal{T}_U &= \mathcal{T}_s \begin{bmatrix} M_0 & \cdots & M_{s-1} \end{bmatrix}, \\ H_U &= \mathcal{H}_{l_1,l_2} \begin{bmatrix} M_0 & \cdots & M_{s-1} \end{bmatrix}, \\ C(\theta) &= O_{l_1}(1:p,:), \quad B(\theta) = C_{l_2}(:,1:m), \end{aligned} \tag{9.40}$$

and

$$H(\Theta) = \begin{bmatrix} H_u & O_{l_1} & \bar{O}_{l_1} \\ C_{l_2} & I & A \\ \bar{C}_{l_2} & A & \bar{A} \end{bmatrix}.$$

It can be seen from this equation that $f(\Theta)$ represents the convex part of the optimization problem (9.39) with respect to the parameter set Θ, and the matrix $H(\Theta)$ is linearly parametrized in terms of Θ. Optimization problem (9.39) can be compactly written as

$$\begin{aligned} \min_{\Theta} \quad & f(\Theta), \\ \text{such that} \quad & \operatorname{rank}[H(\Theta)] = n. \end{aligned} \tag{9.41}$$

The rank constraint in (9.41) implies that the singular values of the matrix $H(\Theta)$ except the n largest ones are zero. In addition, we obtain that $\operatorname{rank}(H(\Theta)) \geq n$ since the matrix $H(\Theta)$ contains an identity matrix of size $n \times n$. As a result, the rank constraint $\operatorname{rank}(H(\Theta)) = n$ is equivalent to $\sigma_i(H(\Theta)) = 0$ for all $i > n$. Define the Ky Fan, Schatten or nuclear norm of the matrix $H(\Theta)$ as

$$g_n(H(\Theta)) = \sum_{i=1}^{n} \sigma_i(H(\Theta)). \tag{9.42}$$

The rank constraint in (9.41) is equivalent to

$$\|H(\Theta)\|_* - g_n(H(\Theta)) = 0. \tag{9.43}$$

As a consequence, the rank-constrained optimization problem can be rewritten as

$$\begin{aligned} \min_{\Theta} \quad & f(\Theta), \\ \text{such that} \quad & \|H(\Theta)\|_* - g_n(H(\Theta)) = 0. \end{aligned} \tag{9.44}$$

Since both the nuclear norm and the Ky Fan norm of a matrix are convex functions (Overton and Womersley, 1993), the constraint in (9.44) is a difference of convex (DC) function. In the above formulation, the function $f(\Theta)$ is differentiable with respect to Θ, whereas the function $\|H(\Theta)\|_* - g_n(H(\Theta))$ may not be differentiable (Overton and Womersley, 1993). In the following, we shall use ∇ to represent the gradient (or subgradient) operator for a differentiable (or non-differentiable) function.

For the constrained non-linear programming problem (9.44), the associated Lagrangian function can be written as (Bertsekas, 1999):

$$L(\Theta, \gamma) = f(\Theta) + \gamma\left[\|H(\Theta)\|_* - g_n(H(\Theta))\right], \tag{9.45}$$

where γ is the Lagrange multiplier.

LEMMA 9.3 *Given a local minimum Θ^* to the constrained optimization problem (9.44), if the gradient $\nabla f(\Theta^*) \neq 0$, then there does not exist the Lagrange multiplier γ such that*

$$\nabla f(\Theta^*) + \gamma\left[\nabla\left\|H(\Theta^*)\right\|_* - \nabla g_n(H(\Theta^*))\right] = 0. \tag{9.46}$$

Proof Given the local minimum Θ^*, the constraint in (9.44) should be satisfied, i.e.,

$$\left\|H(\Theta^*)\right\|_* - g_n(H(\Theta^*)) = 0. \tag{9.47}$$

According to the definitions of the nuclear norm and the Ky Fan norm, we can obtain that

$$\|H(\Theta)\|_* - g_n(H(\Theta)) \geq 0, \tag{9.48}$$

for all Θ. As a result, Θ^* is one of the global minima to the function $\|H(\Theta)\|_* - g_n(H(\Theta))$. It then follows that $0 \in \nabla\|H(\Theta^*)\|_* - \nabla g_n(H(\Theta^*))$, where ∇ is the gradient (or subgradient) operator.

If Θ^* satisfies Equation (9.46), it can be established that, for any γ, it has

$$0 \in -\gamma\left[\nabla\left\|H(\Theta^*)\right\|_* - \nabla g_n(H(\Theta^*))\right] = \nabla f(\Theta), \tag{9.49}$$

which contradicts the condition that $\nabla f(\Theta^*) \neq 0$. This completes the proof of the lemma. □

Lemma 9.3 indicates that there exist Lagrangian multipliers for a local minimum if the objective function $f(\Theta)$ and the DC function $\|H(\Theta)\|_* - g_n(H(\Theta))$ have the same local minimum. In practice, this condition is difficult to satisfy. In order to cope with this problem, the penalty approach (Bertsekas, 1999, chapter 3) will be adopted. Here, we approximate the original constrained optimization problem (9.44) by an unconstrained optimization as follows

$$\min_{\Theta} \quad f(\Theta) + \rho\left[\|H(\Theta)\|_* - g_n(H(\Theta))\right], \tag{9.50}$$

where ρ is the regularization parameter to balance the two terms in the objective function. The regularization term $\rho\left[\|H(\Theta)\|_* - g_n(H(\Theta))\right]$, that is always non-negative, represents the violation of the equality constraint $\|H(\Theta)\|_* - g_n(H(\Theta)) = 0$. As a result, when the regularization parameter ρ is large enough, the global minimum of (9.50) will be close to that of (9.44).

To address the difference of convex programming problem (9.50), the sequential convex programming method (Lipp and Boyd, 2016) will be used. For each optimization iteration, it is necessary to convexify the objective function (or to linearize the concave part of the objective function). Given the estimate $\hat{\Theta}^j$ at the jth iteration and the SVD decomposition as follows

$$H(\hat{\Theta}^j) = \begin{bmatrix} U_1^j & U_2^j \end{bmatrix} \begin{bmatrix} \Sigma_1^j & \\ & \Sigma_2^j \end{bmatrix} \begin{bmatrix} V_1^j \\ V_2^j \end{bmatrix}, \tag{9.51}$$

where Σ_1^j is a diagonal matrix consisting of the n largest singular values. When

$$\sigma_n\left(H(\hat{\Theta}^j)\right) > \sigma_{n+1}\left(H(\hat{\Theta}^j)\right),$$

the subgradient of $g_n\,(H(\Theta))$ with respect to $H(\Theta)$ at the point $\Theta = \Theta^j$ can be written as (Qi and Womersley, 1996):

$$U_1^j V_1^{j,T} \in \nabla g_n\left(H(\hat{\Theta}^j)\right). \tag{9.52}$$

Note that, in order to simplify the notation, the above subgradient is calculated with respect to $H(\Theta)$ instead of Θ. This does not affect the validity of the convex-concave procedure for the DC problem. Due to the convexity of the function $g_n(\cdot)$, it has that

$$g_n(H(\Theta)) \geq g_n\left(H(\hat{\Theta}^j)\right) + \left\langle U_1^j V_1^{j,T}, H(\Theta) - H(\hat{\Theta}^j)\right\rangle, \tag{9.53}$$

where j is the iteration index.

The sequential convex programming procedure (Lipp and Boyd, 2016) for the DC problem (9.50) is given as

$$\Theta^{j+1} := \arg\min_{\Theta} \quad f(\Theta) + \rho\left(\|H(\Theta)\|_* - \operatorname{tr}\left[U_1^{j,T} H(\Theta) V_1^j\right]\right). \tag{9.54}$$

It can be verified that the above iterative optimization will result in decreasing objective values along with the increase of the iteration index k (Lipp and Boyd, 2016).

As analyzed previously, the regularization parameter ρ should be large enough that the optimal solution to (9.54) approaches that to (9.44). As a result, the value of ρ is increased with the increase of the iteration index k. In order to avoid the divergence problem that is caused by the increment of ρ, the maximum value is set to ρ_{max} so that the objective function becomes (9.50) after the maximum value of ρ being reached.

As described in Algorithm 9.2, the first iteration inherently solves a nuclear-norm optimization problem, which provides a good initial parameter estimate for the sequential iterations. This can illustrate the better performance of the difference of convex programming method with relation to the nuclear norm method.

Algorithm 9.2: Difference of convex programming method (or COSMOS method). ©2020 IEEE Reprinted, with permission, from Yu et al. (2020)

Input : $U_1^0, V_1^0, \rho_{max}, \mu, \rho, \epsilon, iter_{max}$
Output: Θ

```
/* Default values */
```
1 $U_1^0 = 0, V_1^0 = 0, \rho = 0.1, \mu = 1, \rho_{max} = 1.05, \epsilon = 10^{-6}, iter_{max} = 100$
2 **while** $j \leq iter_{max}$ and $|f(\hat{\Theta}^j) - f(\hat{\Theta}^{j-1})| > \epsilon$ **do**
3 | Obtain the estimate Θ^{j+1} by solving (9.54)
4 | Compute U_1^{j+1} and V_1^{j+1} by the SVD in (9.51)
5 | $\rho \leftarrow \min\{\mu\rho, \rho_{max}\}$ where $\mu \geq 1$ and $\rho_{max} > 0$
6 | $j \leftarrow j + 1$
7 **end**

9.5.5 Analysis of the Algorithm

The proposed COSMOS method that deals with the rank-constrained optimization problem does not require any prior knowledge of the initial parameter estimate, which is produced by solving a nuclear-norm regularized optimization problem. In addition, the sequential relaxing operation enables the parameter estimate to get around local minima so that the global optimal solution can be obtained in most of the simulation trials. Here, in order to show the performance of the developed COSMOS method without the influence of measurement noise, the matrix completion problem is tested, where the nuclear norm method (Recht et al., 2010), the log-det method (Fazel et al., 2003) and the manifold optimization method (Boumal et al., 2013) are simulated together with the COSMOS method for comparison purposes.

For the matrix completion problem, a low-rank matrix of size 10×10 is generated randomly with its entries following the standard Gaussian distribution, where the rank is set to 5. By randomly selecting a number of matrix entries as unknown parameters, the four different methods aim to recover these matrix entries such that the recovered matrix has a fixed rank 5. To test the performance of these four methods, the number of unknown matrix entries varies from 1 to 50.

Let the generated low-rank matrix be denoted by Ω and the index set of known entries be denoted by $\mathcal{D}$. Then, the optimization problem of the nuclear norm method (Recht et al., 2010) is formulated as

$$\begin{aligned} \min_{\Psi \in \mathbb{R}^{10\times 10}} \quad & \|\Psi\|_* \\ \text{such that} \quad & \Psi(i,j) = \Omega(i,j), \quad (i,j) \in \mathcal{D}. \end{aligned}$$

The log-det method (Fazel et al., 2003) is an iterative optimization approach, and the associated optimization problem at the kth iteration is given as

$$[X^{k+1}, Z^{k+1}, \Psi^{k+1}] = \min_{X, Z, \Psi} \quad \mathrm{tr}\left[\left(\begin{bmatrix} X^k & \\ & Z^k \end{bmatrix} + \sigma I\right)^{-1} \begin{bmatrix} X & \\ & Z \end{bmatrix}\right],$$

$$\text{such that} \quad \begin{bmatrix} X & \Psi \\ \Psi^T & Z \end{bmatrix} \geq 0,$$

$$\Psi(i,j) = \Omega(i,j), \quad (i,j) \in \mathcal{D},$$

where the regularization parameter σ is empirically set to $\sigma = 10^{-6}$. The initial parameter estimate is set to $X^0 = 0$ and $Z^0 = 0$, and the stopping criterion is set to

$$\frac{\left\|\Psi^{k+1} - \Psi^k\right\|_F}{\left\|\Psi^k\right\|_F} \leq 10^{-10}.$$

The log-det method will stop at the 100th iteration even if the stopping criterion is not satisfied.

The manifold optimization method (Boumal et al., 2013) aims to solve the original rank-constrained matrix completion problem which is formulated as

$$\min_{\Psi} \quad \|\Psi(\mathcal{D}) - \Omega(\mathcal{D})\|_F^2,$$

$$\text{such that} \quad \mathrm{rank}(\Psi) = 5.$$

The initial parameter is obtained by setting the unknown parameters to zero and computing the best rank-5 approximation of Ω in the sense of the Frobenius norm. The stopping criterion is set to be the same as the log-det method.

For the proposed COSMOS method, the regularization parameter ρ is set to $\rho = 0.1$, and the stopping criterion in terms of the relative estimation error is set to $\epsilon = 10^{-10}$. Also, the COSMOS method will stop at the 100th iteration even if the stopping criterion is not satisfied.

For the same number of unknown matrix entries, the four methods individually perform 100 Monte Carlo trials by randomly generating the low-rank matrix. The performance is evaluated by the success rate, which counts the number of simulation trials whose relative estimation errors are smaller than 10^{-6}, i.e.,

$$\frac{\|\hat{\Psi} - \Omega\|_F}{\|\Omega\|_F} \leq 10^{-6}.$$

Figure 9.1 shows the matrix completion performance of the four methods in terms of the success rates, where we can observe that COSMOS outperforms the other three methods. COSMOS can fulfill the matrix completion problem perfectly when the number of parameters is less than 15. Although the performance of COSMOS degrades when the number of missing entries increases, it is still the best among these four tested methods. The performance comparison in Figure 9.1 can be explained in detail as:

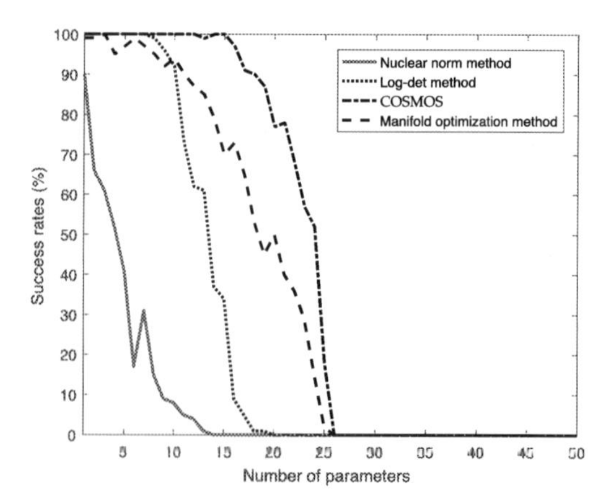

Figure 9.1 Success rates for the nuclear norm method, the log-det method, the manifold optimization method and the COSMOS method. ©2020 IEEE Reprinted, with permission, from Yu et al. (2020)

- The worse performance of the nuclear norm method is because the COSMOS method is carried out by treating the nuclear-norm solution as its initial parameter estimate.
- Both COSMOS and log-det methods rely on iterative optimization. However, COSMOS performs better than the log-det method. For instance, the success rates of COSMOS and the log-det methods are, respectively, 77% and 0% when the number of missing matrix entries is 20. This might be caused by the exact rank information that is made use of in the COSMOS method.
- Both the COSMOS and manifold optimization methods aim to solve the same rank-constrained optimization problem. However, the COSMOS method performs better than the manifold method. This is caused by the sequential relaxing technique of the COSMOS method which can avoid being stuck in local minima.
- COSMOS and manifold optimization methods generally perform better than the other two methods as the number of missing entries is larger than 10. This might be caused by the exact rank information utilized in COSMOS and manifold optimization methods.

From this, it can be concluded that the COSMOS and manifold optimization methods can handle rank-constrained optimization problems better, while the nuclear norm and log-det method can be applied for low-rank optimization problems where the rank is not known as a priori knowledge. As the concerned structured state-space model identification problem is a rank-constrained optimization problem, COSMOS and manifold optimization methods will be tested and compared in the next section.

The numerical examples are simulated on a laptop with a 2.9 GHz processor and 8.0 GB RAM. To deal with the same matrix completion problem with ten missing entries, the associated computation times of four methods are given in Table 9.1, where it can be observed that:

Table 9.1 Comparison of computation times. ©2020 IEEE Reprinted, with permission, from Yu et al. (2020).

Method	Times (s)	No. Iter.	Time/Iter. (s)
Nuclear norm method	0.3437	1	0.3437
Log-det method	1.2290	3	0.4097
COSMOS method	1.3818	3	0.4606
Manifold optimization method	0.0006	12	5×10^{-5}

- COSMOS is more computationally demanding than the other three methods since it needs to solve a nuclear norm optimization problem plus an extra SVD decomposition at each iteration;
- the manifold optimization method is the least computationally demanding as it is inherently a gradient descendent method.

9.5.6 Numerical Simulation Example

Since the proposed COSMOS method can estimate finite-length Markov parameters as well as the system parameters simultaneously, its effectiveness will be demonstrated through two examples: one is for Markov parameter estimation and the other is for parametrized state-space model identification.

9.5.6.1 Markov Parameter Estimation

To address the Markov parameter estimation, the COSMOS method boils down to solving the rank-constrained optimization problem (9.34). The tested state-space model is described by the following system matrices:

$$A = \begin{bmatrix} 1.5610 & -0.6414 \\ 1.0000 & 0 \end{bmatrix}, B = \begin{bmatrix} 1 \\ 0 \end{bmatrix},$$
$$C = \begin{bmatrix} 0.0715 & 0.0072 \end{bmatrix}.$$

The input signal $u(k)$ is generated as a white-noise signal such that it is persistently exciting. The number of input-output data pairs is 300. The first 15 Markov parameters of the state-space model will be identified. To show the identification performance under different levels of noise perturbation, the signal-to-noise ratio adopted is

$$\text{SNR} = 10 \log \frac{\text{var}[y(k) - e(k)]}{\text{var}[e(k)]},$$

and performance is evaluated as the relative estimation error

$$\text{Relative error} = \frac{\sum_{i=1}^{15} \|\hat{M}_i - M_i^*\|_F}{\sum_{i=1}^{15} \|M_i^*\|_F}, \tag{9.55}$$

where $\hat{M}_i$ and M_i^* are the estimated and the true Markov parameters, respectively.

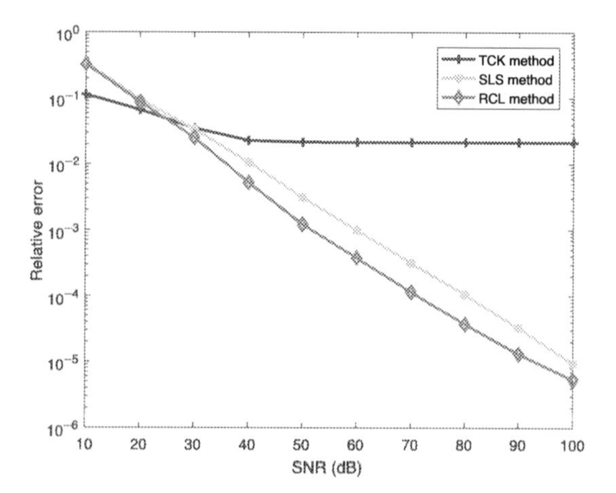

Figure 9.2 Mean relative errors of the 'TC' Kernel (TCK) method, the structured least-squares (SLS) method and the rank-constrained least-squares (RCL) method at different SNRs. ©2020 IEEE Reprinted, with permission, from Yu et al. (2020)

To deal with the Markov parameter estimation problem, the presented rank-constrained least-squares (RCL) method (9.34), the structured least-squares (SLS) estimation method (Mercère et al., 2016) and the 'TC' kernel (TCK)-based FIR model identification method (Pillonetto et al., 2014) will be tested and compared.

The RCL method presented is a sequentially convex programming method, as described in Algorithm 9.2, where the associated parameters are given as

$$\rho = 0.01, \ \mu = 1.02, \ \rho_{max} = 10, \ \epsilon = 10^{-10}.$$

In the simulation, the algorithm stops running at the 100th iteration if the stopping criterion is not satisfied.

For TCK-based FIR model identification method, the associated Matlab commands are given as (Ljung et al., 2015, subsection 3.2):

```
aropt=arxRegulOptions;
aropt.RegulKernel='TC';
[L,R]=arxRegul(data,[0 14 0],aropt);
aopt=arxOptions;
aopt.Regularization.Lambda=L;
aopt.Regularizatoin.R=R;
mest=arx(data,[0 14 0],aopt);
```

The three identification methods are simulated under different noise levels. At each SNR, 30 Monte Carlo trials are carried out to show the distribution of the relative estimation errors, which are plotted in Figure 9.2. In addition, the box plots of the relative estimation errors are provided in Figure 9.3. From these two figures, we can observe that:

1. The estimation errors of the TCK method change slightly with the increase of SNR. This is caused by inappropriate tuning of the involved regularization parameter.

Table 9.2 Computation times of the TCK, SLS and RCL methods. ©2020 IEEE Reprinted, with permission, from Yu et al. (2020).

Method	Times (s)	No. Iter.	Time/Iter. (s)
TCK	0.4672	1	0.4672
SLS	0.5402	1	0.5402
RCL	6.9855	9	0.7762

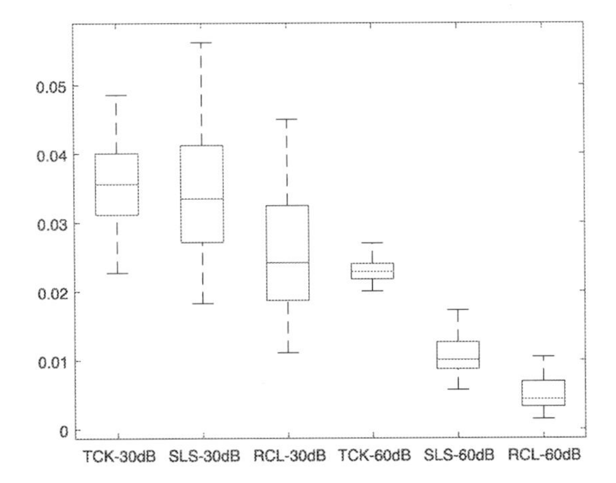

Figure 9.3 Box plots of the TCK method, the SLS method and the RCL method through 50 Monte Carlo trials. ©2020 IEEE Reprinted, with permission, from Yu et al. (2020)

2. The estimation errors of the SLS and RCL methods are quite close to zero when the SNR are large enough, indicating that these two methods can handle the Markov parameter identification perfectly in the absence of measurement noise.
3. The TCK method performs better than the SLS method due to the fact the exact low-rank information of the Markov-parameter stacked Hankel matrix is utilized.

Since the proposed RCL method needs to solve a nuclear norm optimization problem at each iteration, the corresponding computational burden is much heavier than other two methods, as illustrated in Table 9.2.

9.5.6.2 Identification of a Linearly Parametrized State-Space Model

To validate the proposed COSMOS for identifying the parametrized state-space models, a three-compartment model (see Figure 9.4) is considered that can be described by the parametrized system matrices as follows:

$$A(\theta) = \begin{bmatrix} -\theta_1 & \theta_2 & 0 \\ \theta_1 & -(\theta_2+\theta_3) & \theta_4 \\ 0 & \theta_3 & -\theta_4 \end{bmatrix}, \quad B = \begin{bmatrix} 0 \\ 0 \\ 1 \end{bmatrix},$$

$$C = \begin{bmatrix} 0 & 0 & 1 \end{bmatrix}.$$

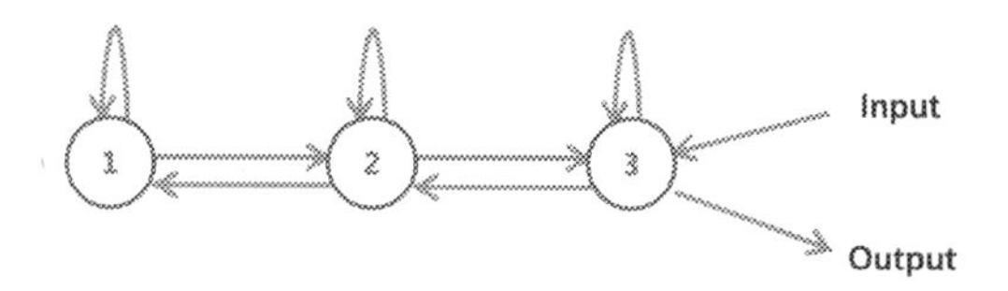

Figure 9.4 Network structure of the simulation example.

According to the identifiability analysis in Bellman and Astrom (1970), the system parameters in this three-compartmental model are identifiable, i.e. they can be uniquely determined from the input and output observations.

In the numerical simulation, the true parameter vector θ is set to

$$\theta = [0.10 \;\; 0.32 \;\; 0.21 \;\; 0.45].$$

The input signal $u(k)$ and measurement noise $e(k)$ are randomly generated following the standard Gaussian distribution. The number of input and output data pairs used for the identification task is set to 50. To show the identification performance of the COSMOS method with relation to the traditional gray box identification methods, the prediction-error method (PEM) and the expectation-maximization (EM) method are tested for comparison purposes. Since the PEM and EM methods rely on the initial parameter estimates, they are randomly produced in the absence of any prior knowledge. The stopping criterion of these three method is the relative estimation error being less than 10^{-10}, and they stop at the 100th iteration if the stopping criterion is not satisfied.

In order to comprehensively evaluate the identification performance, the *normalized estimation error* and the *output relative error* are adopted; these are explicitly defined below.

The *Normalized estimation error* (NEE) is defined as

$$\text{NEE}^r = \frac{\|\hat{\theta}^r - \theta^*\|}{\|\theta^*\|},$$

where $\hat{\theta}^r$ and θ^* are, respectively, the estimated and true values of θ at the rth simulation trial. This criterion can directly demonstrate the estimation accuracy of the system parameters.

The *Output relative error* (ORE) is defined as

$$\text{ORE}^r = \frac{\left\| \mathbf{Y}_{s,N_t} - \Phi_s^r \mathbf{Y}_{s,N_t}^p - \Psi_s^r \mathbf{U}_{s-1,N_t}^p - \mathcal{T}_U^r \mathbf{U}_{s-1,N_t} \right\|_F}{\|\mathbf{Y}_{s,N_t}\|_F},$$

where $\Phi_s^r, \Psi_s^r, \mathcal{T}_U^r$ are the estimates of $\Phi_s, \Psi_s, \mathcal{T}_U$ at the rth simulation trial, respectively. This criterion can demonstrate the estimation accuracy of the input-output mapping, so that it works well whether the parametrized system model is identifiable or not.

The identification performance of the PEM, EM and COSMOS methods in terms of NEE and ORE are shown in the scattering plots; see Figures 9.5 and 9.6. It can be

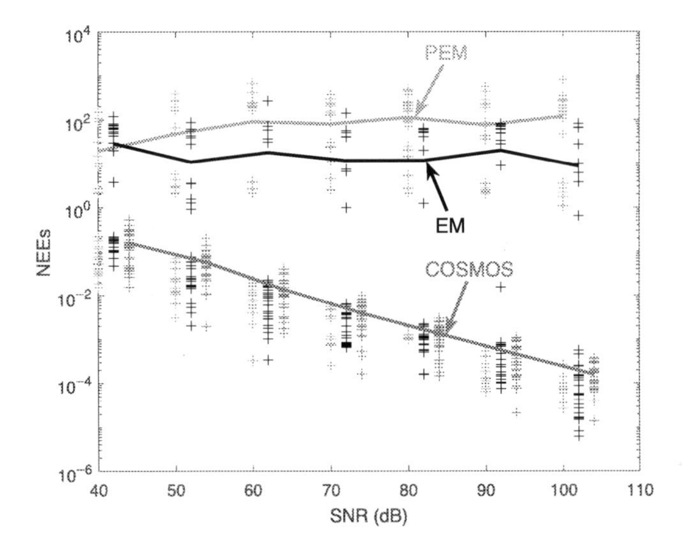

Figure 9.5 Performance of identifying the linearly parametrized state-space model in terms of the NEE criterion. The NEE values are represented by crosses, while their mean values are shown using solid curves. Note that the NEE values of these three methods are computed at the same SNRs; however, they are slightly separated for better comparison. ©2020 IEEE Reprinted, with permission, from Yu et al. (2020)

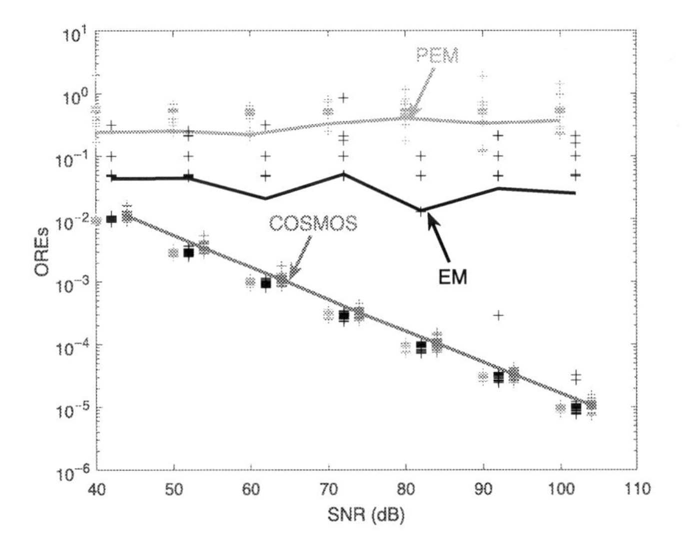

Figure 9.6 Performance of identifying the linearly parametrized state-space model in terms of the ORE criterion. The ORE values are represented by crosses, while their mean values are shown using solid curves. Note that the ORE values of these three methods are computed at the same SNRs; however, they are slightly separated for better comparison. ©2020 IEEE Reprinted, with permission, from Yu et al. (2020)

found that, in contrast to the PEM and EM methods, the COSMOS method does not produce large NEE and ORE values, indicating that the COSMOS method gets stuck at local minima less often than the other two methods. In addition, according to the decaying trends of the NEE and ORE curves with respect to SNR, it can be conjectured that the COSMOS can handle the concerned identification problem perfectly in the absence of measurement noise.

Since the COSMOS method solves nuclear norm optimization problems iteratively, it is more computationally demanding than the PEM and EM methods. To run 100 iterations, the PEM takes about 4.03s, the EM takes about 5.67s, while the COSMOS takes about 94.53s.

9.6 Extensions

The structured state-space models are usually introduced for describing practical physical systems or interconnected network systems. However, many practical state-space models have their system parameters coupled together such that the presented COSMOS cannot work. To deal with this challenge, the COSMOS will be modified slightly to address the polynomially parametrized state-space models. For notational simplicity, we consider a state-space model where the system matrix A is parametrized by a third-order polynomial:

$$A(\theta) = \sum_{i=1}^{l}\sum_{j=1}^{l}\sum_{k=1}^{l} A_{i,j,k}\theta_i\theta_j\theta_k, \tag{9.56}$$

where the matrix coefficients $A_{i,j,k}$ are assumed to be known. The above polynomial parametrization can be rewritten as

$$\begin{aligned} A(\theta) &= \sum_{i=1}^{l}\sum_{j=1}^{l}\sum_{k=1}^{l} A_{i,j,k}\Omega_{i,j,k}, \\ vec(\Omega) &= \theta \otimes \theta \otimes \theta, \end{aligned} \tag{9.57}$$

where Ω denotes a third-order tensor and $vec(\Omega)$ represents the vectorization of the tensor Ω. As a matter of fact, the tensor Ω in Equation (9.57) is a symmetric rank-one tensor, so that it can be equivalently expressed as

$$\begin{aligned} &\Omega_{i,j,k} = \Omega_{i,k,j} = \Omega_{j,i,k} = \cdots = \Omega_{k,j,i}, \quad i,j,k \in \{1,\ldots,l\}, \\ &rank\begin{bmatrix}\Omega(:,1,1) & \cdots & \Omega(:,1,l) & \cdots & \Omega(:,l,l)\end{bmatrix} = 1, \end{aligned} \tag{9.58}$$

where the first equality represents the symmetrical structure of Ω and the second equality represents the rank-one constraint. Therefore, the above third-order polynomial parametrization of system matrices will pose more rank constraints to the COSMOS framework.

When facing the identification problem for large-scale networks, it is computational (or storage) prohibitive to identify the whole network. If we consider the local identification of network clusters, it will introduce the unknown input signals to the local identification problem so that the identification problem becomes a challenging (semi-) blind identification problem. As investigated in Chapter 8, by exploiting the special topological property of the 1D network, the state-evolution equations corresponding to the unknown inputs can be removed without sacrificing the observability of the local network cluster. In addition, when the dimension of the input signal is much less than the dimension of the output signal, a rank minimization formulation can be provided to deal with this identification problem. Moreover, using the strong observability condition is another way to deal with unknown input signals, for which the state sequences are estimated first following the identification of parametrized system matrices. To sum up, the above mentioned approaches enable the identification method presented in this chapter for the local identification of large-scale networks.

9.7 Conclusions

This chapter has looked at an old problem in system identification, namely the estimation of parameters in a given model. Even for linear models, like the standard linear state-space model, the estimation of parameters turns out to be a challenging nonlinear least-squares problem.

Before estimating these parameters it is important to derive the cost function to be optimized with respect to these parameters. For that purpose, two methodologies have been reviewed, namely the maximum likelihood method and expectation minimization. The cost functions of these methods have been derived along with a numerical optimization scheme that is either gradient-based (maximum likelihood) or based on alternating optimization (expectation maximization).

In this chapter a new hybrid method of parameter optimization and subspace identification is presented that is known as COSMOS. This integration resulted in a rank-constrained optimization problem that is solved via difference of convex programming. The numerical experience with this novel model parameter estimation method is that it is less sensitive to the problem of local minima.

Finally, a number of extensions have been presented that demonstrate the applicability of this new methodology for the estimation of parameters in DNS.

Part III

Illustrating with an Application to Adaptive Optics

10 Towards Control of Large-Scale Adaptive Optics Systems

Some of the methods developed in this book are relevant to the application of adaptive optics systems of (extreme) large telescopes. For that reason, a number of results are summarized and presented for that case study. Sections 10.2 and 10.3 provide an introduction to the field of adaptive optics. Results based on laboratory experiments conducted in the Smart Optics lab of the Delft Center for Systems and Control[1] are then presented in Section 10.4. In Section 10.5, the performance of the controllers is evaluated based on telemetry data collected at the William Herschel and Isaac Newton Telescopes in July 2019. The William Herschel and Isaac Newton telescopes on the island of La Palma are part of the the Isaac Newton Group of Telescopes at the Spanish Observatorio del Roque de los Muchachos operated by the Instituto de Astrofí-sica de Canarias. The adaptive optics bench at the William Herschel Telescope (WHT) is developed and maintained by Observatoire de Paris and Durham University (Gendron et al., 2016).

The scientists involved in the third part of this chapter are: *Baptiste Sinquin,*[1] *Léonard Prengère,*[1,2] *Caroline Kulcsàr,*[1] *Henri-François Raynaud,*[1] *Eric Gendron,*[3] *James Osborn,*[4] *Alastair Basden,*[5] *Jean-Marc Conan,*[2] *Nazim Bharmal,*[4] *Lisa Bardou,*[4] *Lazar Staykov,*[4] *Tim Morris,*[4] *Tristan Buey,*[3] *Fanny Chemla,*[6] *Matthieu Cohen*[6]

[1]*Université Paris-Saclay, Institut d'Optique Graduate School, CNRS, Laboratoire Charles Fabry, Palaiseau, France*
[2]*ONERA, DOTA, Université Paris-Saclay, F-92322 Châtillon, France*
[3]*Laboratoire d'Etudes Spatiales et d'Instrumentation en Astrophysique, Observatoire de Paris, Meudon, France*
[4]*Centre for Advanced Instrumentation, Durham University, South Road, Durham, United Kingdom*
[5]*Institute for Computational Cosmology, Durham University, South Road, Durham, United Kingdom*
[6]*Galaxies Etoiles Physique et Instrumentation, Observatoire de Paris, Meudon, France*

[1] www.youtube.com/watch?v=rnfeUuN9xV0

10.1 Introduction

The principle of adaptive optics (AO) is outlined in the schematic drawing in Figure 10.1. The wavefront of light coming from space is planar. However, it reaches the ground distorted after crossing several turbulent layers of atmosphere. An AO system aims to cancel out the wavefront distortions as if there were no atmosphere.

An AO system is typically composed of the following components: a wavefront sensor (WFS), a wavefront corrector element to influence the optical path length differences or phase and a feedback controller. In most systems, like the one depicted in Figure 10.1, the wavefront corrector element is a (set of) deformable mirror(s) (DM).

The light that enters the telescope first reaches the DM that applies a phase correction $\phi_m(r,t)$. The commands sent to this DM are the output of the controller that is fed by closed-loop sensor measurements. Several ways to design the feedback controller, model-based or not, that will be reviewed in this chapter. After applying the wavefront correction, a beam splitter divides the reflected light beam in two parts: one part of the

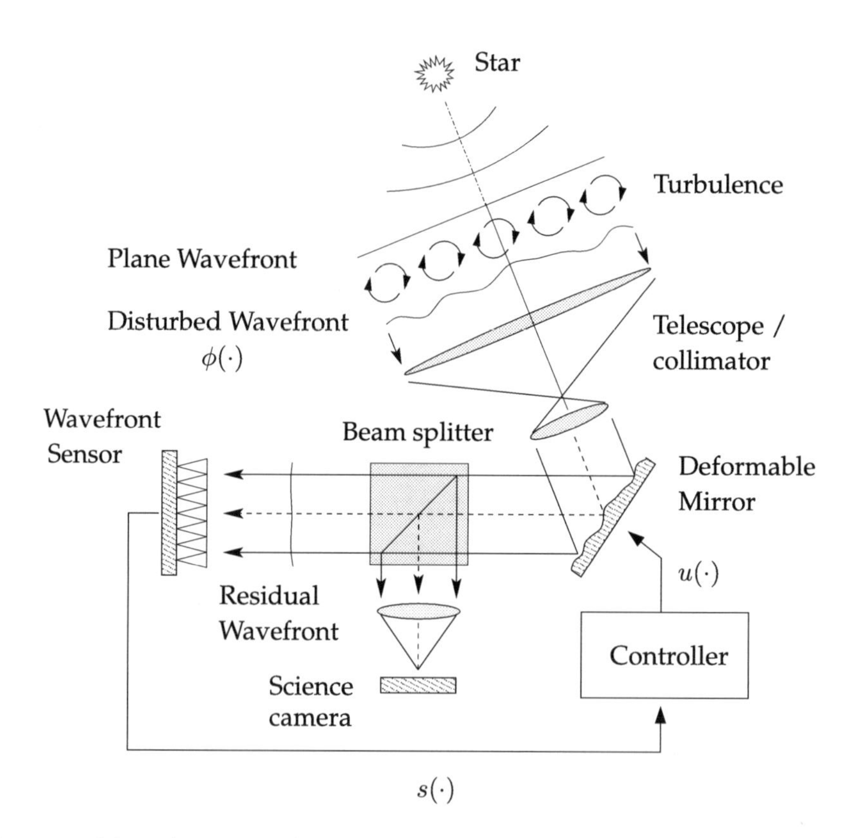

Figure 10.1 Schematic representation of an AO system and its main components: a wavefront sensor (WFS), a wavefront corrector element depicted as a deformable mirror (DM) to influence the optical path length differences or phase and a (feedback) controller. The science camera is used to image the star and should be as diffraction-limited as possible. © 2008 IEEE Reprinted, with permission, from Hinnen et al. (2008)

corrected light beam leaves the AO system and is used by the science camera for imaging the object of interest, the second part is directed to the wavefront sensor. The wavefront sensor provides quantitative information about the residual wavefront that is used by the controller to compute the actuator inputs $u(\cdot)$ to the DM. The performance indicator is the residual phase error, that is the difference between the turbulence induced wavefront and the applied correction, i.e. $\epsilon = \phi - \phi_m$.

AO systems reduce the nefarious effect of atmospheric turbulence on the science image. If the AO system is working properly, the light reaching the science camera should have an almost flat wavefront such that images taken with a long exposure time still have a high resolution. Large ground-based based telescopes are then able to reach close to diffraction limited performance in the near infrared (Roddier, 1999). For an example of the real improvement AO can achieve of a starburst galaxy as recorded with the Canada-France-Hawaii Telescope (CFHT), atop of the Mauna Kea volcano, Hawaii, we refer to `www2.keck.hawaii.edu/realpublic/ao/ngc7469.html`.

In astronomy, a distinction is made between adaptive optics and active optics. Active optics refers to the technique of compensating static and low (temporal) frequency errors in the primary mirror geometry of the telescope itself. For example, compensated error sources include mechanical errors introduced by gravitational sag and wind forces at different telescope inclinations. Active optics is also used to align the different segments in telescopes with a segmented primary mirror. In contrast to adaptive optics, which operates at few hundred Hertz, active optics only requires a fairly low temporal frequency in the order of 0.05 Hertz.

Before introducing control strategies that have been developed for AO, as done in Section 10.3, we will (in Section 10.2) describe a number of important ideas specific to the field.

10.2 Key Terminology for AO

10.2.1 Seeing-Limited and Diffraction-Limited Imaging Systems

Let $\rho \in \mathbb{R}^2$ represent the spatial coordinates (x, y) in a two-dimensional plane orthogonal to the direction of propagation of the light, and j the complex number satisfying $j^2 = -1$. The monochromatic light is modelled by the electromagnetic field,

$$u(\rho, t) = a(\rho)\mathrm{Re}\left(e^{j(wt-\phi(\rho))}\right), \tag{10.1}$$

where $a(\rho)$ is the (real) amplitude, $w = 2\pi\nu$ the temporal pulsation associated to the frequency ν and $\phi(\rho)$ is the phase expressed in radians. The amplitude of the lightwave is generally assumed to be unmodified when propagating through the atmosphere, such that $a(\rho) = a$. The term $U(\rho) = ae^{-j\phi(\rho)}$ is the complex amplitude of the wave. Assuming Fraunhofer diffraction, the intensity I in the focal plane of the telescope is derived from the squared modulus of the Fourier transform of the complex amplitude $U(\rho)$,

$$I(x,y) \propto \left| \iint_{x_0,y_0 \in \Omega} e^{-j\phi(x_0,y_0)} e^{-j\frac{2\pi}{\lambda f}(xx_0+yy_0)} dx_0 dy_0 \right|^2, \tag{10.2}$$

where Ω is a circular aperture of diameter D, λ is the wavelength and f is the focal length of the imaging lens. If the phase $\phi(x_0, y_0)$ is independent of the spatial coordinates, Equation (10.2) reduces to,

$$I(x,y) \propto \left| \iint_{x_0,y_0 \in \Omega} e^{-j\frac{2\pi}{\lambda f}(xx_0+yy_0)} dx_0 dy_0 \right|^2. \tag{10.3}$$

After integrating Equation (10.3), and introducing θ, the angular coordinate in the focal plane, equal to $\sqrt{x^2+y^2}/f$ for small angles, the image of a point source is written with,

$$p_0(\theta) = \frac{\pi D^2}{4\lambda^2} \left(\frac{2J_1(\pi D|\theta|/\lambda)}{\pi D|\theta|/\lambda} \right)^2, \tag{10.4}$$

where J_1 is the Bessel function of the first kind. Equation (10.4) is the point spread function (PSF). It represents a bright spot surrounded by alternating dark and bright rings. This *Airy* pattern is a function of the pupil aperture D of the optical system. The first dark ring occurs at,

$$\sin(\theta) \approx 1.22\frac{\lambda}{D}. \tag{10.5}$$

An imaging system with pupil size D cannot distinguish objects separated with angular distances smaller than θ. Widening the pupil increases the resolving power of the telescope as much as it increases the signal-to-noise ratio by collecting more photons within the pupil.

The wavefront is defined as a surface of equal phase. If the phase is constant on a parabola, then the wavefront is the surface defined from the set of coordinates $(x, y, \frac{2\pi}{\lambda}\delta)$ where $\delta = \sqrt{x^2+y^2}$. The distance δ is denoted in optics as the difference of path length.

In astronomy, the star is assumed infinitely far away and therefore the wavefront is well approximated as a plane before crossing the atmosphere layer, i.e. the phase is independent of the spatial coordinates ρ. It is only when propagating through the atmosphere that optical path length differences are introduced resulting in a distorted wavefront and a PSF that does not resemble the Airy disk. Collecting photons over long exposure times yields a blurred PSF image that has lost much of the resolution that the diameter could have permitted. The imaging system is then said to be seeing-limited rather than diffraction-limited.

10.2.2 Components of an Adaptive Optics System

When a single star is used as a reference, the optical system is said to be single conjugated. Figure 10.1 illustrates a single-conjugate adaptive optics system.

Objects that are of interest to astronomers may not be single stars, but rather extend over a wider field of view. The wavefront should then be corrected over a large field of view. This is achieved using multiple guide stars; this scheme is not detailed further here although some of the identification methods proposed in this book could be studied to make a difference in this scope.

From the key components of an AO system listed in Figure 10.1, we are now going to zoom in briefly on the following ones:

1. The *atmospheric turbulence* that induces wavefront aberrations, as depicted in Figure 10.1.
2. The *deformable mirror*, which is used to compensate for these wavefront aberrations.
3. The *wavefront sensor* that measures a quantity from which the wavefront aberrations can be reconstructed.

These components are discussed next.

10.2.2.1 Atmospheric Turbulence

The wavefront aberrations that are seen on ground-based telescopes are caused by inhomogeneities in the refraction index, mainly due to local variations of the temperature, densities and water vapor content, and that are essentially driven by the wind. The Kolmogorov model is a milestone in modelling these wavefront aberrations. Kolmogorov (1941) proposed modelling the dynamics of the turbulence as large eddies that collapse one onto another into smaller structures that do not sustain but rather dissipate by viscous friction. This is a step toward deriving covariance equations from first principles.

In general, the wavefront of the light travelling through a time-varying and heterogeneous medium is modified. Let z denote the height coordinate along the line of sight, which is the vertical altitude for sources located azimuthally at $z = h$. The phase difference that is created because of crossing the atmosphere is a linear function of the refraction index integrated over the line of sight,

$$\phi(\rho) - \Phi = \frac{2\pi}{\lambda} \underbrace{\int_0^h n(\rho, z) dz}_{\delta(\rho)} \tag{10.6}$$

where Φ is a constant. The phase difference is a function of the wavelength (although this is not the case for the optical path difference). The larger the wavelength, the smaller the wavefront distortions and the larger the diffraction-limited angular resolution, as seen in Section 10.2.1. This explains why a good AO correction is more easily achieved in the infra-red spectrum.

The turbulent atmopshere is modelled mathematically with a linear combination of infinitely thin layers, independent and stationary, each driven by a wind blowing at a specific speed and direction. Each of these layers, or phase screens, is modelled by a zero-mean Gaussian signal with a covariance function depending on physical parameters related to the size of the eddies,

$$C_\phi(r) = \left(\frac{L_0}{r_0}\right)^{5/3} \frac{\alpha}{2} \left(\frac{2\pi r}{L_0}\right)^{5/6} K_{5/6}\left(\frac{2\pi r}{L_0}\right), \tag{10.7}$$

where r is the distance between two phase points $\rho_1 = (x_1, y_1)$ and $\rho_2 = (x_2, y_2)$, $K_{5/6}$ is the modified Bessel function of the third type and α a constant such that:

$$\alpha = \frac{2^{1/6}\Gamma(11/6)}{\pi^{8/3}} \left(\frac{24}{5}\Gamma(6/5)\right)^{5/6}. \tag{10.8}$$

The Fried parameter r_0 and the outer scale L_0 are of particular use in describing how much the wavefront is distorted, and how much the turbulence is spatially correlated. The Fried parameter r_0 is defined as the diameter of a circular area over which the root-mean-square of the wavefront is equal to one radian. The smaller r_0, the stronger the turbulence. The larger the outer scale L_0, the more coupling between far away wavefront values, that is, the denser the wavefront covariance matrix. Realistic values for L_0 and r_0 are 20 m and 5–20 cm, respectively. Equation (10.7) reflects the isotropic property of the turbulence as the spatial statistical properties reduce to a one-dimensional function. For a wavefront discretized on a regular square grid and lifted into a vector, its covariance matrix cannot be approximated by a sparse matrix with good accuracy, although some other properties may be exploited. It is Toeplitz with Toeplitz blocks and has a low-Kronecker rank. Moreover, its inverse is (approximated as) sparse, as has been used for deriving computationally efficient control algorithms (e.g. Correia et al., 2010) that will be reviewed in Sections 10.4 and 10.5.

Such description in Equation (10.7) relies on a zonal representation of the wavefront. A modal representation, however, shows that only few modes represent most of the disturbance. The wavefront root mean square error for Kolmogorov turbulence associated with each of the wavefront modes is, without AO, equal to $1.029(D/r_0)^{5/3}$ compared to $0.134(D/r_0)^{5/3}$ when correcting the tip and tilt modes only (Noll, 1976). The lowest order modes, such as tip and tilt, contribute the most to the turbulence, even when considering vibrations and wind-shake on the mechanical structure of the telescope. Another advantage of this modal decomposition is to reconstruct and correct only for the modes that can be corrected with the DM. However, a drawback for large-scale AO is the large number of modes that need to be corrected without possibilities to exploit sparsity, or spatial invariance, for example. The Karhunen–Loeve basis is an alternative basis for large-scale wavefront control.

10.2.2.2 The Deformable Mirror

Reshaping the distorted wavefront usually requires two deformable mirrors; a tip-tilt with large stroke and another one with many more actuators to correct for the higher spatial frequencies. Membrane mirrors with continuous face sheets are a common choice for this second type of mirror. Each actuator is coupled to its closest neighbours: the interaction is limited to its close neighbourhood and is modelled with an identical two-dimensional Gaussian influence function for each actuator. Typical values of coupling are such that the four closest neighbours of the ith actuator reach 15–30% of the value set on the ith actuator. Mirrors relying on the micro-electromechanical technology have their first resonant frequency much larger than the usual sampling frequencies of the sensor, and settle sufficiently fast to its steady state to neglect its temporal dynamics. This is the assumption for a infinitely fast mirror.

A one-step delay is assumed between the time at which the control inputs are applied and the time at which the wavefront is actually induced. Between two consecutive sampling times, the input is maintained to the same value using a zero-th order hold. When applying $u(k)$ at time instant kT_s, the relationship between the control inputs and the wavefront induced by the mirror only, $\phi_m(t)$, is:

$$\phi_m(t) = Hu(k-1), \quad \text{for all} \quad t \in [kT_s, (k+1)T_s]\,. \tag{10.9}$$

For a wavefront ϕ_m expressed in a zonal basis, the matrix H is sparse if the coupling between the actuators does not exceed 15–20%. If the wavefront is evaluated on a square sampling grid and the influence functions are Gaussians, then the matrix H has a Kronecker rank of one. This matrix was illustrated in Figure 3.7 and was used as a core example to explain the interest in using Kronecker products for modelling 2D dynamics.

The mirror cannot take any arbitrary shape; there remains a residual error between the wavefront that we would like to correct based on what the sensor has seen and the applied correction. The variance of this fitting error is evaluated as follows (Roddier, 1999):

$$\sigma^2_{\text{fit}} = \kappa_f \left(\frac{d}{r_0}\right)^{5/3}, \tag{10.10}$$

where d is the inter-actuator spacing projected on the primary aperture, and $\kappa_f = 0.28$ for membrane mirrors.

10.2.2.3 The Wavefront Sensor

The Shack–Hartmann sensor is composed of a two-dimensional array of micro-lenses that each focuses the wavefront located over its aperture on a camera placed at the back focal point. If the wavefront is planar, an Airy pattern is observed behind each lenslet. The centre of this bright spot sets a reference location. Local tilts of the wavefront deviate the point where light rays focus on the CCD plane. The difference with respect to the reference in both the horizontal and vertical directions define a local slope. A schematic is presented in Figure 10.2.

The SH sensor measures the local gradients averaged over the respective subaperture, which is defined by corner coordinates $\{(h_i, v_i), (h_{i+1}, v_i), (h_i, v_{i+1}), (h_{i+1}, v_{i+1})\}$ in the horizontal and vertical directions,

$$\begin{cases} y_{h_{i,j}}(k+1) = \frac{1}{T_s}\int_{kT_s}^{(k+1)T_s} \alpha_h \Big(\int_{v_i}^{v_{i+1}} \phi_{th_i,v} dv - \int_{v_i}^{v_{i+1}} \phi_{th_{i+1},v} dv\Big) dt + w_{h_{i,j}}(k), \\ y_{v_{i,j}}(k+1) = \frac{1}{T_s}\int_{kT_s}^{(k+1)T_s} \alpha_v \Big(\int_{h_i}^{h_{i+1}} \phi_{th,v_i} dh - \int_{h_i}^{h_{i+1}} \phi_{th,v_{i+1}} dh\Big) dt + w_{v_{i,j}}(k), \end{cases} \tag{10.11}$$

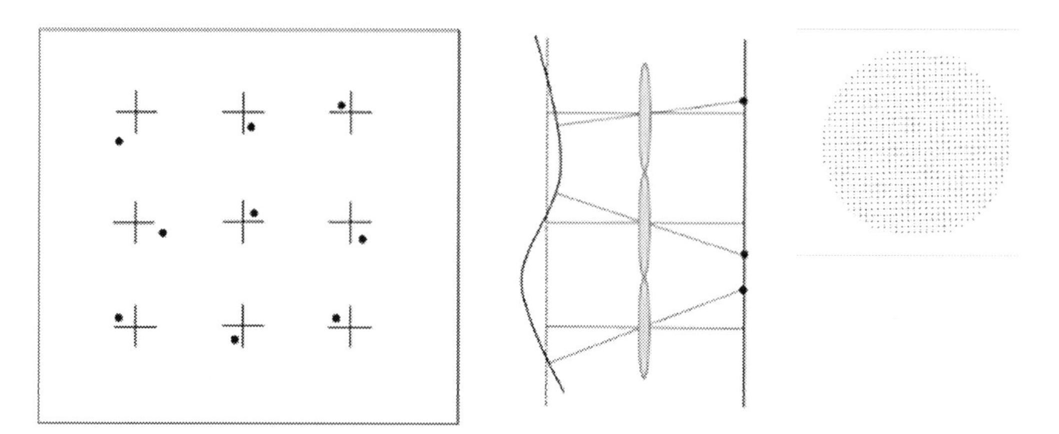

Figure 10.2 (Left) Schematic representation of a SH two-dimensional array of lenses projecting the wavefront onto a CCD plane. The crosses indicate the reference location of centroids induced by each lens. The dots correspond to the measured centroids when the wavefront is not flat. (Middle) One-dimensional view of an abberrated wavefront with lenslets focusing the local wavefront on their back focal plane. The local displacements are measured with respect to the reference obtained when the wavefront is planar. (Right) Reading of a medium-size Shack–Hartmann sensor used in the laboratory testbed for the experiments reported in Section 10.4. Each black dot represents a reference position for the centroids. Reprinted/adapted from Sinquin (2019) with permission from TU Delft

where α_h, α_v are geometrical properties and $w(k) = \left[w_x(k)^T \quad w_y(k)^T\right]^T$ is a zero-mean white noise with covariance matrix $\sigma_w^2 I$. The measurements are integrated over the sampling period T_s (assumed equal to the exposure time). The additive noise w on the wavefront sensor accounts for Poisson noise due to the arrival of photons on the camera, the read-out and thermal noise, non-linearities in the sensor among which the spatial discretization of the CCD with pixels and discretization of CCD intensity values. It is, however, approximated as zero-mean white, Gaussian and stationary. All channels are uncorrelated. A delay of one step is assumed to account for the time required for collecting photons and reading out the frames from the camera.

Equation (10.11) is re-written between the discretized wavefront $\phi(k)$ (or $\epsilon(k)$ in closed-loop) and $y(k)$ as follows,

$$y(k+1) = D\phi(k) + w(k). \tag{10.12}$$

The measurement matrix D is rank-deficient due to the presence of unseen modes by the sensor such as the piston and waffle modes (Roddier, 1999). The piston mode does not affect the image quality on the PSF and the waffle mode corresponds to a very large spatial frequency (Roddier, 1999). The set of the slopes allow to reconstruct a wavefront e.g. using finite differencing, or stochastic least-squares when using the covariance information.

10.3 Control Strategies for Medium-Size AO

Before delving into the control methodology, we review the main error sources that should be kept in mind when designing regulators. The few controllers that are mentioned in this paragraph do not represent an exhaustive list of what has been proposed although they allow us to illustrate the challenges at stake.

The most significant wavefront error sources include the non-common path errors, the fitting error, the aliasing error, the noise propagation error and the temporal error. The non-common path errors result from a different wavefront reaching the sensor and the science camera: the distortions that only affect the science camera cannot be compensated for when closing the loop from the sensor. The DM cannot take any arbitrary shape due to a finite number of actuators hence causing a fitting error. Both sources are independent of the controller. When the wavefront sensor is a Shack–Hartmann sensor, the first derivative of the wavefront is measured although with a finite sampling that is due to the spacing of the micro-lenslets. The aliasing error is theoretically evaluated as,

$$\sigma_{alias}^2 = 0.08 \left(\frac{D}{r_0 N} \right)^{5/3}, \tag{10.13}$$

where N is the number of lenslets in one dimension. Further, the measurement noise propagates in the feedback loop. The temporal error is related to the time delay between the measure and actually applying the commands. The turbulence layers above the telescope are driven by the wind and therefore the control commands become quickly outdated if the wavefront aberrations have significantly evolved during a sampling period. The mean-square wavefront distortion due to this time delay is (Roddier, 1999),

$$\sigma_{\text{temp}}^2 = \left(\frac{v}{r_0 f_S} \right)^{5/3}. \tag{10.14}$$

The larger the control frequency f_S, the smaller the temporal error. There are, however, limitations in increasing the control frequency of AO systems when using model-free controllers such as the integrator. One of these is the implied requirement for a bright guide star which limits the sky coverage. The temporal wavefront error due to the bandwidth limitations of a system using a first-order feedback controller at a sampling frequency f_S is,

$$\sigma_{temp}^2 \propto \left(\frac{f_G}{f_S} \right)^{5/3}, \tag{10.15}$$

where the Greenwood frequency f_G (see Hinnen et al., 2007) is equal to $0.427 \frac{v}{r_0}$. The dependency with the power 5/3 shows that predictive control may have a significant impact with respect to the integrator when the turbulent wavefront quickly evolves over time with respect to the sampling frequency. A model-based alternative to the integrator is the linear-quadratic-Gaussian (LQG) controller which relies on a quadratic cost function in order to minimize the variance of the residual wavefront $\epsilon(k) = \phi(k) - \phi_m(k)$ over an infinite time horizon (Kulcsár et al., 2012; Roux et al., 2004). Using the mirror model (10.9), the optimal solution in terms of control inputs minimizes the cost function,

$$\min_{u(k)} \lim_{K\to\infty} \frac{1}{K} \sum_{k=1}^{K} \|\phi(k+1) - Hu(k)\|_2^2, \tag{10.16}$$

and is further addressed using the separation principle. The separation principle states that the spatio-temporal prediction of the disturbance wavefront can be carried out independently of the projection onto the mirror's space via the following equation:

$$u(k) = \left(H^T H\right)^{-1} H^T \widehat{\phi}(k+1|k), \tag{10.17}$$

where $\widehat{\phi}(k+1|k)$ is the unbiased minimum variance estimate of the wavefront at time instant $k+1$ using all the data up to time k. A dynamical model of the stochastic (open-loop) disturbance is required to derive the wavefront prediction, which is in the form of a linear time-invariant stochastic state-space,

$$\begin{cases} x(k+1) = Ax(k) + \eta(k), \\ \phi(k) = Mx(k), \\ y(k) = D\phi(k-1) + w(k), \end{cases} \tag{10.18}$$

where the extra delay in the output equation is introduced to take the time delay in the wavefront sensing operation (10.12) into account. This model may be based on priors (wind speed, seeing) or telemetry data (open-loop slopes) to describe the dynamics of the distortion induced by the atmospheric turbulence, but it may also describe additional wavefront perturbations induced by windshake and telescope vibrations. This latter source of perturbation is all the more nefarious for the integrator as the vibrations frequently appear at frequencies that correspond to its overshoot. This led to further developments of the LQG framework for AO. It was first shown in laboratory experiments (Petit et al., 2008), then on-sky at the WHT (Sivo et al., 2014), that a data-driven model of the tip-tilt combined with priors for the other modes in a LQG controller outperforms the classical integral feedback solution. The temporal dynamics of the wavefront (decomposed in a Zernike basis; see Roddier (1999)) is then an AR with temporal order two:

$$\begin{cases} \phi(k+1) = A_1\phi(k) + A_2\phi(k-1) + Q^{1/2}v(k), \\ y(k) = D\phi(k-1) + w(k). \end{cases} \tag{10.19}$$

The state in (10.18) is restricted to two consecutive time occurences of the wavefront. Although reaching very promising performance on sky, such a model essentially represents a boiling behaviour of the atmospheric disturbance but fails to represent e.g. frozen-flow or dome turbulence. This model does not scale to large sizes because of the extremely large number of Zernike modes that would need to be considered, along with numerical issues with high-order Zernike modes and the absence of structure in the system matrices for efficiently solving the difference algebraic Riccati equation (DARE).

An alternative is to use a model derived only from data that would include all possible behaviours of the disturbance including vibrations. The state-space matrices in (10.18) are then identified from open-loop data with a subspace algorithm and further used to formulate an $\mathcal{H}_2$ optimal controller (Hinnen et al., 2007). The subspace identification method SSARX (Jansson, 2003) on which this controller relies on is not able to handle the large sensors in the next generation of extremely large telescopes.

The main challenges for scalable controllers are twofold. First, to derive a predictor offline by efficiently computing the Kalman gain. This can be done either directly from data, via the solution of a Riccati equation, or via an autoregressive model. Second, this filter should be data sparse to pave the way for efficient online computations.

In this respect, the wavefront is been expressed in a zonal basis. This also helps to use the frozen-flow hypothesis: the phase screen above the telescope at the next time instant is a shifted version of the one that is currently seen. Autoregressive models of temporal order one are a first step toward the design of more elaborate models used to focus on the essential building blocks for solving the DARE. A distributed control approach approximates the phase screen over the aperture as a cropped version of an infinitely long screen (Massioni et al., 2011), so that the state-space is transformed into the Fourier domain. It has a structure that is adequate for carrying out parallel calculations and reminds us of the developments on decomposable and circulant systems mentioned in Chapter 4. The Kalman gain is efficiently computed in the Fourier domain, and its transformation back to the spatial domain reveals that the prediction of each wavefront spatial sample relies on a large spatial neighbourhood. This kernel is, however, spatially-invariant. This differs from the work in (Poyneer et al., 2007) where, at each time instant, the slopes (extrapolated to a square array) are transformed into the Fourier domain in which the Kalman filter is formulated. Other works tackling the problem of large-scale filtering for AO include Correia et al. (2010) and Massioni et al. (2015). The former relies on a sparse approximation of the inverse of the covariance matrix and the DARE is solved using iterative methods. The latter assumes that the measurement noise is small such that solving the DARE is non-iterative.

10.4 Laboratory Experiments

This paragraph compares the structures of VAR models that we have described in Chapters 3 and 6 in the context of AO laboratory experiments.

The temporal evolution of open-loop slopes $y(k)$ is modelled with the VAR,

$$y(k+1) = \sum_{i=0}^{p-1} M_i y(k-i) + \xi(k), \tag{10.20}$$

where $\xi(k)$ is a zero-mean white Gaussian noise with covariance matrix σ_ξ^2. The coefficient matrices are sums-of-Kronecker products, $M_i = \sum_{j=1}^{r} M_{i,j,d} \otimes \cdots \otimes M_{i,j,1}$ given a Kronecker rank r and a tensor order d. The factors of the coefficient matrices are estimated solving the Quarks using data collected in open-loop. This model is then used in closed-loop to derive a prediction of the slopes vector. The control loop is summarized in Algorithm 10.1.

Algorithm 10.1: Control algorithm minimizing the residual sensor measurement with a tensor-based prediction of the measurements.

Input : $\{M_{i,j,n}\}_{i=0..p-1, j=1..r, n=1..d}, M_{int}, \{y(k-i)\}_{i=1..p-1}, y_{res}(k), u(k-1)$
Output: $u(k)$

1 $y(k) = y_{res}(k) + M_{int}u(k-1)$
2 Reshuffle $y(k)$ into $\mathcal{Y}(k)$
3 $\widehat{\mathcal{Y}}(k+1|k) = \sum_{i=0}^{p-1} \sum_{j=1}^{r} \mathcal{Y}(k-i) \times_1 M_{i,j,1} \times_2 \ldots \times_d M_{i,j,d}$
4 Reshuffle $\widehat{\mathcal{Y}}(k+1|k)$ into $\widehat{y}(k+1|k)$
5 Solve the sparse least-squares, $\min_{u(k)} \|\widehat{y}(k+1|k) - M_{int}u(k)\|_2^2$ possibly with box-bounded constraints to get $u(k)$

During closed-loop operation, at each time sample the signal $y(k)$ is computed by removing the influence of the previous inputs. It uses the sparse structure of the interaction matrix M_{int} to achieve $\mathcal{O}(N^2)$ complexity. Tensorizing the vector of measurements scales with $\mathcal{O}(N^2)$ whereas computing the prediction with the n-mode matrix product scales with $prdN^{\frac{2(d+1)}{d}}$. The control inputs are computed solving a large-least squares with coefficient matrix M_{int} nonetheless sparse (or at least with spatial invariance and Kronecker structures to be exploited if the coupling between actuators is large). Optimization algorithms such as the conjugate gradient or interior-point methods are relevant all the more if box-bounded constraints on the inputs limit the stroke of each actuator.

One limitation of this regulator that minimizes the residual slopes rather than the wavefront is the inapplicability to multi-conjugate AO that consists of using multiple guide stars and DMs to extend the corrected field of view; see Kulcsár et al. (2012) for an overview of the challenges at stake. Another limitation is that the wavefront is not reconstructed and the controller might suffer from aliasing all the more if the sensor sampling in terms of lenslets per r_0 is not sufficient. A possible extension of this algorithm to go beyond SCAO and correct for these drawbacks is to formulate a tensor-based VAR model from a temporal sequence of reconstructed wavefronts. This structured model may then be used for computing an approximate Kalman filter using, for example, doubling algorithms (Rice and Verhaegen, 2009), or low-measurement noise assumptions (Massioni et al., 2015) for non-iterative schemes.

10.4.1 The Experimental Testbed

The testbed is represented in the schematic in Figure 10.3 and the hardware is shown in Figure 10.4.

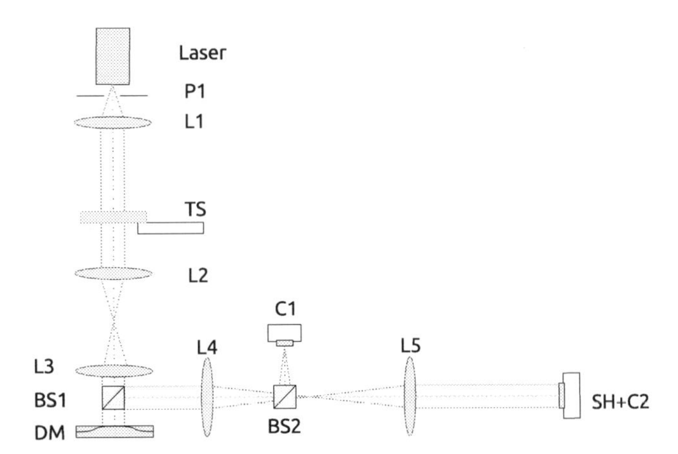

Figure 10.3 Schematic view of the laboratory testbed. P1 is a pinhole, L1–L5 are lenses, TS is a rotating disk for simulating the turbulence, BS1 and BS2 are beam splitters, DM is the deformable mirror, C1 is the point-spread-function (PSF) camera, SH+C2 is the wavefront sensor. Reprinted/adapted with permission from Sinquin and Verhaegen (2018) © The Optical Society

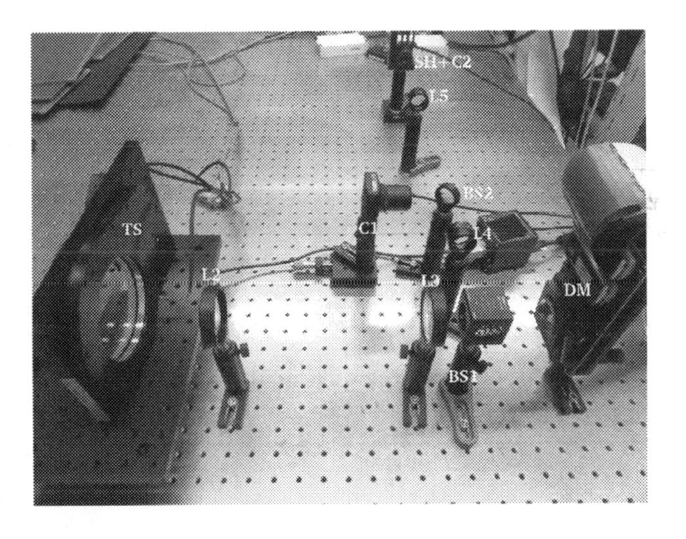

Figure 10.4 Annotated picture of the laboratory testbed in the Smart Optics Lab of Delft Center for Systems and Control used in the experiments of Section 10.4.1. Reprinted/adapted with permission from Sinquin and Verhaegen (2018) © The Optical Society

A reference guide star is simulated using a laser source at a wavelength $\lambda = 635$nm. This light is collimated into a beam of size $D = 9$mm with the lens L_1 and goes through a turbulence disk TS placed at the focal plane of the lens L2 of length $f_2 = 10$cm. This represents a frozen layer with statistics following Kolmogorov distribution and whose windspeed varies with the rotating speed of the disk. The phase pattern printed on the disk creates temporally varying seeing conditions as the local value of the Fried parameter r_0 varies as $(1 + 1/5 \cdot \sin\theta)$, from 1.2 to 1.8 mm.

The plane of the disturbance is conjugated with the one of the deformable mirror using the lenses L2 and L3. A total of 952 actuators (of which 5 are failing) are positioned within the circular aperture of a regular grid of size 34×34. The stroke is 1.8 μm, the actuator pitch is 300μm and there is no hysteresis. The response time from 10% to 90% of the mechanical response is below 20μs. The coupling between the actuators is 15% with an error margin of 5%. The light beam is reflected by the mirror to the lens L4 of focal length 150mm and divided in two using the beam splitter BS2. One part of the beam travels to the science camera C1, a Thorlabs-CMOS camera DCC1545M with 1280×1024 pixels, that is used to record the PSF. The second part goes to the wavefront sensor SH+C2. An OKOtech Shack–Hartmann wavefront sensor, 1-inch optical format, with a lenslet array pitch of 300μm and focal length 18.6mm, is placed perpendicular to the optical beam path at the focal point of L5. An array of 30×30 lenslets is selected, of which 689 are illuminated within a circular aperture and considered as active, hence providing 1378 slope measurements.

The loop runs at a sampling frequency $f_S = 10$Hz. During one time period, the DM commands are first sent, a PSF image and a WFS image are then recorded, the slopes measurements are calculated and the control commands are computed. The PSF and WFS camera do not record simultaneously.

10.4.2 Analysis from Open-Loop Data

A total of $3 \cdot 10^3$ temporal samples are collected in open-loop to identify a model. One part of the dataset is used for estimating the coefficient matrices, and the second part for checking the accuracy of the model before closing the loop. From the measurement estimate $\widehat{y}(k+1|k)$, the wavefront $\widehat{\phi}(k+1|k)$ is reconstructed using stochastic least-squares. The actual wavefront at time $k+1$ is similarly estimated from the actual measurement $y(k+1)$. The variance of the residual wavefront is evaluated with a mean-square error criteria, over a temporal sequence,

$$\sigma_{res}^2 = \frac{1}{N_{val}} \sum_{k=0}^{N_{val}-1} \|\widehat{\phi}(k+1|k) - \phi(k+1)\|_2^2. \tag{10.21}$$

The different models that will be presented correspond to the following structural assumptions:

1. quasi-static turbulence, $\widehat{y}(k+1|k) = y(k)$. The AR-1 model is such that M_0 equals the identity.
2. sparse coefficient matrices in a VAR model with a temporal order of 5. For each lenslet, only the neighbouring ones have influence; the choice of the bandwidth of the coefficient matrices is part of a trade-off that is made between sparsity and robustness of the model with respect to varying wind conditions.
3. low-Kronecker rank coefficient matrices for a VAR with varying temporal order. The three tensor decompositions analyzed are displayed in Table 10.1.

Table 10.1 Partitions in the SH sensor associated with the Quarks model. Reprinted/adapted with permission from Sinquin and Verhaegen (2018) © The Optical Society.

Tensor order, d	Size of factor matrices, J
2	$(60, 30)$
3	$(12, 6, 25)$
4	$(12, 5, 6, 5)$

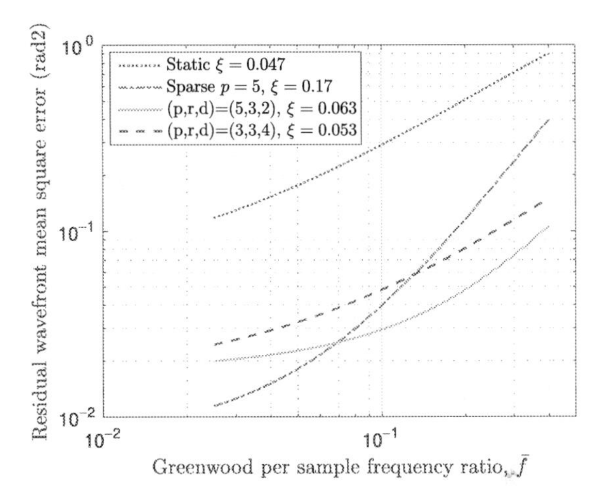

Figure 10.5 Mean-squared-error on validation data as a function of the Greenwood per sample frequency ratio. The index of performance ξ is the relative root-mean-square error (RMSE) between the interpolation with a second-order polynomial and the dataset to evaluate how the experimental points fit with the polynomial. Reprinted/adapted with permission from Sinquin and Verhaegen (2018) © The Optical Society

The performance of the models is evaluated as a function of the rotating speed of the disks that simulate varying wind speeds. More precisely, the mean-squared-error as a function of the Greenwood per sample frequency ratio $\bar{f}$ is shown in Figure 10.5. For low values of $\bar{f}$, the turbulence is quasi-static. For larger values, performance improvement using predictive methods is expected.

For low wind speeds, all data-driven models result in small residual errors and simple models relying on the quasi-static assumption have comparable performance with the data-driven models. When the wind speed increases, the quasi-static assumption is not valid anymore, which explains why the model relying on this assumption performs worse. On the other hand, the banded pattern on the coefficient matrices is well suited for values of $\bar{f}$ below 0.1, after which the mean-squared-error increases significantly. Kronecker structures are especially interesting for large wind speeds. The set of parameters (p, r, d) tunes the trade-off between data compression (and computational efficiency for online purposes) and prediction error. When the tensor

Table 10.2 Relative improvement on σ^2 when increasing either the temporal order or the Kronecker rank while $d = 2$; $(r_a, p_a) \to (r_b, p_b) := |\sigma^2_{(p,r)=(p_a,r_a)} - \sigma^2_{(p,r)=(p_b,r_b)}| = \sigma^2_{(p,r)=(p_a,r_a)}$.
Reprinted/adapted with permission from Sinquin and Verhaegen (2018) © The Optical Society.

$(r_a, p_a) \to (r_b, p_b)$	$\bar{f} \in [0.026, 0.061]$	$\bar{f} \in [0.069, 0.10]$	$\bar{f} \in [0.11, 0.15]$	$\bar{f} \in [0.15, 0.22]$	$\bar{f} \in [0.24, 0.31]$	$\bar{f} \in [0.33, 0.40]$
$(3,1) \to (3,3)$	0.23	0.22	0.29	0.27	0.29	0.28
$(3,3) \to (3,5)$	0.067	0.10	0.11	0.12	0.13	0.14
$(1,3) \to (3,3)$	0.48	0.30	0.35	0.30	0.28	0.24
$(3,3) \to (5,3)$	0.16	0.14	0.13	0.12	0.10	0.11

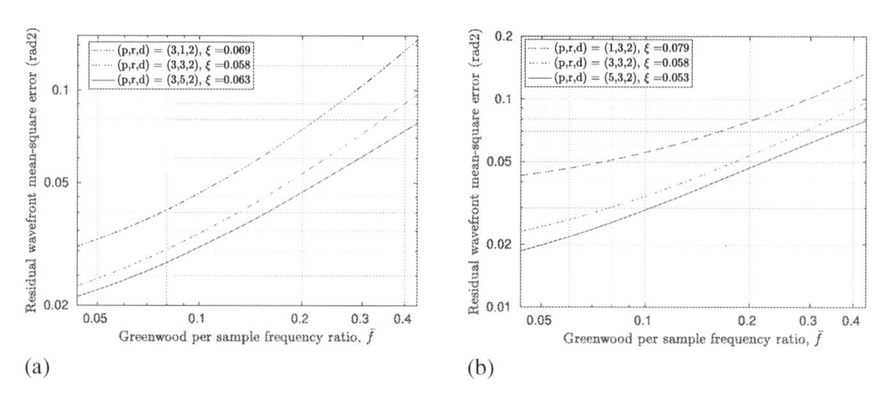

Figure 10.6 Mean-squared-error on validation data as a function of the Greenwood per sample frequency ratio $\bar{f}$ for (a) increasing Kronecker rank, and (b) increasing temporal order for $d = 2$. Reprinted/adapted with permission from Sinquin and Verhaegen (2018) © The Optical Society

order d is set larger than 2, the data fitting capability of the structure decreases all the more as we are identifying dynamics on a 2D grid. A tensor representation with $d = 4$ reaches a prediction error slightly larger than the case $d = 2$ while the number of stored entries in the factor matrices is $pr \times 122$ contrary to $pr \times 4500$. Increasing the Kronecker rank parameter decreases the prediction error as shown in Figure 10.6: the model fit to the slopes data is better while the data compression is less efficient. Conclusions are very similar to those made in Chapter 6.

Relative improvements that occur when increasing the Kronecker rank in the case $d = 2$ are shown in Table 10.2. It is averaged in different ranges of $\bar{f}$ to highlight the main trends. The larger the wind speed, the larger $\bar{f}$, and the more useful it is to increase r for accurate estimations. The relative improvement is less significant from $r = 3$ to $r = 5$ than from $r = 3$ to $r = 1$. Approximating the spatial-temporal dynamics of the sensor data with a function separable in d-dimensions is a valid assumption, even for low values of the Kronecker rank. Increasing the temporal order

p only leads to decreased prediction-error where the decrease becomes smaller and smaller with increasing p, as can be seen in Figure 10.6 and in Table 10.2. Similar conclusions are drawn for model order selection in system identification.

10.4.3 Closed-Loop Performance

AO systems aim to re-shape the incoming wavefront so that it is as flat as possible after correction. The variance of the residual wavefront over the pupil aperture Ω and averaged over infinitely long exposures is,

$$\sigma_{res}^2 = \lim_{T_s \to \infty} \frac{1}{T_s} \int_0^{T_s} \left(\iint_{\rho\in\Omega} \epsilon(\rho)^2 d\rho - \left(\iint_{\rho\in\Omega} \epsilon(\rho) d\rho \right)^2 \right) dt. \tag{10.22}$$

This residual phase variance translates into sharpness of the long exposure PSF. As the optical flux that reaches the PSF camera is identical whatever the aberration, the more concentrated over a central core the intensities are, the better. The main performance index is the Strehl ratio. It is defined as the ratio between the maximum value of the PSF during closed-loop and the value without turbulence (that is, diffraction-limited). Its evaluation from experimental data is as follows. Let $\bar{I}(p)$ denote the intensity of the long exposure image at the pixel coordinate p in the CCD frame $\mathcal{I}$, and $\underline{I}(p)$ denote the intensity of the diffraction-limited PSF image obtained with no turbulence at position p. All values below 2% of the CCD range are first thresholded to zero. Let the maximum intensity be denoted by $\bar{I}_0$ and its position by p_0. Let q_0 denote the radius between the centre of the theoretical Airy disk and the position where the first minimum occurs. Let $\mathcal{B}_r(p_0) = \{p \in \mathcal{I}, \|p - p_0\|_2 < r\}$. Summing the intensities in the neighborhood $\mathcal{B}_{2q_0}(p_0)$ yields the flux $\bar{I}_{2q_0}$. The Strehl ratio is computed using,

$$S = \frac{\bar{I}_0}{\bar{I}_{2q_0}} \frac{\underline{I}_{2q_0}}{\underline{I}_0}. \tag{10.23}$$

Another performance index is the normalized encircled energy that measures the flux entering a circle of radius r centered around the position of the maximum value, p_0 in the PSF image I,

$$EE(r) = \frac{\lim_{T_s\to\infty} \frac{1}{T_s} \int_0^{T_s} \iint_{\rho\in\mathcal{B}_r(p_0)} I(\rho,t) d\rho dt}{\lim_{T_s\to\infty} \frac{1}{T_s} \int_0^{T_s} \iint_{\rho\in\mathcal{I}} I(\rho,t) d\rho dt}. \tag{10.24}$$

The closed-loop experiment lasts $5 \cdot 10^2 \times f_S$ seconds. We compare with a proportional integrator (PI),

$$u(k) = \frac{c_1}{1 - c_2 z^{-1}} R y_{res}(k), \tag{10.25}$$

where R is a command matrix.

Figure 10.7 depicts the Strehl ratio as a function of the Greenwood per sample frequency ratio $\bar{f}$.

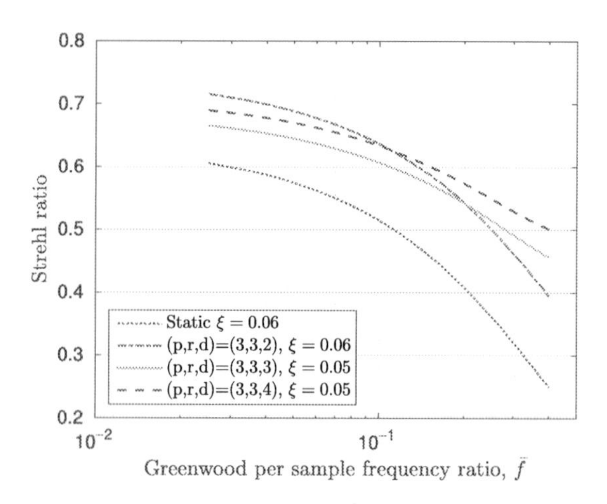

Figure 10.7 Strehl ratio as a function of the Greenwood per sample frequency ratio, $\bar{f}$. Reprinted/adapted with permission from Sinquin and Verhaegen (2018) © The Optical Society

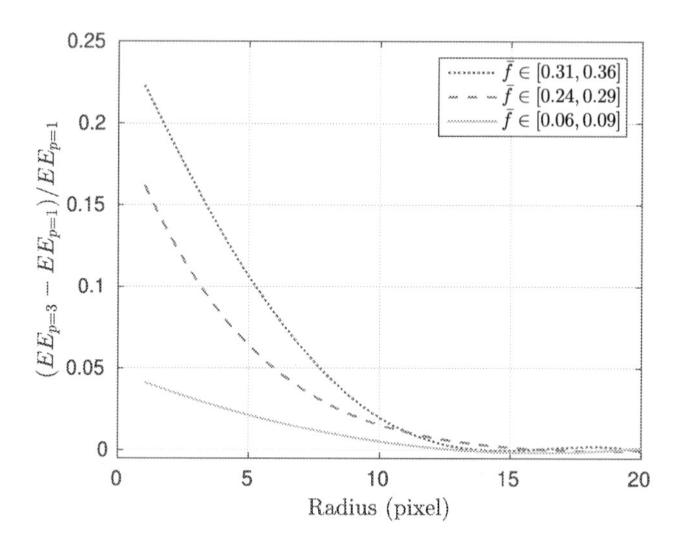

Figure 10.8 Encircled energy as a function of the Greenwood per sample frequency ratio. The relative improvement brought by the case $(p,r,d) = (3,3,4)$ over $(p,r,d) = (1,3,4)$ is shown. Reprinted/adapted with permission from Sinquin and Verhaegen (2018) © The Optical Society

The performance of the PI controller degrades with increasing windspeed whereas the use of the predictive models presented in Algorithm 10.1 reduces the temporal error. Better disturbance rejection is observed when $d = 4$, especially for large $\bar{f}$. One reason is the robustness to noise of a parametrization with fewer coefficients. When gradually increasing the radius for computing the encircled energy, the normalized value converges to one. The impact of increasing the temporal order is shown in Figure 10.8. The relative improvement that the model with $(p,r,d) = (3,3,4)$ brings over the structure $(p,r,d) = (1,3,4)$ when increasing the radius used for computing the encircled energy increases with the windspeed. For quasi-static turbulence however, large temporal orders marginally increase the performance over a simple tensor model with $p = 1$.

To summarize, tensorizing the sensor measurements for identifying the spatial-temporal dynamics of the turbulence from data and applying subsequent predictive control is most advantageous when the wind speed is large and when the sensor array has many lenslets. The aliasing is then low, and the wavefront is well sampled, hence allowing use of the slopes measurements directly without wavefront reconstruction. A banded structure may be further parametrized on the coefficient matrices of the VAR model to further improve the scalability. It requires knowledge of the (overall) wind speed that is usually available in practice.

10.5 Replaying On-Sky Data

In this section, we replay AO telemetry data that has been collected on sky to further assess the identification algorithms presented in this book in the context of disturbance rejection.

The AO bench CANARY (Gendron et al., 2016) is setup on a Nasmyth platform of the 4.2m WHT. The single-conjugate AO mode at the WHT features an on-axis Shack–Hartmann sensor with 14×14 lenslets, for which the angle viewed on-sky for each pixel is $0.428''$. There are 288 active lenslets. Both a tip-tilt and a deformable mirror (coupling of 0.45) with 243 active actuators are used to correct the incoming wavefront disturbance. The control frequency is set to 200 Hz and a loop delay of 2.25 frames was measured.

The closed-loop data that was stored throughout two nights in July 2019 with very different wind speeds is re-used after the experiment to evaluate the estimated models. The residual slopes are evaluated by subtracting the slopes induced by the deformable mirror to the open-loop ones,

$$y_{\text{res}}(k) = y(k) - \delta(z) M_{\text{int}} u(k), \tag{10.26}$$

where $\delta(z) = 0.75z^{-2} + 0.25z^{-3}$. Different controllers in closed-loop yielding commands $u(k)$ and then residual slopes $y_{\text{res}}(k)$ using Equation (10.26) are tested. The residual wavefront is reconstructed with a weighted least-squares,

$$\epsilon(k) = \left(D^T D + \Sigma_w\right)^{-1} D^T y_{\text{res}}(k), \tag{10.27}$$

where Σ_w is the (diagonal) covariance of the measurement noise. We evaluate the performance as the sum over the variance of the first 197 modes, which are the mirror modes, at the wavelength of the imaging camera. This criterion is such that the fitting and aliasing errors are under-evaluated. The aliasing error has a very low contribution to the error budget for the current configuration of CANARY thanks to the sampling of the atmosphere with a 14×14 Shack–Hartmann sensor for a 4.2 m-telescope.

Because the sensor delivers only 288 outputs at each time sample, we may compare data-driven and structured controllers with state-of-the-art methods devised for medium-scale AO; see, for example, Sivo et al. (2014) and Sinquin et al. (2020). Both of these include a model built from priors on 495 Zernike modes to which is combined a data-driven model for the tip-tilt modes for the former, and for the low-orders in general for the latter. Criterion (10.16) is reformulated to minimize the residual slopes rather than the residual wavefront. This allows us to rely on the very good knowledge of the interaction matrix and not deal with unknown transformations between the sensor and actuator grid (magnifications, rotations). The commands are then expressed with,

$$u(k) = M_{\text{com}} D \widehat{\phi}(k+1|k), \tag{10.28}$$

where M_{com} is the command matrix. The method described in Sivo et al. (2014) is denoted with LqgZer-TT (N4SID) whereas the one in Sinquin et al. (2020) with LqgZer-LOn_{LO}, where n_{LO} indicates the number of modes whose temporal dynamics are described from both data and priors.

The data-driven controllers derive a prediction of the disturbance-induced slopes $y(k+1|k)$ with either an AR or a state-space model in innovation form, from which the commands are deduced,

$$u(k) = M_{\text{com}} \widehat{y}(k+1|k). \tag{10.29}$$

The subspace algorithm N4SID (Overschee and Moor, 1994) is used as a reference for identifying a model using a databatch with 15×10^3 points. The model order is set equal to the first index of the singular values vector whose corresponding value is larger than 99% of the accumulated sum of the singular values.

Three algorithms presented in this book have been used for deriving $\widehat{y}(k+1|k)$: a sparse AR model with multi-banded coefficient matrices, the Quarks, and K4SID for using matrix state-space models. A time sequence of length only 5×10^3 samples is used for identifying the models.

The first 50 seconds of pseudo open-loop data are used for the identification. Datasets that are used for identifying models are not used for simulating the closed-loop.

10.5.1 Case 1: Large Wind Speeds

The first dataset was collected during the night of July 18, 2019, between 3h08 and 4h21. The guide star had a magnitude of 5.2 in the near-infrared. The main turbulent layers at altitudes of 0, 2 and 8 km were measured with a wind speed of, respectively, 12, 15 and 13m/s using the stereo-SCIDAR (Osborn, 2018). Wind profiles measured by James Osborn at the Isaac Newton Telescope are shown in Figure 10.9.

Table 2 of Sinquin et al. (2020) displays the residual phase variance in nm root-mean-square,

$$\delta_{\text{res}} = \frac{\lambda_{\text{PSF}}}{2\pi} \sigma_{\text{res}} \times 10^9. \tag{10.30}$$

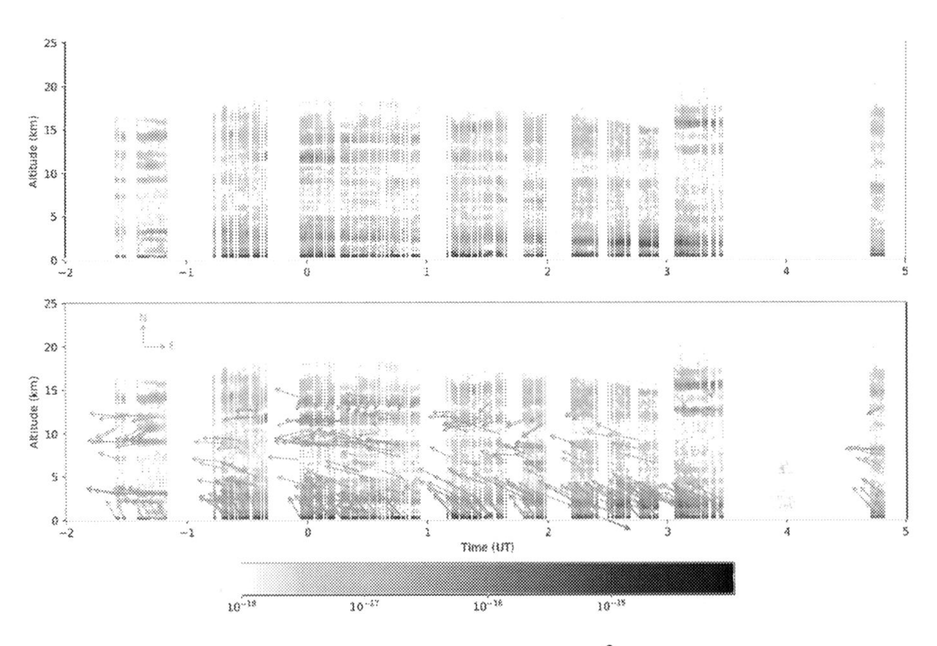

Figure 10.9 Altitude profile of the turbulence strength factor $C_n^2 dh$ as a function of time for the night from July 17th to the 18th. The length of the arrow is equivalent to 10 m/s for scale. Reprinted/adapted from Sinquin et al. (2020) with permission from © MNRAS

The method referenced in this table with LqgZer-LO that combines a data-driven model for low orders (e.g. shown for 9 or 65 Zernike modes) and priors on 495 modes to reject in a least-squares sense the related disturbance modes. This method does not scale to large sizes of the wavefront sensor. The controller using a state-space identified with N4SID for deriving the prediction of the slopes shows lower residual than the integrator and LqgZer-TT (Sivo et al., 2014). The sparse AR model performs better than the Quarks for these examples. It would be required to further increase the Kronecker rank of the Quarks (here set to two) in order to converge to such a value of residual variance at the expense of a larger computational complexity. Similarly, K4SID restricts us to the identifiation of Kronecker rank-one state space matrices, which is one reason why the performance is worse than for controllers relying on AR models. Another is the non-globally convergent algorithm for identifying the unknown scaling parameters in the factor matrices. The controller relying on K4SID shows, nonetheless, a significant improvement with respect to the integrator.

10.5.2 Case 2: Low Wind Speed

The second dataset was collected during the night of July 21st, from 4h31 to 6h00. The guide star has magnitude 5.8 in near-infrared. The windspeed at the ground was no more than 3 m/s. No significant turbulent layer was observed in altitude, as shown in Figure 10.10.

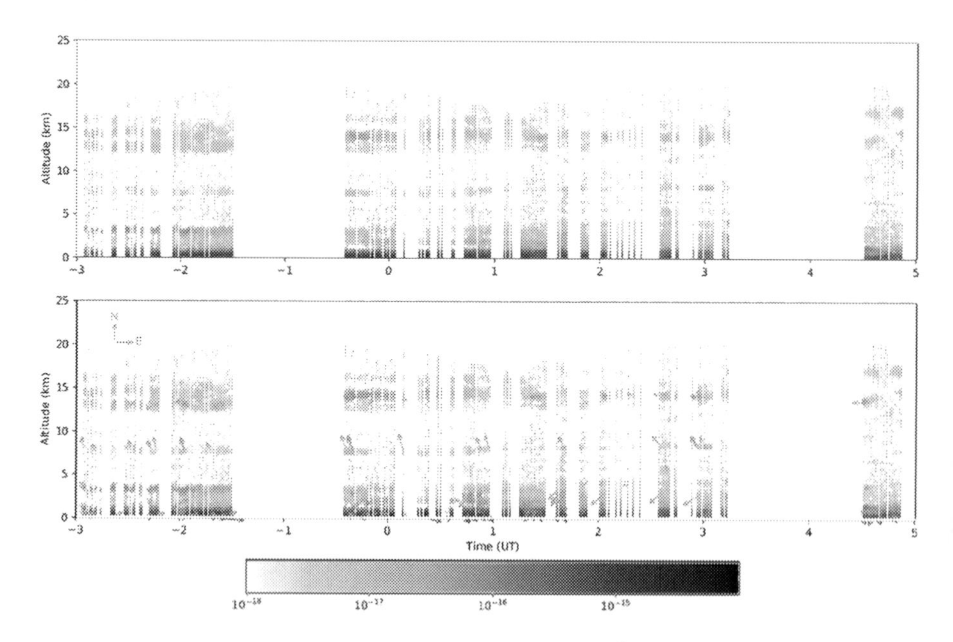

Figure 10.10 Altitude profile of the turbulence strength factor $C_n^2 dh$ as a function of time for the night from July 20th to the 21st. The length of the arrow is equivalent to 10m/s for scale. Reprinted/adapted from Sinquin et al. (2020) with permission from © MNRAS

For lower wind speeds, the temporal error reduces and there is less to gain by using LQG controllers. The controller LqgZer-LO9 achieves the best performance. The controller relying on a full data-driven stochastic model (N4SID-based) performs similarly to LqgZer-TT. For all controllers that rely on data only, and as expected, the performance is best when N4SID is applied for identifying unparametrized system matrices. Enforcing a matrix structure increases the residual phase variance although it ensures scalability to larger sizes of sensor. The three methods that rely either on low-Kronecker rank structures or on a banded parametrization of coefficient matrices in auto-regressive models have comparable performance on this dataset of AO telemetry.

10.6 Conclusions

Large-scale adaptive optics in astronomy is a potential application of the system identification methods derived in the previous chapters. Using the terminology that we have developed in this book, the system identification for the disturbance has the following features: the topology is unknown, the subsystems are homogeneous, the sensor grid is regular. In view of these elements, the framework of multi-dimensional modelling that uses sums-of-Kronecker matrices to parametrize the coefficient or state-space matrices appeared relevant. For single-conjugate adaptive optics, and when the aliasing error is negligible compared to the temporal error, reconstructing the wavefront may be bypassed allowing the use of the signal flow models in the form of either (or both) multi-banded or low-Kronecker rank matrices.

This is no longer applicable in multi-conjugate AO for which a much larger difference is observed between state-of-the-art LQG and integrators than in SCAO. Structured state-space models such as MSSM or TSSM come into the picture for deriving controllers that rely on a cost function that minimizes the residual wavefront (and not the residual slopes). Despite showing interesting performance replaying on-sky data collected on a medium-size telescope, K4SID is dedicated to Kronecker rank-one matrices that is rather conservative for real-life applications.

In the on-sky experiments presented in this chapter, the control frequency is only 200 Hz. As a consequence, first, the temporal error due to large wind speed is significant with respect to the other sources of error; and second, the vibrations are amplified by the integrator. When AO systems operate at frequencies of about 500 Hz or more, the performance gap between the integrator and LQG narrows down; this calls for even better structured identification algorithms that would approach the performance of N4SID in all wind situations.

The target for AO as far as system identification is concerned is to be able to reproduce the results in Hinnen et al. (2007) for large-scale systems. The derivation of K4SID was a step toward this, although is incomplete because of Kronecker rank-one restrictions, and the requirements to estimate a structured matrix relating state and wavefront. Moreover, the benefits of using entirely data-driven controllers on-sky has never been demonstrated: whether first principles models for few modes with high impact on the disturbance should be merged with this zonal, data-driven and structured approach is still an open question. This highlights the need for a more flexible matrix structure than the low-Kronecker rank. For example, all computational assets for standard matrix operations would remain when working with low-Kronecker rank plus low-rank matrices.

Acknowledgements

CANARY was supported by Agence Nationale de la Recherche (ANR) program 06-BLAN-0191, CNRS/INSU, Observatoire de Paris, and Universitíl’ Paris Diderot Paris 7 in France, Science and Technology Facilities Council (Grants PP/E007651/1, ST/I002871/1, ST/K003569/1, ST/M007669/1 and ST/N002660/1), and the University of Durham in the UK and European Commission Framework Programme 7 (E-ELT Preparation Infrastructure Grant 211257 and OPTICON Research Infrastructures Grant 226604). Horizon 2020: This project has received funding from the European Union's Horizon 2020 Research and Innovation Programme under grant agreement number 730890. The work of James Osborn was supported by the Science and Technology Funding Council (UK) (ST/P000541/1), UKRI Future Leaders Fellowship (UK) (MR/S035338/1) and Horizon 2020.

11 Conclusions

The identification of dynamic systems operating in large networks is in its infancy. In this book, two general approaches that are under full development have been described. Many extensions are underway. Some of these naturally follow from the lines of development outlined in this book. Many methodologies that have been introduced in this book have been formulated for *core* identification problems. This means that the problems have often been simplified to focus on essential and challenging new problems, leaving extensions for further research. Such extensions may be relevant for future applications.

One such example is the often used simplifying assumption on the additive perturbations. To address more general noise scenarios, VARMAX, or Box-Jenkins models for example, have to be considered instead of the simple VARX models used in this book. An interesting research question is whether the VARX solutions presented can be used as a building block for the more general noise scenarios such as in the case for lumped parameter systems (Ljung, 1999; Verhaegen and Verdult, 2007). On the other hand, the simple innovation noise that was assumed to be additive to the output in the considered state-space model examples can, however, be used directly to consider much more general noise model classes. This insight stems from the fact that a more general noise model structure gives rise to an innovation model, which in its predictor or observer format exactly fits in the considered model class however now having as input, both the measured input and output. A key assumption for a number of methods is that this predictor or observer form preserves the assumed structure of the system matrices, as in, for example, the class of SSS systems in Rice and Verhaegen (2011). For the latter class of systems, it was explicitly shown that under mild conditions the predictor inherits the SSS model structure when the original system matrices are within this class. Such structure-preserving property does not hold in general.

An important research question is to strive not only for compact models but for models whose compactness or sparsity pattern in the system or coefficient matrices can be exploited in subsequent synthesis steps, such as in designing predictors or controllers. For large-scale systems, this means that the model structures allow us, for example, to solve the discrete algebraic Riccati equation (DARE) in a scalable manner. A challenge of equal importance is to integrate (some) of the identification methods with data driven synthesis methodologies as considered for fault estimation for lumped parameter dense low complexity models in Wan et al. (2016). Taking such (additional) objectives into consideration gives rise to *identification for distributed control*. This topic is due full exploration in the near future.

The identification algorithms presented in the book all use a data batch of temporal samples. For non-stationary systems, or when the computing facilities for data storage are very limited, use can be made of recursive variants that have been derived for unstructured autoregressive or state-space models in Houtzager et al. (2009b); Sayed and Kailath (1994). Developments in this area for low-Kronecker rank autoregressive models have just started. A first exploration in this direction is the work presented in Sinquin et al. (2018) for the recursive identification of Quarks models.

On a more theoretical level, postulating a matrix structure for a state-space model implies properties of observability, controllability and stability. For example, the dynamics identified for a string of interconnected subsystems guarantee these properties to hold on a local scale, but not for the global system (when lifting the local states into a global state vector). Similarly, for Kronecker structures theoretical guarantees hold for the factors but it fails to extend further. It has not been investigated to what extent this hampers further derivation of distributed controllers that yield good performance and robustness.

Recent special sessions, for example at the Conference on Decision and Control in Nice, 2019, on '*Theoretical Foundations for the Representation and Identification of Dynamic Networks I and II*', organized by Prof. P. van den Hof and Prof. Sean Warnick, indicate that the field is in full development. These developments depend partly on new theoretical contributions, such as those elaborated on in this book, but also depend, in a crucial manner, on the availability of high-performance software implementations. Such implementations can be prototyped on standard Matlab® or Python®. However, taking the large-scale nature into account for potential applications, dedicated efforts are needed to develop software solutions that make use of distributed computing possibilities like GPU. An example of such a development is the work on Quarks identification methods in Sinquin et al. (2018), on the control of multilevel SSS systems (Qiu, 2013) and the MORLab toolbox for the control and model reduction of sparse state space (SSS)[1](Castagnotto et al., 2017). But here as well the developments have just started.

To conclude this book, a final comment is made about applications. In this book, one motivating application that has been discussed is the use of system identification in adaptive optics. Chapter 10 provided some 'food for thought' to demonstrate the potential of system identification to adaptive optics. The increasing size of telescopes, such as the Thirty Meter Telescope (Ellerbroek, 2011) or the European Extremely Large Telescope currently under construction in Chile, are two examples where distributed identification may become very useful. Other applications using large-scale sensor and actuator arrays, some of which were illustrated in Chapter 1, may naturally start making use of the developed methodologies. This will result in extending the methodologies presented in this book. In particular, there is a lack of real-world datasets to further confront the algorithms and evaluate the potential

[1] Also here the acronym SSS is used, but as this is the only reference here no further distinction is made.

impact that system identification can make. An interesting potential area where both applications and methodology development for distributed system identification will come together is in *data driven control* of large-scale networked systems. In such a context, the availability of real-world datasets allows us to compare the methods with more standard controllers such as integrators that are often still preferred in many engineering environments unless more elaborate controllers based on (identified) models show a significant performance improvement and are also relatively easy to tune, update and maintain.

Appendices

A Some Properties of the Kronecker Product

In this appendix some properties of the Kronecker product, which are used in this book, are summarized. For more details we refer to Golub et al. (2000).

First, we start with the definition of the operation of the Kronecker product. Denoting the entry in matrix A in the i row and j column by a_{ij} and given two matrices $B \in \mathbb{R}^{m \times n}$ and $\mathbb{C} \in \mathbb{R}^{p \times q}$, then the Kronecker product between them is given in the matrix $A \in \mathbb{R}^{mp \times nq}$ as:

$$A = B \otimes C = \begin{bmatrix} b_{11}C & b_{12}C & \dots & b_{1n}C \\ b_{21}C & b_{22}C & \dots & b_{2,n}C \\ \vdots & \vdots & \ddots & \vdots \\ b_{m1}C & b_{m2}C & \dots & b_{mn}C \end{bmatrix}. \tag{A.1}$$

When the matrix A is given, the factors B and C in its Kronecker product are not unique. The ambiguity is a consequence of the fact that we can multiply B by the scalar γ and scale C by its reciprocal value (considering $\gamma \neq 0$), such that the Kronecker product is preserved. This is denoted as:

$$A = B \otimes C = \gamma B \otimes \frac{1}{\gamma} C. \tag{A.2}$$

PROPERTY A.1 (Distributivity and associativity) *The Kronecker product is a distributive and associative operation, expressed as:*

$$D \otimes (B + C) = D \otimes B + D \otimes C, \tag{A.3}$$

$$D \otimes (B \otimes C) = (D \otimes B) \otimes C. \tag{A.4}$$

As a direct consequence of the associative property is the following:

$$(B_1 \otimes C_1)(B_2 \otimes C_2) = (B_1 B_2) \otimes (C_1 C_2). \tag{A.5}$$

PROPERTY A.2 (Transpose) *The transpose of a Kronecker product of two matrices satisfies:*

$$(B \otimes C)^T = B^T \otimes C^T. \tag{A.6}$$

PROPERTY A.3 (Trace) *The trace of a Kronecker of two matrices satisfies:*

$$Trace(A \otimes B) = Trace(A)Trace(B). \tag{A.7}$$

PROPERTY A.4 (Inverse) *If the factor matrices B and C are invertible (and hence also square), the inverse of their Kronecker product A exists and is given by:*

$$(B \otimes C)^{-1} = B^{-1} \otimes C^{-1}. \tag{A.8}$$

For the case where B and C are not square and (only) have a pseudo-inverse, a similar relationship holds replacing the inverse by the pseudo-inverse.

PROPERTY A.5 (Inner product) *Let the standard inner product between square matrices B and C be defined as:*

$$\langle B, C \rangle = Trace\left(BC^T\right),$$

then the inner product between two Kronecker products is given as:

$$\langle B_1 \otimes C_1, B_2 \otimes C_2 \rangle = \langle B_1, B_2 \rangle \langle C_1, C_2 \rangle. \tag{A.9}$$

Proof This property can be derived from Property A.3. □

The above properties can be generalized when the Kronecker product of more than two matrices is considered. The proof of these generalizations follows by induction. This is illustrated below for the inverse of the Kronecker product between d factor matrices.

COROLLARY A.1 (Inverse Kronecker product of d factor matrices) *Let all factor matrices A_i be square and invertible (for $i = 1 : d$). Then the following holds:*

$$(A_d \otimes \cdots \otimes A_1)^{-1} = A_d^{-1} \otimes \cdots \otimes A_1^{-1}.$$

Proof Based on the property of associativity and Property A.4, we can write:

$$\begin{aligned}(A_d \otimes A_{d-1} \otimes \cdots \otimes A_1)^{-1} &= (A_d \otimes (A_{d-1} \otimes \cdots \otimes A_1))^{-1} \\ &= A_d^{-1} \otimes (A_{d-1} \otimes \cdots \otimes A_1)^{-1}.\end{aligned}$$

Continuing via induction completes the proof. □

The Kronecker product in a number of cases preserves the structure of its factor matrices. Table A.1 summarizes a number of such properties. In general, these properties are one-sided, meaning that the reverse is not true.

PROPERTY A.6 (Vectorization) *Given the matrices $A \in \mathbb{R}^{m \times n}$, $B \in \mathbb{R}^{n \times p}$, $C \in \mathbb{R}^{p \times q}$, and using the notation $vec(A)$ to denote the column-wise vectorization of the matrix A, then the following relationship holds:*

$$vec(ABC) = (C^T \otimes A)vec(B). \tag{A.10}$$

Table A.1 Some cases where the Kronecker product preserves the matrix structure of its factor matrices. Reprinted/adapted from Sinquin (2019) with permission from TU Delft.

if B and C are both ...	then $B \otimes C$ is ...
banded	multi-banded
Toeplitz	two-level Toeplitz
sequentially semi-separable	two-level sequentially semi-separable
symmetric	symmetric
positive definite	positive definite
orthogonal	orthogonal

PROPERTY A.7 (Khatri–Rao product) *The Khatri–Rao product $\odot$ between two matrices B and C, which have an equal number of n columns, each column denoted respectively as b_i and c_i, is expressed by the following column-wise Kronecker product:*

$$A = B \odot C = \begin{bmatrix} b_1 \otimes c_1 & \dots & b_n \otimes c_n \end{bmatrix}. \tag{A.11}$$

B Matrices and Tensors

B.1 Introduction

Matrices represent a mapping between vector spaces. However, when these vectors contain structure, for example due to lifting of a 2D object (like an image), these matrices may perceive a two-level structure. Equivalently, the matrices are composed of blocks that are matrices themselves and have a certain matrix partitioning. This partitioning may be exploited to efficiently store and do calculations with such matrices especially when their sizes are large. We illustrate this concept with the nearest Kronecker product (NKP) problem (Loan and Pitsianis, 1992). This problem is an introduction to tensors.

The NKP problem reads as follows. Given a matrix $A \in \mathbb{R}^{m \times n}$ with $m = m_1 m_2$ and $n = n_1 n_2$ then the problem is,

$$\min_{B \in \mathbb{R}^{m_1 \times n_1}, C \in \mathbb{R}^{m_2 \times n_2}} \|A - B \otimes C\|_F. \tag{B.1}$$

The solution to this problem is shown to be equivalent to finding a nearest rank-1 matrix to a reshuffled matrix version of A (Loan and Pitsianis, 1992). This reshuffling and the rank-1 approximation is best seen via an example.

EXAMPLE B.1 (Reshuffling a partitioned matrix) Consider the following matrix A that is partitioned with $m_1 = 3$ and $m_2 = n_1 = n_2 = 2$,

$$A = \left[\begin{array}{cc|cc} a_{11} & a_{12} & a_{13} & a_{14} \\ a_{21} & a_{22} & a_{23} & a_{24} \\ \hline a_{31} & a_{32} & a_{33} & a_{34} \\ a_{41} & a_{42} & a_{43} & a_{44} \\ \hline a_{51} & a_{52} & a_{53} & a_{54} \\ a_{61} & a_{62} & a_{63} & a_{64} \end{array}\right] = \left[\begin{array}{c|c} A_{11} & A_{12} \\ \hline A_{21} & A_{22} \\ \hline A_{31} & A_{32} \end{array}\right].$$

Then the reshuffling of A, denoted by $\mathcal{R}(A)$ equals,

$$\mathcal{R}(A) = \begin{bmatrix} \text{vec}(A_{11})^T \\ \text{vec}(A_{21})^T \\ \text{vec}(A_{31})^T \\ \text{vec}(A_{12})^T \\ \text{vec}(A_{22})^T \\ \text{vec}(A_{32})^T \end{bmatrix}.$$

It was shown in Loan and Pitsianis (1992) that the NKP problem in (B.1) is equivalent to the following rank-1 approximation problem:

$$\min_{B \in \mathbb{R}^{m_1 \times n_1}, C \in \mathbb{R}^{m_2 \times n_2}} \| R(A) - \text{vec}(B)\text{vec}(C)^T \|_F.$$

This concept can be generalized to multi-level structured matrices that are represented as a sum of Kronecker products. This is outlined using tensor calculus that is reviewed briefly in this appendix. This material is based on appendix B of Sinquin (2019) and a Matlab® tutorial can be found in Sinquin et al. (2018). More details can be found in Kolda (2006). Numerical operations with tensors are well documented and accessible in a user-friendly manner in the TensorLab toolbox (Vervliet et al., 2016).

B.2 Definitions

Let d, I, J be three integers. Let $(I_1, \ldots, I_d)$ and $(J_1, \ldots, J_d)$ be two tuples of integers such that $\prod_{j=1}^{d} I_j = I$ and $\prod_{j=1}^{d} J_j = J$.

DEFINITION B.1 (Tensorization of a vector) For $j \in \{1, \ldots, d\}$, let i_j an integer such that $1 \leq i_j \leq I_j$. A vector $x \in \mathbb{R}^I$ is tensorized into $\mathcal{X} \in \mathbb{R}^{I_1 \times \cdots \times I_d}$ using the element-wise relationship:

$$x_{i_1,\ldots,i_d} = x_{\overline{i_1 \ldots i_d}}, \tag{B.2}$$

where $\overline{i_1 \cdots i_n} = i_1 + (i_2 - 1)I_1 + \cdots + (i_N - 1)I_1 \cdots I_d$.

This tensorization of a vector is done via re-indexing the elements of the vector in a tensor with d indices.

EXAMPLE B.2 (Tensorization of a vector) The tensorization is illustrated with a small-scale example. Let us choose $I = 16$, $d = 3$ and consider the vector $x \in \mathbb{R}^{16}$ with $\mathrm{x}_i = i$ for $i \in \{1, \ldots, 16\}$. Then the tensorization yields a tensor $\mathcal{X} \in \mathbb{R}^{2 \times 4 \times 2}$ with slices given as:

$$\mathcal{X}_{:,:,1} = \begin{bmatrix} 1 & 3 & 5 & 7 \\ 2 & 4 & 6 & 8 \end{bmatrix}, \quad \mathcal{X}_{:,:,2} = \begin{bmatrix} 9 & 11 & 13 & 15 \\ 10 & 12 & 14 & 16 \end{bmatrix}. \tag{B.3}$$

Such tensorization (into a tensor of order three) is not unique. With the given data we can also create the tensor $\mathcal{X} \in \mathbb{R}^{4 \times 2 \times 2}$, with slices:

$$\mathcal{X}_{:,:,1} = \begin{bmatrix} 1 & 2 & 3 & 4 \\ 5 & 6 & 7 & 8 \end{bmatrix}^T, \quad \mathcal{X}_{:,:,2} = \begin{bmatrix} 9 & 10 & 11 & 12 \\ 13 & 14 & 15 & 16 \end{bmatrix}^T. \tag{B.4}$$

Turning tensors into a matrix is called matricization. We define this next in a linear algebra manner.

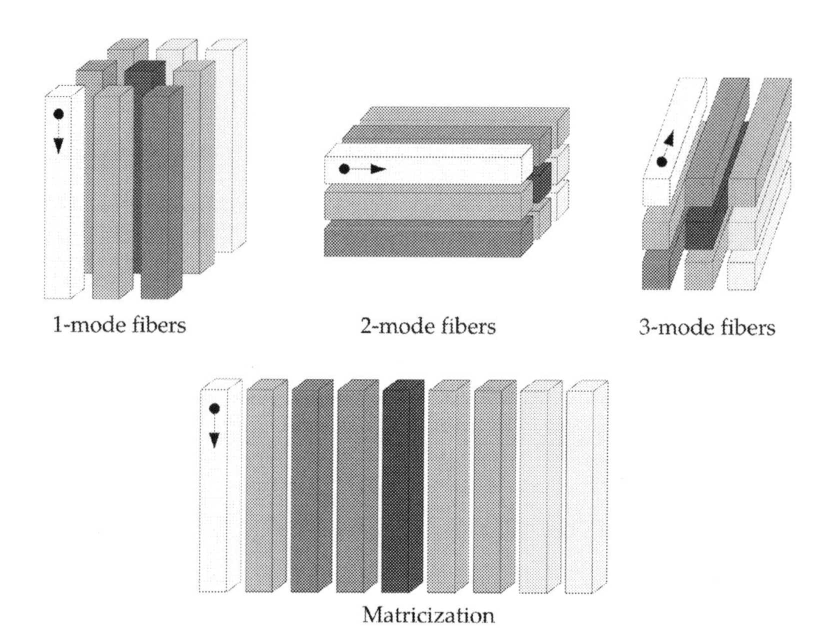

Figure B.1 (Top) A 3D plot is given for a $3 \times 3 \times 3$ tensor is represented in terms of 1-, 2- and 3-mode fibers, respectively. (Bottom) The matricization of that tensor. The arrows in each subplot indicate the direction of the fibers used for matricizing. Reprinted/adapted with permission from Sinquin and Verhaegen (2018) ©The Optical Society

DEFINITION B.2 (n-mode matricization (Kolda, 2006)) The n-mode matricization of a tensor $\mathcal{X} \in \mathbb{R}^{J_1 \times \cdots \times J_d}$, denoted as $X_{(n)}$, is a matrix in $\mathbb{R}^{J_n \times J_1 \cdots J_{n-1} J_{n+1} \cdots J_d}$ and given as:

$$X_{(n)} = \begin{bmatrix} x_{1,\ldots,1,1,1,\ldots,1} & x_{2,\ldots,1,1,1,\ldots,1} & \cdots & x_{J_1,\ldots,J_{n-1},1,J_{n+1},\ldots,J_d} \\ x_{1,\ldots,1,2,1,\ldots,1} & x_{2,\ldots,1,2,1,\ldots,1} & & x_{J_1,\ldots,J_{n-1},2,J_{n+1},\ldots,J_d} \\ \vdots & \vdots & & \vdots \\ x_{1,\ldots,1,J_n,1,\ldots,1} & x_{2,\ldots,1,J_n,1,\ldots,1} & \cdots & x_{J_1,\ldots,J_{n-1},J_n,J_{n+1},\ldots,J_d} \end{bmatrix}. \tag{B.5}$$

For a more intuitive way of defining the matricizing of a tensor, the concept of ‘fibers’ is introduced.

DEFINITION B.3 (n-mode fiber) Using the notation above, an n-mode fiber is a column vector that contains the elements $\mathcal{X}_{j_1,\ldots,j_{n-1},:,j_{n+1},\ldots,j_d}$.

For a second-order tensor, which is a matrix, a 1-mode fibre is a column of that matrix while a 2-mode fiber is a row of that matrix. The n-mode matricization is formed by reshuffling the n-mode fibers to be the columns of the matrix $X_{(n)}$, as is illustrated in Figure B.1.

EXAMPLE B.3 (n-mode matricization) For the tensor $\mathcal{X} \in \mathbb{R}^{2\times 4\times 2}$ mentioned in Example B.2, we have:

$$X_{(1)} = \begin{bmatrix} 1 & 3 & 5 & 7 & 9 & 11 & 13 & 15 \\ 2 & 4 & 6 & 8 & 10 & 12 & 14 & 16 \end{bmatrix}, \tag{B.6}$$

$$X_{(2)} = \begin{bmatrix} 1 & 2 & 9 & 10 \\ 3 & 4 & 11 & 12 \\ 5 & 6 & 13 & 14 \\ 7 & 8 & 15 & 16 \end{bmatrix}, \tag{B.7}$$

$$X_{(3)} = \begin{bmatrix} 1 & 2 & 3 & 4 & 5 & 6 & 7 & 8 \\ 9 & 10 & 11 & 12 & 13 & 14 & 15 & 16 \end{bmatrix}. \tag{B.8}$$

The n-mode matricization can be used to compute n-mode matrix products. Before outlining this procedure, we provide the following definition:

DEFINITION B.4 (n-mode matrix tensor product) Let $I \in \mathbb{N}$. The n-mode matrix product of a tensor $\mathcal{X} \in \mathbb{R}^{J_1 \times \cdots \times J_d}$ with a second-order tensor (or a matrix) $M \in \mathbb{R}^{I \times J_n}$ is denoted as $\mathcal{X} \times_n M$. The resulting operation defines a tensor of size $J_1 \times \cdots J_{n-1} \times I \times J_{n+1} \times \cdots \times J_d$ with elements given as:

$$(\mathcal{X} \times_n M)_{j_1,\ldots,j_{n-1},i,j_{n+1},\ldots,j_d} = \sum_{j_n=1}^{J_n} x_{j_1,\ldots,j_d} m_{i,j_n}. \tag{B.9}$$

The n-mode product between a matrix and a tensor is illustrated in the following example.

EXAMPLE B.4 (n-mode matrix tensor product) Consider a 3D tensor $\mathcal{Y}$ with entries as illustrated in Figure B.2. Let $M_1 \in \mathbb{R}^{3\times 3}$, then,

$$(\mathcal{Y} \times_1 M_1)_{j,j_2,j_3} = \sum_{j_1=1}^{J_1} \mathcal{Y}_{j_1,j_2,j_3} M_1(j,j_1).$$

If we call this resulting tensor $\mathcal{Z}$, then the 1-mode matricization of this tensor equals:

$$\mathcal{Z}_{(1)} = M_1 \mathcal{Y}_{(1)}.$$

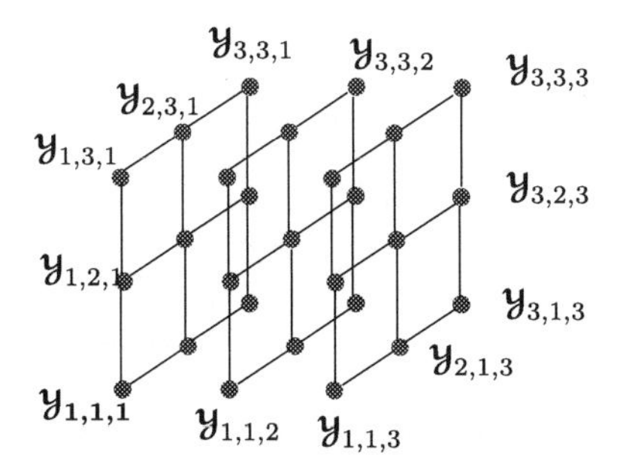

Figure B.2 A 3D tensor $\mathcal{Y} \in \mathbb{R}^{J_1 \times J_2 \times J_3}$ with $J_i = 3$ for $i = 1,2,3$.

Further, let $M_2 \in \mathbb{R}^{3\times 3}$, then,

$$(\mathcal{Z} \times_2 M_2)_{j_1,j,j_3} = \sum_{j_2=1}^{J_2} \mathcal{Z}_{j_1,j_2,j_3} M_2(j,j_2).$$

The 2-mode matricization of this tensor equals:

$$M_2 \mathcal{Z}_{(2)}.$$

It is often more intuitive to view the n-mode matrix product by highlighting its relationship to the n-mode matricization.

B.3 Multi-level Decompositions

PROPOSITION B.1 Let $(\mathcal{X},\mathcal{Y}) \in \mathbb{R}^{J_1\times\cdots\times J_d} \times \mathbb{R}^{I_1\times\cdots\times I_d}$. If $M \in \mathbb{R}^{I_n\times J_n}$, then $\mathcal{Y} = \mathcal{X} \times_n M$ can equivalently be written as $Y_{(n)} = MX_{(n)}$.

Moreover, for $j \in \{1,\ldots,d\}$ and a sequence of matrices $M_j \in \mathbb{R}^{I_j\times J_j}$, we define: $\bar{M} := M_d \otimes \cdots \otimes M_{n+1} \otimes M_{n-1} \otimes \cdots \otimes M_1$. Then, with this matrix sequence, we have the following equivalent relationships:

$$\mathcal{Y} = \mathcal{X} \times_1 M_1 \times_2 \cdots \times_d M_d, \tag{B.10}$$

$$Y_{(n)} = M_n X_{(n)} \bar{M}^T, \tag{B.11}$$

$$\text{vec}(Y_{(n)}) = \left(\bar{M} \otimes M_n\right)\text{vec}(X_{(n)}). \tag{B.12}$$

These relationships show that it is possible to recast the n-mode matrix product into matrix-matrix products with two matricized tensors. For $d = 2$, $n = 1$, this becomes the well-known relationship between the vectorization of the expression $Y_{(1)} = M_1 X_{(1)} M_2^T$ and $\text{vec}\left(Y_{(1)}\right) = (M_2 \otimes M_1)\text{vec}\left(X_{(1)}\right)$ as highlighted in Property A.6. When the matrix $M_2 \otimes M_1$ may be large, it is more efficient from a memory point of view, as well as from a computational point of view, to use n-mode matrix products.

EXAMPLE B.5 (1-mode product) The 1-mode matrix-tensor product for the case $d = 3$, the tensor $\mathcal{X}$ defined in Example B.2 and the matrix M given as:

$$M^T = \begin{bmatrix} 1 & 3 & 5 \\ 2 & 4 & 6 \end{bmatrix}, \tag{B.13}$$

is given in terms of the following two slides of $\mathcal{X}$, related to $\mathcal{X}$ via Equation (B.10):

$$\mathcal{Y}_{:,:,1} = \begin{bmatrix} 5 & 11 & 17 & 23 \\ 11 & 25 & 39 & 53 \\ 17 & 39 & 61 & 83 \end{bmatrix}, \mathcal{Y}_{:,:,2} = \begin{bmatrix} 29 & 35 & 41 & 47 \\ 67 & 81 & 95 & 109 \\ 105 & 127 & 149 & 171 \end{bmatrix}. \tag{B.14}$$

To illustrate the computational and memory advantages, consider the following. For the matrix $M \in \mathbb{R}^{I\times J}$ defined by d Kronecker products as in Proposition B.1 and $x \in \mathbb{R}^J$, the multiplication of M and x using n-mode matrix products instead of the ordinary matrix-vector product reduces the computational cost. This is because the matrix-vector product Mx can be written as $\mathcal{X} \times_1 M_1 \times_2 \cdots \times_d M_d$. The cost for computing a n-mode matrix product $\mathcal{X} \times_n M_n$ is $I_n J$. As a consequence, the computational cost of $\mathcal{X} \times_1 M_1 \times_2 \cdots \times_d M_d$ equals $J \sum_{j=1}^{d} I_j$. For the speical case $I = J$, where I_j is independent of j, the matrix-vector multiplication using a n-mode matrix product costs $dI^{(d+1)/d}$. The exponent of the integer I converges to one for increasing values of d.

In general, the polyadic decomposition of a tensor is a decomposition of a tensor into a sum of rank-1 tensors. For example a third-order tensor $\mathcal{X} \in \mathbb{R}^{I \times J \times K}$, a polyadic decomposition of $\mathcal{X}$ is given as follows:

$$\mathcal{X} = \sum_{i=1}^{R} x_i \circ y_i \circ z_i, \tag{B.15}$$

for $x_i \in \mathbb{R}^I, y_i \in \mathbb{R}^J, z_i \in \mathbb{R}^K$ non-zero vectors and where $\circ$ denotes the outer product of vectors. The latter is explicitly denoted component-wise as:

$$\mathbf{x}_{rst} = \sum_{i=1}^{R} x_i(r) y_i(r) z_i(r),$$

with $x_i(r)$ denoting the rth component of the vector x_i.

DEFINITION B.5 (Rank of a tensor (Hitchcock, 1927)) The rank of a tensor $\mathcal{X} \in \mathbb{R}^{I \times J \times K}$ is the minimum number of rank-1 tensors in a polyadic decomposition of $\mathcal{X}$.

DEFINITION B.6 (Canonical polyadic decomposition (CPD) of a tensor (Carroll and Chang, 1970; Harshman, 1970)) A canonical polyadic decomposition (CPD) of a third-order tensor $\mathcal{X}$ expresses $\mathcal{X}$ as a polyadic decomposition with a minimal number of terms in its sum.

An illustration of this definition is given in Figure B.3 for a third-order tensor. The lines in this figure represent non-zero vectors.

LEMMA B.1 (CPD decomposition) *Any d-order tensor in $\mathbb{R}^{I \times \cdots \times I}$ can be decomposed* exactly *into a CPD.*

Proof Let $\mathcal{X} \in \mathbb{R}^{I \times \cdots \times I}$. Let $e_i \in \mathbb{R}^I$ with only zero entries except a one at position i. The set $\mathcal{S} = \{e_{i_1} \circ \cdots \circ e_{i_d} / i_j \in \{1, \ldots, I\}, j \in \{1, \ldots, d\}\}$ is the basis of

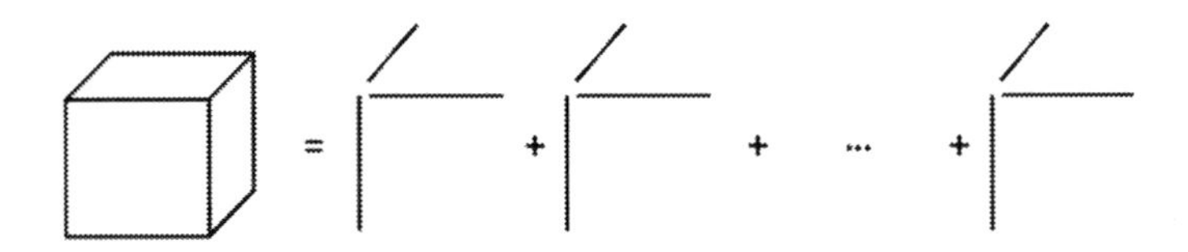

Figure B.3 Schematic of a canonical polyadic decomposition (CPD) of a third-order tensor. Reprinted/adapted from Sinquin (2019) with permission from TU Delft

the space $\mathbb{R}^{\overbrace{I \times \cdots \times I}^{d\text{-times}}}$. Alternatively, let us work in $\mathbb{R}^{I^d}$ and replace the outer product in Equation (B.15) with the Kronecker product. The rank is equal to the minimum dimension of the subspace containing $\mathcal{X}$ and formed from a linear combination of vectors in $\mathcal{S}$. Let $\alpha \in \mathbb{R}^{I \times \cdots \times I}$, then $\mathcal{X}$ can be written in terms of the following linear combination of rank-one tensors:

$$\mathcal{X} = \sum_{i_1=1}^{I} \cdots \sum_{i_d=1}^{I} \alpha_{i_1,\ldots,i_d} e_{i_1} \circ \cdots \circ e_{i_d}. \tag{B.16}$$

□

This lemma is only an existence result and does not provide a numerical procedure to find the optimal rank approximation. Finding the rank of a CPD of a tensor in general is an NP-hard problem (Håstad, 1990).

The problem of the best rank approximation for matrices is given by the well-known Eckart–Young theorem (Eckart and Young, 1936). This theorem states that for a matrix $X \in \mathbb{R}^{m \times n}$, its SVD is given as:

$$A = U \Sigma V = \sum_{i=1}^{rank(A)} \sigma_i u_i \circ v_i, \tag{B.17}$$

with singular values ordered in descending order, then a best rank-r approximation is given by the first r terms in the above sum. Moreover, when r equals the rank of the matrix, the matrix is exactly represented by its SVD decomposition. For tensors a different situation arises. The factors in the rank-2 CPD of a third-order tensor may not be factors of the rank-3 CPD, as was illustrated in Harshman (2004). Furthermore, it may happen that a rank-r tensor can be approximated arbitrarily well with a tensor of rank strictly less than r, meaning that the set of tensors of rank at most r is not closed (de Silva and Lim, 2008). In such cases, the numerical procedure for computing a low-rank tensor approximation may yield factors in the summation that cancel each other, while the cost function that is used in the numerical calculation keeps on decreasing (Vervliet et al., 2016). Such degeneracy was shown in de Silva and Lim (2008) and occurs when the CPD is not a proper way to represent the tensor. A situation may arise when the tensor is determined from measurements and hence is noisy.

References

Baldi, P. and Hornik, K. Neural networks and principal component analysis: Learning from examples without local minima. *Neural Networks*, 2(1):53–58, 1989.

Bamieh, B. and Filo, M. An input-output approach to structured stochastic uncertainty. Technical Report arXiv:1806.07473v1, University of California, Santa Barbara, 2018.

Bamieh, B. and Voulgaris, P. G. A convex characterization of distributed control problems in spatially invariant systems with communication constraints. *Systems & Control Letters*, 54(6):575–583, 2005.

Bellman, R. and Astrom, K. J. On structural identifiability. *Mathematical Biosciences*, 7(3):329–339, 1970.

Bertsekas, D. P. *Nonlinear Programming*. Athena Scientific, Nashua, NH, 1999.

Beylkin, G. and Mohlenkamp, M. Algorithms for numerical analysis in high dimensions. *SIAM Journal on Scientific Computing*, 26(6):2133–2159, 2005.

Boersma, S., Doekemeijer, B., Vali, M., Meyers, J. and van Wingerden, J.-W. A control-oriented dynamic wind farm model: WFSim. *Wind Energy Science*, 3(1):75–95, 2018.

Borelli, F. and Keviczky, T. Distributed LQR design for identically dynamically decoupled systems. *IEEE Transactions on Automatic Control*, 53(8):1901–1912, 2008.

Boumal, N., Mishra, B., Absil, P. A. and Sepulchre, R. Manopt, a Matlab toolbox for optimization on manifolds. *Journal of Machine Learning Research*, 15(1):1455–1459, 2013.

Boussé, M. *Explicit and Implicit Tensor Decomposition-Based Algorithms and Applications*. PhD thesis, KU Leuven, 2019.

Bruls, J., Chou, C., Haverkamp, B. and Verhaegen, M. Linear and non-linear system identification using separable least-squares. *European Journal of Control*, 5(1):116–128, 1999.

Carroll, J. D. and Chang, J.-J. Analysis of individual differences in multidimensional scaling via an n-way generalization of Eckart-Young decomposition. *Psychometrika*, 35(3):283–319, 1970.

Castagnotto, A., Cruz Varona, M., Jeschek, L. and Lohmann, B. sss & sssMOR: Analysis and reduction of large-scale dynamic systems in Matlab. *at-Automatisierungstechnik*, 65(2):134–150, February 2017.

Chen, T., Ohlsson, H. and Ljung, L. On the estimation of transfer functions, regularizations and Gaussian processes revisited. *Automatica*, 48(8):1525–1535, 2012.

Chiuso, A. The role of vector autoregressive modeling in predictor-based subspace identification. *Automatica*, 43(6):1034–1048, 2007.

Chiuso, A. and Pillonetto, G. A Bayesian approach to sparse dynamic network identification. *Automatica*, 48(8):1553–1565, 2012.

Choudhary, S. and Mitra, U. Identifiability scaling laws in bilinear inverse problems. *arXiv preprint arXiv:1402.2637*, 2014.

Correia, C., Conan, J.-M., Kulcsár, C., Raynaud, H.-F. and Petit, C. Adapting optimal LQG methods to ELT-sized AO systems. Paper presented at the first Adaptive Optics for Extremely Large Telescopes conference (1st AO4ELT), 2010.

Cuthill, E. and McKee, J. Reducing the bandwidth of sparse symmetric matrices. In *Proceedings of the 1969 24th National Conference of ACM*, pp. 157–172, New York, 1969.

Dahlhaus, R. and Eichler, M. Causality and graphical models in time series analysis. In Green, P., Hjort, N. and Richardson, S., eds. *Highly Structured Stochastic Systems*, pp. 115–137. Oxford University Press, Oxford, 2003.

D'Andrea, R. Software for modeling, analysis, and control design for multidimensional systems. In *Proceedings of the 1999 IEEE International Symposium on Computer Aided Control System Design* (Cat. No. 99TH8404), pp. 24–27, IEEE, 1999.

D'Andrea, R. and Dullerud, G. Distributed control for spatially interconnected systems. *IEEE Transactions on Automatic Control*, 48(9):1478–1495, 2003.

De Moor, B. *Mathematical Concepts and Techniques for Modelling of Static and Dynamic Systems*. PhD thesis, KU Leuven, 1988.

Dempster, A., Laird, N. and Rubin, D. *Maximum Likelihood from Incomplete Data via EM Algorithm*. Prentice Hall, Englewood Cliffs, NJ, 1977.

de Silva, V. and Lim, L. Tensor rank and the ill-posedness of the best low-rank approximation problem. *SIAM Journal on Matrix Analysis and Applications*, 30(3):1084–1127, 2008.

Dewilde, P. and van der Veen, A. *Time-Varying Systems and Computations*. Kluwer Academic Publisher, Norwell, MA, 1998.

Diestel, R. *Graph Theory*. Springer-Verlag, New York, 1997.

Ding, F. and Chen, T. Iterative least-squares solutions of coupled Sylvester matrix equations. *Systems & Control Letters*, 54(2):95–107, 2005.

Doekemeijer, B. M., Boersma, S., Pao, L. Y., Knudsen, T. and van Wingerden, J.-W. Online model calibration for a simplified LES model in pursuit of real-time closed-loop wind farm control. *Wind Energy Science*, 3(2):749–765, 2018.

Doelman, R. and Verhaegen, M. Sequential convex relaxation for convex optimization with bilinear matrix equalities. In *2016 European Control Conference (ECC)*, pp. 1946–1951, 2016.

Dullerud, G. and D'Andrea, R. Distributed control of heterogeneous systems. *IEEE Transactions on Automatic Control*, 49(12):2113–2128, 2004.

Eckart, C. and Young, G. The approximation of one matrix by another of lower rank. *Psychometrika*, 1(3):211–218, 1936.

Efron, B., Hastie, T., Johnstone, I. and Tibshirani, R. Least angle regression. *Annals of Statistics*, 2(32):407–499, 2004.

Ellerbroek, B. The TMT adaptive optics program. Paper presented at the second Adaptive Optics for Extremely Large Telescopes international conference (2nd AO4ELT), 2011.

Epperlein, J. P. and Bamieh, B. Spatially invariant embeddings of systems with boundaries. Paper presented at the *American Control Conference*, 2016.

Fax, J. and Murray, R. Information flow and cooperative control of vehicle formations. *IEEE Transactions on Automatic Control*, 49(9):1465–1476, 2004.

Fazel, M., Hindi, H. and Boyd, S. P. Log-det heuristic for matrix rank minimization with applications to Hankel and Euclidean distance matrices. Paper presented at the *American Control Conference*, 2003.

Fazel, M., Pong, T. K., Sun, D. and Tseng, P. Hankel matrix rank minimization with applications to system identification and realization. *Siam Journal on Matrix Analysis and Applications*, 34(3):946–977, 2012.

Fraanje, R., Massioni, P. and Verhaegen, M. A decomposition approach to distributed control of dynamic deformable mirrors. *International Journal of Optomechatronics*, 4(3):269–284, 2010.

Fritzson, P. *Introduction to Modeling and Simulation of Technical and Physical Systems with Modelica*. Wiley-IEEE, New Jersey, 2011.

Gebraad, P. *Data-driven wind plant control*. PhD thesis, Delft Center for Systems and Control, Delft University of Technology, 2014.

Gendron, E., Morris, T., Basden, A., et al. Final two-stage MOAO on-sky demonstration with CANARY. In Marchetti, E., Close, L. M., and Véran, J.-P., eds., *Adaptive Optics Systems V*, vol. 9909, pp. 126–142. International Society for Optics and Photonics, SPIE, 2016.

Gevers, M., Bazanella, A. and Pimentel, G. Identifiability of dynamical networks with singular noise spectra. *IEEE Transactions on Automatic Control*, 64(6):2473–2479, 2019.

Giraldi, L., Nouy, A. and Legrain, G. Low-rank approximate inverse for preconditioning tensor-structured linear systems. *SIAM Journal on Scientific Computing*, 36(4):A1850–A1870, 2014.

Golub, G. H. and Van Loan, C. F. *Matrix Computations*. The Johns Hopkins University Press, third edition, 1996.

Golub, G., Kolda, T., Nagy, J. and Loan, C. V. Workshop on tensor decompositions. *SIAM J. Matrix Analysis and Applications*, 21(2):1253–1278, 2000.

Gonçalves, J. and Warnick, S. Necessary and sufficient conditions for dynamical structure reconstruction of lti networks. *IEEE Transactions on Automatic Control*, 53(7):1670–1674, 2008.

Gragg, W. B. and Lindquist, A. On the partial realization problem. *Linear Algebra and its Applications*, 50:277–319, 1983.

Guyon, O. Extreme adaptive optics. *Annual Review of Astronomy and Astrophysics*, 56(1):315–355, 2018.

Halko, N., Martinsson, P. G. and Tropp, J. A. Finding structure with randomness: Probabilistic algorithms for constructing approximate matrix decompositions. *SIAM Review*, 53(2):217–288, 2011.

Harshman, R. A. Foundations of the PARAFAC procedure: Models and conditions for an explanatory multimodal factor analysis. UCLA working papers in phonetics, 16:1–84, (University Microfilms, Ann Arbor, Michigan, No. 10,085), 1970.

Harshman, R. A. The problem and nature of degenerate solutions or decompositions of 3-way arrays. Paper presented at the *American Institute of Mathematics Tensor Decomposition Workshop, Palo Alto, CA*, 2004.

Håstad, J. Tensor rank is NP-complete. *Journal of Algorithms*, 11(4):644–654, 1990.

Henrion, D. and Šebek, M. Polynomial and matrix fraction description. 2000. Available at `https://homepages.laas.fr/henrion/Papers/6-43-13-5.pdf`.

Hinnen, K., Verhaegen, M. and Doelman, N. Exploiting the spatiotemporal correlation in adaptive optics using data-driven H_2-optimal control. *Journal of the Optical Society of America A*, 24(6):1714–1725, 2007.

Hinnen, K., Verhaegen, M. and Doelman, N. A data-driven H_2-optimal control approach for adaptive optics. *IEEE Transactions on Control Systems Technology*, 16(3):381–395, 2008.

Hippe, P. and Deutscher, J. *Design of Observer-based Compensators: From the Time to the Frequency Domain*. Springer-Verlag, London, 2009.

Hitchcock, F. L. The expression of a tensor or a polyadic as a sum of products. *Journal of Mathematics and Physics*, 6(1–4):164–189, 1927.

Ho, B. and Kalman, R. Effective construction of linear state-variable models from input/output data. *Regelungstechnik*, 14:545–548, 1966.

Hoff, P. D. Multilinear tensor regression for longitudinal relational data. *Annals of Applied Statistics*, 9(3):1169–1193, 2015.

Houtzager, I., van Wingerden, J. and Verhaegen, M. VARMAX-based closed-loop subspace model identification. In *Proceedings of the 48th IEEE Conference on Decision and Control*, 2009a.

Houtzager, I., van Wingerden, J.-W. and Verhaegen, M. Fast-array recursive closed-loop subspace model identification. In *15th IFAC Symposium on System Identification*, volume 42, pp. 96–101, 2009b.

Inigo, G. *Estimation and control of noise amplifier flows using data-based approaches*. PhD thesis, Fluids mechanics [physics.class-ph], Ecole Polytechnique, France, 2015.

Jansson, M. Subspace identification and ARX modeling, *IFAC Proceedings Volumes*, 36(16), 1585–1590, 2003, https://doi.org/10.1016/S1474-6670(17)34986-8.

Jovanović, M. and Bamieh, B. Lyapunov-based distributed control of systems on lattices. *IEEE Transactions on Automatic Control*, 50(4):422–433, 2005.

Juang, J., Chen, C. and Phan, M. Estimation of Kalman filter gain from output residuals. *Journal of Guidance, Control, and Dynamics*, 16(5):903–6908, 1993.

Kailath, T. *Linear Systems*. Prentice-Hall Information and System Sciences Series. Prentice Hall, 1980.

Kailath, T., Sayed, A. and Hassibi, B. *Linear Estimation*. Prentice Hall, New Jersey, 2000.

Kibangou, A. Y., Garin, F. and Gracy, S. Input and state observability of network systems with a single unknown input. *IFAC-PapersOnLine*, 49(22):37–42, 2016. Paper presented at the 6th IFAC Workshop on Distributed Estimation and Control in Networked Systems NECSYS 2016. www.sciencedirect.com/science/article/pii/S2405896316319565.

Kim, J. and Bewley, T. R. A linear systems approach to flow control. *Annual Review of Fluid Mechanics*, 39(1):383–417, 2007.

Knudsen, T. Consistency analysis of subspace identification methods based on a linear regression approach. *Automatica*, 37(1):81–89, 2001.

Kolda, T. Multilinear operators for higher-order decompositions. Technical Report SAND2006-2081, Sandia National Laboratories, 2006.

Kolda, T. and Bader, B. Tensor decompositions and applications. *SIAM Review*, 51(3):455–500, 2009.

Kolmogorov, A. The local structure of turbulence in incompressible viscous fluid for very large Reynolds' numbers. *Akademiia Nauk SSSR Doklady*, 30:301–305, 1941.

Kulcsár, C., Raynaud, H.-F., Petit, C. and Conan, J.-M. Minimum variance prediction and control for adaptive optics. *Automatica*, 48(9):1939–1954, 2012.

Langbort, C., Chandra, R. and D'Andrea, R. Distributed control design for systems interconnected over an arbitrary graph. *IEEE Transactions on Automatic Control*, 49(9):1502–1519, 2004.

Leskovec, J., Chakrabarti, D., Kleinberg, J., Faloutsos, C. and Ghahramani, Z. Kronecker graphs: An approach to modeling networks. *Journal of Machine Learning Research*, 11:985–1042, 2010.

Li, N., Kindermann, S. and Navasca, C. Some convergence results on the regularized alternating least-squares method for tensor decomposition. *Linear Algebra and its Applications*, 438(2):796–812, 2013.

Li, G., Wen, C. and Zhang, A. Fixed point iteration in identifying bilinear models. *Systems & Control Letters*, 83:28–37, 2015.

Lipp, T. and Boyd, S. Variations and extension of the convex cconcave procedure. *Optimizationa & Engineering*, 17(2):263–287, 2016.

Liu, Q. *Modeling, Distributed Identification and Control of Spatially-Distributed Systems with Application to an Actuated Beam*. PhD thesis, Institute of Control Systems, Hamburg University of Technology, 2015.

Liu, Z., Hansson, A. and Vandenberghe, L. Nuclear norm system identification with missing inputs and outputs. *Systems and Control Letters*, 62(8):605–612, 2013.

Ljung, L. *System Identification – Theory for the User*. Prentice Hall, Upper Saddle River, NJ, 2nd edition, 1999.

Ljung, L. and Parrilo, P. Initialization of physical parameter estimates. Paper presented at the *13th IFAC Symposium on System Identification SYSID*, 2003.

Ljung, L., Singh, R. and Chen, T. Regularization features in the system identification toolbox. *IFAC-PapersOnLine*, 48(28):745–750, 2015. www.sciencedirect.com/science/article/pii/S2405896315028438

Loan, C. V. and Pitsianis, N. Approximation with Kronecker products. In Moonen, M. and Golub, G., eds., *Linear Algebra for Large Scale and Real Time Applications*, pp. 293–314. Kluwer Publications, Dordrecht, 1992.

Massioni, P. Distributed control α-heterogeneous dynamically coupled systems. *Systems and Control Letters*, 72:30–35, 2014.

Massioni, P. *Decomposition Methods for Distributed Control and Identification*. PhD thesis, Delft Center for Systems and Control, Delft University of Technology, 2015.

Massioni, P. and Verhaegen, M. Distributed control for identical dynamically coupled systems: A decomposition approach. *IEEE Transactions on Automatic Control*, 54(1):124–135, 2009.

Massioni, P., Kulcsár, C., Raynaud, H.-F. and Conan, J.-M. Fast computation of an optimal controller for large-scale adaptive optics. *Journal of the Optical Society of America A*, 28(11):2298–2309, 2011.

Massioni, P., Raynaud, H.-F., Kulcsár, C. and Conan, J.-M. An approximation of the Riccati equation in large-scale systems with application to adaptive optics. *IEEE Transactions on Control Systems Technology*, 2(23):479–487, 2015.

Mercère, G., Markovsky, I. and Ramos, J. A. Innovation-based subspace identification in open- and closed-loop. Paper presented at the *2016 IEEE 55th Conference on Decision and Control*, 2016.

Mohan, K. and Fazel, M. Reweighted nuclear norm minimization with application to system identification. In *Proceedings of the 2010 American Control Conference*, pp. 2953–2959. IEEE, 2010.

Mohlenkamp, M. J. Musings on multilinear fitting. *Linear Algebra and its Applications*, 438(2):834–852, 2013.

Moonen, M., Moor, B. D., Vandenberghe, L. and Vandewalle, J. On- and off-line identification of linear state-space models. *International Journal of Control*, 49(1):219–232, 1989.

Motee, N. and Jadbabaie, J. Optimal control of spatially distributed systems. *IEEE Transactions on Automatic Control*, 53(7):1616–1629, 2008.

Neichel, B., Fusco, T., Sauvage, J.-F., et al. The adaptive optics modes for HARMONI: from classical to laser assisted tomographic AO. In *Proceedings of SPIE 9909, Adaptive Optics Systems V*, 990909 (26 July 2016); https://doi.org/10.1117/12.2231681

Nocedal, J. and Wright, S. J. *Numerical Optimization*, 2nd ed. Springer, New York, 2nd edition, 2006.

Noll, R. J. Zernike polynomials and atmospheric turbulence∗. *Journal of the Optical Society of America*, 66(3):207–211, 1976.

Ogata, K. *Modern Control Engineering 4th ed.* Prentice Hall, New Jersey, 4th edition, 2002.

Osborn, J. Characterising atmospheric turbulence using scidar techniques. In *Imaging and Applied Optics 2018*, p. JW5I.5. Optical Society of America, 2018.

Overschee, P. V. and Moor, B. D. N4SID: Subspace algorithms for the identification of combined deterministic-stochastic systems. *Automatica*, 30(1):75–93, 1994.

Overton, M. L. and Womersley, R. S. Optimality conditions and duality theory for minimizing sums of the largest eigenvalues of symmetric matrices. *Mathematical Programming*, 62(1–3):321–357, 1993.

Peternell, K., Scherrer, W. and Deistler, M. Statistical analysis of novel subspace identification methods. *Signal Processing*, 52(2):161–177, 1996.

Petit, C., Conan, J.-M., Kulcsár, C., Raynaud, H.-F. and Fusco, T. First laboratory validation of vibration filtering with LQG control law for adaptive optics. *Optics Express*, 16(1):87–97, 2008.

Pillonetto, G., Dinuzzo, F., Chen, T., Nicolao, G. D. and Ljung, L. Kernel methods in system identification, machine learning and function estimation: A survey. *Automatica*, 50(3):657–682, 2014.

Pircher, M. and Zawadzki, R. J. Review of adaptive optics OCT (AO-OCT): Principles and applications for retinal imaging. *Biomedical Optics Express*, 8(5):2536–2562, 2017.

Piscaer, P. *Sparse VARX Model Identification for Large-Scale Adaptive Optics*. MSc thesis, Delft Center for Systems and Control, Delft University of Technology, 2016.

Poyneer, L. A., Macintosh, B. A. and Véran, J.-P. Fourier transform wavefront control with adaptive prediction of the atmosphere. *Journal of the Optical Society of America A*, 24(9):2645–2660, 2007.

Poyneer, L. A., Palmer, D. W., Macintosh, B., et al. Performance of the Gemini planet imager adaptive optics system. *Applied Optics*, 55(2):323–340, 2016.

Pozzi, P., Quintavalla, M., Wong, A. B., Borst, J. G. G., Bonora, S. and Verhaegen, M. Plug-and-play adaptive optics for commercial laser scanning fluorescence microscopes based on an adaptive lens. *Optics Letters*, 45(13):3585–3588, 2020. doi: 10.1364/OL.396998.

Qi, L. and Womersley, R. S. On extreme singular values of matrix valued functions. *Journal of Convex Analysis*, 3:153–166, 1996.

Qiu, Y. Multilevel sequential separable (MSSS) matrix computation toolbox version 0.7. `http://ta.twi.tudelft.nl/nw/users/yueqiu/software.html`, 2013.

Qiu, Y. *Preconditioning Optimal Flow Control Problems Using Multilevel Sequentially Semiseparable Matrix Computations*. PhD thesis, Delft Center for Systems and Control, Delft University of Technology, 2015.

Recht, B. and D'Andrea, R. Distributed control of systems over discrete groups. *IEEE Transactions on Automatic Control*, 49(9):1270–1283, 2009.

Recht, B., Fazel, M. and Parrilo, P. Guaranteed minimum-rank solutions of linear matrix equations via nuclear norm minimization. *SIAM Review*, 52(3):471–501, 2010.

Rice, J. and Verhaegen, M. Distributed control: A sequentially semi-separable approach for spatially heterogeneous linear systems. *IEEE Transactions on Automatic Control*, 54(6):1270–1283, 2009.

Rice, J. and Verhaegen, M. Efficient system identification of heterogeneous distributed systems via a structure exploiting extended Kalman filter. *IEEE Transactions on Automatic Control*, 56(7):1713–1718, 2011.

Roddier, F. *Adaptive Optics in Astronomy*. Cambridge University Press, 1999.

Roesser, R. A discrete state-space model for linear image processing. *IEEE Transactions on Automatic Control*, 20(1):1–10, 1975.

Rojo, O. A new algebra of Toeplitz-plus-Hankel matrices and applications. *Computers & Mathematics with Applications*, 55(12):2856–2869, 2008.

Roux, B. L., Conan, J.-M., Kulcsár, C., Raynaud, H.-F., Mugnier, L. M. and Fusco, T. Optimal control law for classical and multiconjugate adaptive optics. *Journal of the Optical Society of America A*, 21(7):1261–1276, 2004.

Rudin, W. *Real and Complex Analysis*. McGraw-Hill International Editions, New York, 1986.

Sayed, A. H. and Kailath, T. A state-space approach to adaptive RLS filtering. *IEEE Signal Processing Magazine*, 11(3):18–60, 1994.

Scobee, D., Ratliff, L., Dong, R., Ohlsson, H., Verhaegen, M. and Sastry, S. S. Nuclear norm minimization for blind subspace identification (N2BSID). In *Decision and Control (CDC), 2015 IEEE 54th Annual Conference on*, pp. 2127–2132. IEEE, 2015.

Shamma, J. and Arslan, G. A decomposition approach to distributed control of systems on lattices. *IEEE Transactions on Automatic Control*, 51(4):701–707, 2006.

Shen, X., Diamond, S., Udell, M., Gu, Y. and Boyd, S. Disciplined multi-convex programming. Technical Report arXiv:1609.03285, Stanford University, 2016.

Sinquin, B. *Structured Matrices for Predictive Control of Large and Multi-dimensional Systems*. PhD thesis, Delft University of Technology, 2019.

Sinquin, B. and Verhaegen, M. A subspace like identification method for large-scale LTI dynamical systems. Paper presented at the *2017 Signal Processing Symposium (SPSympo)*, 2017.

Sinquin, B. and Verhaegen, M. Tensor-based predictive control for extremely large-scale single conjugate adaptive optics. *Journal of Optical Society of America A*, 36(9):1612–1626, 2018.

Sinquin, B. and Verhaegen, M. K4SID: Large-scale subspace identification with Kronecker modeling. *IEEE Transactions on Automatic Control*, 64(3):960–975, 2019a.

Sinquin, B. and Verhaegen, M. QUARKS: Identification of large-scale Kronecker vector-autoregressive models. *IEEE Transactions on Automatic Control*, 64(2):448–463, 2019b.

Sinquin, B., Varnai, P., Monchen, G. and Verhaegen, M. Tensor toolbox for identifyin multi-dimensional systems. Available at `https://bitbucket.org/csi-dcsc/t4sid`, 2018.

Sinquin, B., Prengère, L., Kulcsár, C., et al. On-sky results for adaptive optics control with data-driven models on low-order modes. Submitted to *Monthly Notices of Royal Astronomical Society*, 2020.

Sivo, G., Kulcsár, C., Conan, J.-M., et al. First on-sky SCAO validation of full LQG control with vibration mitigation on the CANARY pathfinder. *Optical Express*, 22(19):23565–23591, 2014.

Songsiri, J. and Vandenberghe, L. Topology selection in graphical models of autoregressive processes. *Journal of Machine Learning Research*, 11:2010, 2010.

Southwell, R. *Relaxation Methods in Theoretical Physics*. Oxford University Press, Oxford, 1946.

Strang, G. Fast transforms: Banded matrices with banded inverses. *Proceedings of the National Academy of Sciences of the United States of America*, 107(28):12413–12416, 2009.

Sundaram, S. Fault-tolerant and secure control systems. *University of Waterloo, Lecture Notes*, 2012.

Tibshirani, R. Regression shrinkage and selection via the lasso. *Journal of the Royal Statistical Society. Series B (Methodological)*, 58(1):267–288, 1996.

Timoshenko, S. and Woinowsky-Krieger, S. *Theory of Plates and Shells*. McGraw-Hill, New York, 1959.

Tsiligkaridis, T. and Hero, A. O. Covariance estimation in high dimensions via Kronecker product expansions. *IEEE Transactions on Signal Processing*, 61(21):5347–5360, 2013.

Udell, M., Horn, C., Zadeh, R. and Boyd, S. Generalized low-rank models. *Foundations and Trends in Machine Learning*, 9:1–118, 2016.

van den Hof, J. Structural identifiability of linear compartemental systems. *IEEE Transactions on Automatic Control*, 43(6):800–818, 2002.

Van Overschee, P. and De Moor, B. N4SID: Subspace algorithms for the identification of combined deterministic-stochastic systems. *Automatica*, 30(1):75–93, 1994.

Van Overschee, P. and De Moor, B. *Identification for Linear Systems: Theory - Implementation - Applications*. Springer-Verlag, 2011.

Varnai, P. *Exploiting Kronecker Structures, with Applications to Optimization Problems Arising in the Field of Adaptive Optics*. MSc thesis, Delft Center for Systems and Control, Delft University of Technology, 2017.

Verhaegen, M. Subspace techniques in system identification. In Baillieul, J. and Samad, T., eds., *Encyclopedia of Systems and Control*. Springer-Verlag, London, 2015.

Verhaegen, M. The identification of network connected systems. In Baillieul, J. and Samad, T., eds., *Encyclopedia of Systems and Control*, pp. 1–11. Springer London, London, 2019. doi: 10.1007/978-1-4471-5102-9_100088-1.

Verhaegen, M. and Hansson, A. N2SID: Nuclear norm subspace identification of innovation models. *Automatica*, 72(5):57–63, 2016.

Verhaegen, M. and Verdult, V. *Filtering and System Identification: A Least Squares Approach*. Cambridge University Press, Cambridge, 1st edition, 2007.

Vervliet, N., Debals, O., Sorber, L., Barel, M. V. and Lathauwer, L. D. Tensorlab 3.0. Technical report, KU Leuven, 2016. Available at www.esat.kuleuven.be/stadius/tensorlab/.

Wan, Y., Keviczky, T., Verhaegen, M. and Gustafsson, F. Data-driven robust receding horizon fault estimation. *Automatica*, 71:210–221, 2016.

Weerts, H. *Identifiability and Identification Methods for Dynamic Networks*. PhD thesis, Eindhoven: Technische Universiteit Eindhoven, 2018.

Wen, F., Chu, L., Liu, P. and Qiu, R. C. A survey on nonconvex regularization-based sparse and low-rank recovery in signal processing, statistics, and machine learning. *IEEE Access*, 6:69883–69906, 2018.

Wills, A., Yu, C., Ljung, L. and Verhaegen, M. Affinely parametrized state-space models: Ways to maximize the likelihood function. Paper presented at the *18th IFAC Symposium on System Identification (SYSID 2018)*, 2018.

Woodbury, N. *Representation and Reconstruction of Linear Time-Invariant Networks*. PhD thesis, Brigham Young University, 2019.

Xu, Y. and Yin, W. A block coordinate descent method for regularized multiconvex optimization with applications to nonnegative tensor factorization and completion. *SIAM Journal on Imaging Sciences*, 6(3):1758–1789, 2013.

Yeung, E., Gonçalves, J., Sandberg, H. and Warnick, S. Representing structure in linear interconnected dynamical systems. Paper presented at the *49th IEEE Conference on Decision and Control (CDC)*, 2010.

Yu, C. and Verhaegen, M. Local subspace identification of distributed homogeneous systems with general interconnection patterns. Paper presented at the *17th IFAC Symposium on System Identification (SYSID 2015)*, 2015.

Yu, C. and Verhaegen, M. Blind multivariable ARMA subspace identification. *Automatica*, 66:3–14, 2016.

Yu, C. and Verhaegen, M. Subspace identification of distributed clusters of homogeneous systems. *IEEE Transactions on Automatic Control*, 62(1):463–468, 2017.

Yu, C. and Verhaegen, M. Subspace identification of individual systems operating in a network (si^{2}on). *IEEE Transactions on Automatic Control*, 63(4):1120–1125, 2018.

Yu, C., Chen, J. and Verhaegen, M. Subspace identification of individual systems in a large-scale heterogeneous network. *Automatica*, 109:108517, 2019.

Yu, C., Ljung, L. and Verhaegen, M. Identification of structured state-space models. *Automatica*, 90:54–61, 2018a.

Yu, C., Verhaegen, M. and Hansson, A. Subspace identification of local systems in one-dimensional homogeneous networks. *IEEE Transactions on Automatic Control*, 63(4):1126–1131, 2018b.

Yu, C., Chen, J., Ljung, L. and Verhaegen, M. Subspace identification of continuous-time models using generalized orthodnormal bases. Paper presented at *IEEE 56th Annual Conference on Decision and Control (CDC)*, pp. 5280–5285, 2017, doi: 10.1109/CDC.2017.8264440. 2017.

Yu, C., Ljung, L., Wills, A. and Verhaegen, M. Constrained subspace method for the identification of structured state-space models (COSMOS). *IEEE Transactions on Automatic Control*, 65(10):4201–4214, 2020. doi: 10.1109/TAC.2019.2957703.

Yue, Z., Thunberg, J., Ljung, L. and Gonçalves, J. A state-space approach to sparse dynamic network reconstruction. Technical Report arXiv:1811.08677v1, Université du Luxembourg, 2018.

Index

Printed in the United States
by Baker & Taylor Publisher Services